ENNIS AND NANCY HAM LIBRARY
ROCHESTER COLLEGE
800 WEST AVON ROAD
ROCHESTER HILLS, MI 48307

# FORESTRY AND ITS CAREER OPPORTUNITIES

25123

**McGraw-Hill Series in Forest Resources**

**Henry J. Vaux,** Consulting Editor

*Allen and Sharpe*   An Introduction to American Forestry
*Avery*   Forest Measurements
*Baker*   Principles of Silviculture
*Boyce*   Forest Pathology
*Brockman and Merriam*   Recreational Use of Wild Lands
*Brown and Davis*   Forest Fire: Control and Use
*Brown, Panshin, and Forsaith*   Textbook of Wood Technology
  Volume II—The Physical, Mechanical, and Chemical Properties of the
  Commercial Woods of the United States
*Chapman and Meyer*   Forest Mensuration
*Dana*   Forest and Range Policy
*Davis*   Forest Management: Regulation and Evaluation
*Duerr*   Fundamentals of Forestry Economics
*Graham and Knight*   Principles of Forest Entomology
*Guise*   The Management of Farm Woodlands
*Harlow and Harrar*   Textbook of Dendrology
*Hunt and Garratt*   Wood Preservation
*Panshin and de Zeeuw*   Textbook of Wood Technology
  Volume I—Structure, Identification, Uses, and Properties of
  the Commercial Woods of the United States
*Panshin, Harrar, Bethel, and Baker*   Forest Products
*Rich*   Marketing of Forest Products: Text and Cases
*Shirley*   Forestry and Its Career Opportunities
*Stoddart and Smith*   Range Management
*Trippensee*   Wildlife Management
  Volume I—Upland Game and General Principles
  Volume II—Fur Bearers, Waterfowl, and Fish
*Wackerman, Hagenstein and Michell*   Harvesting Timber Crops
*Worrell*   Principles of Forest Policy

WALTER MULFORD WAS CONSULTING EDITOR OF THIS SERIES FROM ITS INCEPTION IN 1931 UNTIL JANUARY 1, 1952.

*Sculpture formed of wood plasticized by ammonia.* (State University of New York, College of Environmental Science and Forestry, Syracuse.)

# FORESTRY AND ITS CAREER OPPORTUNITIES

Third Edition

**HARDY L. SHIRLEY**

*Dean Emeritus*
*State University of New York*
*College of Environmental Science and Forestry*
*Syracuse*

**McGraw-Hill Book Company**

New York   St. Louis   San Francisco   Düsseldorf
Johannesburg   Kuala Lumpur   London   Mexico
Montreal   New Delhi   Panama   Rio de Janeiro
Singapore   Sydney   Toronto

**Library of Congress Cataloging in Publication Data**

Shirley, Hardy Lomax, 1900–
  Forestry and its career opportunities.

  (McGraw-Hill series in forest resources)
  Includes bibliographies.
  1. Forests and forestry.  2. Forestry as a profession.  I. Title.
  SD371.S5    1973     634.9     72-13420
  ISBN 0-07-056978-9

**Forestry and Its Career Opportunities**

Copyright © 1964, 1973 by McGraw-Hill, Inc. All rights reserved. Copyright 1952 by McGraw-Hill, Inc. All rights reserved. Printed in the United States of America. No part of this publication may be reproduced, stored in a retrieval system, or transmitted, in any form or by any means, electronic, mechanical, photocopying, recording, or otherwise, without the prior written permission of the publisher.

67890 KPKP 79876

This book was set in Times Roman by Black Dot, Inc. The editors were William P. Orr and David Damstra; the production supervisor was John A. Sabella. The drawings were done by Black Dot, Inc.
The printer and binder was Kingsport Press, Inc.

# Contents

|  |  | *Preface* | ix |
|---|---|---|---|
| **PART ONE** |  | **Introduction** |  |
|  | 1 | Forests and Man's Homeland | 3 |
|  | 2 | Forestry throughout the Ages | 12 |
| **PART TWO** |  | **United States and World Forestry Overview** |  |
|  | 3 | The Development of Forestry in the United States | 27 |
|  | 4 | Major Forest Regions of the World and of the United States | 53 |
| **PART THREE** |  | **Forest Biology** |  |
|  | 5 | Forest Ecology | 79 |
|  | 6 | Forest Care and Use | 97 |

|  |  |  |
|---|---|---|
| | 7 Protecting Forests | 119 |
| | 8 Chemical Ecology | 143 |
| | 9 Protecting Soil and Water | 156 |
| | 10 Wildlife and Range Management | 174 |

### PART FOUR  Forest Management

| | | |
|---|---|---|
| 11 | Appraising Forest Resources | 207 |
| 12 | Harvesting Timber Crops | 224 |
| 13 | The Business of Growing Timber | 238 |
| 14 | Forest Administration | 251 |

### PART FIVE  Forest Products

| | | |
|---|---|---|
| 15 | Wood: Its Nature and Uses | 267 |
| 16 | Wood Processing and Secondary Products | 284 |
| 17 | Wood Chemistry | 307 |
| 18 | Forest Products in World and National Economy | 327 |

### PART SIX  Forestry and Society

| | | |
|---|---|---|
| 19 | Social Benefits of Forestry | 357 |
| 20 | Forests and Outdoor Recreation | 371 |
| 21 | Forest Policy | 386 |

### PART SEVEN  Forestry Education, Research, and Employment

| | | |
|---|---|---|
| 22 | Education in Forestry | 405 |
| 23 | Research in Forestry | 421 |
| 24 | Employment in Forestry | 436 |

*Index*  453

# Preface

The decade of the 1970s may become known as the time when literate man, worldwide, came to appreciate his place in the life systems of the earth. He came to recognize as never before the roles played by the sun, oceans, soil, and biota in making the earth habitable by man. He saw in the forest more than timber to cut, game to pursue, water to tap, forage to be grazed, and outings to be enjoyed, important though he deems these to be. In addition, the forest became in his mind a vast dynamic system for transforming energy, building and conserving soil, perpetuating life forms, filtering, cooling, and humidifying air, masking noise, regulating stream flow, challenging man's scientific curiosity and spirit of adventure, and appealing to his esthetic sensibilities. Many people began to question whether these broad service functions of forests did not require priority over timber use. Some even advocated use of substitute materials for wood to preserve the forest.

But substitute materials are mostly finite and exhaustible, not renewable through growth. Most of them require large amounts of power to extract, refine, and process, and these operations usually result in polluting the air and water. Few substitute materials are biologically degradable when their service life is spent. Wood, aside from being a superb construction material and the most plentiful fiber for paper manufacture, has two very desirable properties—it burns and it decays, and its basic chemical substance thereby becomes available for recycling.

The purpose of this book is to present a view, in balanced perspective, of man's uses of his forests. The flow of energy and chemicals through forest ecosystems is described, the extent to which products may be removed from the forest with negligible cost in long-term forest benefits is brought out, and some measures for monitoring changes are indicated.

This revision has been extensive. Chapters on forest ecology, forest measurements, and chemical ecology have been added to help the reader appreciate the dynamics of forestry as applied quantitative ecology. Quantitative ecology is relatively new. Foresters have dealt in quantitative terms since ancient times, but spectacular recent advances offer promise of eventual global monitoring of forest biomass. It is such possibilities as these that add new excitement to forestry and place it in a position to make indispensable contributions to environmental science.

The author wishes to reaffirm his gratitude to the many people who contributed significantly to the first two editions of this book. Several of them, as well as many others, have given indispensable help on this edition. The author wants to express his gratitude also to foresters throughout the world who have served with him on work assignments, accompanied him on field studies, and participated with him in seminars, discussions, committee assignments, and world congresses. Outstanding among these men have been the foresters of the UN Food and Agriculture Organization who have personally acquainted him with the achievements of foresters in the Americas, Eurasia, and Africa. It is through this help that the author has been able to try—in this edition—to give worldwide perspective to forestry.

Chief Edward P. Cliff and H. R. Josephson, Director of Economics and Marketing Research of the Forest Service graciously permitted use of preliminary information from the current study of nationwide forest resources and the adequacy of these resources to meet future needs. Other Forest Service officers with whom the author conferred in person or in writing include members of the Denver Regional Office, and of the Northeastern, North Central, and Rocky Mountain Forest and Range Experiment stations. He has conferred with faculty members in forest, range, recreation, and wildlife management of Colorado State University. Mr. David Rice, Executive Secretary of the Colorado Livestock Association, was most helpful in acquainting the author with recent changes in the livestock industry and in the use of the western range. George Schroeder of Crown Zellerbach Corporation and industrial foresters of the South have furnished helpful information on the impact of mechanized operations on silviculture. The author has also conferred with a number of forestry and related experts whom he has met at national and international meetings concerning subjects dealt with in this book.

It would have been impossible to deal with new developments in the several fields without the help of department chairmen and many of their associates at State University of New York, College of Environmental Science and Forestry, Syracuse. Professor Robert M. Silverstein reviewed and gave helpful comments on Chap. 8. Adjunct Professor Leon S. Minckler has reviewed and furnished comments on the entire manuscript.

Photographs were arranged for by Edward Cliff of the Forest Service and John Farrell of the Department of Interior. Others who furnished indispensable material were Neil Stout of the Bureau of Outdoor Recreation and Donald R. Theoe of the Society of American Foresters.

To all of these the author is deeply grateful.

*Hardy L. Shirley*

PART ONE

# Introduction

Chapter 1

# Forests and Man's Homeland

Man's home is the finite earth made life-supporting by the energy received from the sun. Half the solar energy that reaches the earth's surface is used to evaporate water and redistribute it over sea and land. Green plants use somewhat less than 1 percent of this solar energy to make from the air, water, and soil minerals the food and fiber that support life. Animals feed on plants and other animals. They burrow in soil, drill holes in trees, carry seed and pollen, and help to break down plant and animal remains. Completing the energy cycle are the decomposers, the bacteria and fungi that release the last of the energy stored in plant and animal remains, thus freeing the basic chemical components of living things. At this point a new infusion of energy from sunlight enables green plants to start the wheel of life on another revolution through vegetation, herbivores, carnivores, and decomposers. This is the cycle of life on the earth. Its participants, together with their environment, make up the earth's ecosystem.

Since man appeared on earth he has been a part of the earth's ecosystem, and he will continue to be part of it until he becomes extinct as a biological organism. Whether he lives well or meagerly depends on how effectively he modifies the ecosystem to make it serve his needs.

The earth's ecosystem is complex. It embraces all life and all that supports life. Although man can never hope to discover all the earth's secrets he can, through systematic observation and research, learn much about his earthly home and how it has responded to his manipulations of it. Not all his actions have been beneficial. Discharge of polluted effluents into streams has often overtaxed decomposing organisms. Smoke and automobile exhausts have made the air toxic to man and plants. Unchecked demands such as these on the earth's ecosystem can lead to disaster.

## SURFACE FEATURES OF THE EARTH

Oceans, seas, and lakes cover 71 percent of the earth's surface. The land surface consists of:

| Type of surface | Area, millions of sq mi |
|---|---|
| Polar ice caps and wastelands | 6.5 |
| Deserts | 18.1 |
| Croplands | 5.4 |
| Grasslands | 12.0 |
| Forests | 16.4 |
| Total land area | 58.4 |

Man can support highly organized societies in desert and polar regions only by drawing heavily on resources of croplands, grasslands, and forests. These constitute his homeland.

**Cropland** Cropland is man's most precious land resource. It produces his cereals, vegetables, and fruits; his cotton and flax. It should be guarded jealously for, once impaired, it cannot readily be restored. Cropland need not be fragile if properly used. Wheat fields of Anatolia that supplied grain for the Roman legions before the time of Christ yield as generous crops now to their modern Turkish farmers as they did long ago to their pre-Christian ones. But many former wheatlands are now useless. Their topsoils silted up the harbors of Ephesus, Nicomedia, and Tarsus. In America, also, tilling steep slopes has led to eroded soils and impoverished people. A growing threat to cropland are the sprawling cities and suburbs with their supporting industries, airports, and highways. These spread first onto the deep-soiled cropland before moving onto the forested hills.

**Grasslands** Herding cattle, sheep, and goats was one of man's earliest efforts to assure himself a dependable supply of food and clothing. Nomadic people still follow their herds and flocks in the Mediterranean area, Afghanistan, Pakistan, India, and Africa, living much as their forebears did 2,000 and more years ago. The rock-protruding slopes they overgraze react, as stones, loosed by cloven hoofs, go bounding down the mountain. But for all the idyllic charm of the nomadic way of life, modern society offers less onerous ways of making a living than that of following sheep in solitude over rock-strewn mountainsides. Modern range and herd management practices have increased by severalfold the number of domestic stock one man can care for. Such practices make possible continued use of the best grasslands by domestic livestock. Those grasslands ill-suited for livestock may support populations of wild free-ranging herbivores. Where it is not overgrazed by livestock or wildlife, the grass cover can protect the soil, assure a usable water supply in springtime, and yield a harvest of animal protein.

**Forests** Almost one-half of the earth's habitable land space is tree-covered. Where trees grow closely spaced with a continuous crown canopy, they form forests. Where they grow openly spaced without a continuous canopy, they form woodlands. Forests and woodlands once covered most of what is now cropland and much of what is now pastureland as well. Forests generally provide the greatest degree of soil and watershed protection of any plant cover. They are the most effective transformers of sunlight, air, and water into carbohydrates, waxes, and fats. Their leaf surface acts as a great outdoor air conditioner, absorbing heat as tree leaves give up moisture to the atmosphere. Leaves filter dust particles and gaseous pollutants from the air and reduce apparent noise levels. Trees' sturdy trunks are the most common source of building materials for single family dwellings. Wood in its natural state has thousands of uses as man pursues his highly diversified activities throughout the world. Wood supplies the chief paper-making fiber that fills hundreds of special needs in modern society (Fig. 1-1).

Forests preserve the seed stock of many plants and animals: flowers, birds, small mammals, and the decomposing organisms that perform an indispensable service in the life processes of ecosystems. Forests also add to life's amenities by softening and beautifying the landscape.

## FORESTRY

Forestry is the art, the science, and the practice of managing the natural resources that occur on and in association with forest land for the benefit

**Figure 1-1** The Pristine Forest, Nooya Creek Flat Tongass National Forest. *(U.S. Forest Service.)*

of mankind. Through the ages the practice of forestry has been an ever-expanding task. At the time of King Solomon foresters were concerned mainly with harvesting and extracting timber. The Roman Emperor Hadrian used them to establish forest boundaries and to protect the forests from overcutting and overuse by herdsmen. Foresters of the Middle Ages were charged with protecting and looking after the game on lands owned by kings and nobles. Modern foresters engage in such diverse activities as avalanche and torrent control; and forest planting to arrest migration of dunes, to shelter field crops and farmsteads from wind, to protect municipal water supplies, to stabilize eroding slopes, to protect soil, and to reduce damage caused by floods. Decade by decade new tasks are added as man appreciates more completely the benefits forests confer.

Forests provide products, services, and amenities. Forest products include wood, nuts, fruits, mushrooms, litter for cattle bedding, pharmaceutical plants, gums, resins, leaf extractives, greenery, forage, and many other plant products. They include also the game, fish, furbearers, and other useful wildlife. Potable water is one of the most valuable products of forests, especially in regions where rainfall is meager except on the forested mountains. Forest services include recreational use, wilderness use, scientific use, soil and watershed protection, and environmental protection and enhancement. Amenities include esthetic landscapes, the fascinating fall colors of hardwoods, the forest-dwelling birds, the wild flowers, and the many other features that entice people to visit the forest to enjoy natural beauty.

Foresters have the responsibility of protecting the forests from fire, insect pests, diseases, and damage from animals and storms. They have the broader task of protecting the environment that is greatly influenced by the presence of forests. The forester is in fact a land manager responsible for all the goods and services that flow from the forest.

This land manager's task embraces both production and marketing, or utilization. Production involves growing timber and forage, producing usable water, providing food and cover for wildlife, developing facilities for recreational use, and ensuring that the forest makes a positive contribution to environment. Marketing involves harvesting timber, forage, wildlife, and water in such quantities as are currently needed—with the proviso that present harvesting not jeopardize future supplies. It also means opening the forest to recreational use and acquainting the public with the opportunities afforded. It is the revenue received from sale of timber, forage, hunting, and camping permits and other marketable products and services that enables the private forest operator to meet his costs for producing them. It is the public benefits that the people get from forest products, services, and amenities that induce governments to incur the costs of forest protection and management. The forester must be a good salesman in the best meaning of the term if the people at large are to benefit to the greatest degree from the nation's forests.

The physical tasks involved in multiple resource use are so varied and diverse that specialties have developed in the forestry profession. Wood use has become the province of the wood technologist, range use of the range manager, soil protection and water management of the forest hydrologist, habitat improvement for game of the wildlife manager, and recreation facilities of the landscape architect. All share the common responsibility, along with forest entomologists, pathologists, meteorologists, fire control experts, and forest ecologists, of explaining to the public the nature of forest ecosystems and how they benefit man.

The forester's aim is to manage forests so that they can continually provide the optimum combination of products, services, and amenities for man's use and enjoyment. To do this successfully he must employ the mathematical, physical, biological, and social sciences and their technologies. He must have the artist's eye for esthetics, the jurist's sense of equity, the politician's sensitivity to human reactions, and the administrator's skill in making wise decisions amid varying degrees of complexity and uncertainty. It is a demanding task, a challenging task, and at times a frustrating task. It is important that it be well performed if man is to live well on this earth.

## THE TREND IN FOREST USE IS UPWARD

Since 1945, *wood use* in developing nations has increased markedly and the rate of increase has also accelerated. In 1968, wood use in these nations remained far below that of nations with developed economies, although the latter, too, were increasing their wood use. By the year 2000 the Forest Service projects that the United States may require double the 1960–1970 use of lumber, triple that of veneer and plywood, and quadruple that of paper and fiberboard. At the same time, needs for food, dwelling space, transportation, and industrial production will add to the pressure on the nation's land resources. Planned action is urgent, for the wood to be harvested in the year 2000 will come largely from the seedlings, saplings, and poles in the forests of the 1970s (Fig. 1-2).

*Water use* seems likely to increase more rapidly than wood use. Mainly, it offers a means of substantially increasing the food supply while only slightly impairing the environment. Industrial and municipal use of water is due to increase markedly in both developing and more economically advanced nations as both industrialize rapidly, with an accompanying growth in population and consumption. Water use for power development is certain to increase also. This will mean added hydroelectric

**Figure 1-2** Frame house construction using plywood sheets, Santa Barbara, Calif. *(U.S. Forest Service.)*

developments and increased use of water as a coolant in fossil- and atomic-fueled power plants. Man must rely largely on forest lands for such water.

*Forage* for domestic livestock comes mostly from the world's grasslands. In many nations, these grasslands have been overused for centuries. Domestic flocks have been pastured in forests also. A worldwide, intensive program of rangeland management is needed and is being supported by the United Nations Food and Agriculture Organization. Improvement will be difficult to achieve, for practices followed for centuries can rarely be quickly changed no matter how destructive they may be.

Grazing by domestic livestock in the forests of the western United States seemed to have reached a peak by the 1960s but it was still expanding in the South. Because of the marked improvement of the western range since 1937, and especially since 1950, grazing seemed to be causing only limited damage to watersheds in the 1960s. Elsewhere in the world grazing damage varied from region to region. In western Europe few livestock graze in the forest. In Latin America, Africa, and Asia, grazing linked with mountain agriculture was causing considerable erosion and silting of reservoirs during the 1960s and early 1970s. The task of rehabilitating overgrazed forests and grasslands with rapidly increasing demands for food will tax the best efforts of modern forest and range managers.

*Wildlife* adds to forest amenities enjoyed by both sportsmen and recreationists. Both would like to see it increase. A well-balanced wildlife population performs valuable services in the forest that would cost man considerable effort to perform by himself. Unbalanced populations, on the other hand, may deplete their food supply, destroy tree seedlings and saplings, and even damage timber of merchantable size. With growing demands for timber and other forest uses, the need increases for managing wildlife to protect its health, services, and contribution to forest amenities.

*Recreationists* yearly seek the forest in growing numbers. Facilities provided for their use are frequently overtaxed even though extensive programs of expansion have been completed. Forest recreation is enjoyed by people of many nations. It accounts for substantial revenue to countries that have outstanding recreational features, such as the fjords of Norway, the Alps of Central Europe, and the game parks of Africa. Most recreational use of the forest is made by people who live within motoring distance of public and private forests that are open to them. Providing picnic areas and camp grounds in state and national forests is now expected by the American public. Forest managers who have gone beyond this to interpret the forest to the visitor have experienced favorable public response. As forest recreation seems destined to increase it becomes urgent to harmonize such use with the other demands made upon the forest.

## PROTECTING THE ENVIRONMENT

A responsibility of foresters is to regulate forest use so that little impairment of environment occurs and so that what does occur is soon repaired. The four most common sources of damage to forests are (1) wildfire, (2) wind and ice storms, (3) outbreaks of disease and insect pests, and (4) logging. The forester can minimize damage from the first three causes by keeping the forest in a vigorous growing condition. He can repair damage by salvaging dead timber and promptly reforesting the area affected. New approaches to pest control in the forest were being rapidly developed in the early 1970s.

Logs are heavy, bulky products that require powerful machinery for their handling and removal. Careless logging can cause considerable damage to residual trees and can result in new channels for erosion and runoff. By careful operation followed by slash disposal, stabilizing of skidways and logging roads, and prompt reforesting, damage is minimized. Timber harvesting, if properly done, is less disturbing to the forest ecosystem than wildfire or serious outbreaks of disease or insect pests that may otherwise occur.

Clearly the care given our nation's forests will have a significant effect on environment. Protecting the environment is a task to be shared by foresters with associated scientists, landowners, timber operators, and other forest users. The public at large must help by recognizing that there is a limit to how much human consumption can be increased without causing irreparable environmental impairment. There are two possible courses of action. One is to slow the increase of population and consumption. The other is to mobilize man's scientific talent to work creatively with ecosystems. Both should be followed if human privation is to be minimized. The tasks ahead will not be met effectively by improvisation. They require sophisticated research on a teamwork basis matched by courageous action. Man need not despair, for never have tools for meeting crises been more varied or effective than they are today.

## SCOPE OF TREATMENT

The foregoing pages have described in broad outline the space occupied by forests on the earth, the contributions they make to man's welfare, and the care man must exercise if he is to enjoy the benefits of forests during his tenure as a life form on the earth. The following chapters expand on the magnitude and nature of this task. How forestry came about and developed in various countries, including the United States, is sketched with some emphasis on why forestry must necessarily have the backing of government to attain stature and permanence in any land. Examples of substantial social benefits that nations, states, and communities have enjoyed from applied forestry are given to indicate what forests can contribute to the welfare of society and to the prosperity of industries and people.

The technology of forestry is introduced mainly to acquaint the reader with its scope and with the many scientific disciplines from which this technology is drawn. Some of the contributions that foresters themselves have made to basic science are mentioned to show the close relationship that prevails between the forest investigator and his colleagues in many branches of basic and applied science. The technology of wood processing is outlined to call attention to the impact of such processing on the environment and to suggest some of the opportunities for new developments in the future. The role played by forest industries in national and world economy is dealt with to indicate the measure of their contribution to mankind. Effort is made in the chapters dealing with silviculture, forest protection, wildlife and range management, forest research and chemical ecology, to introduce the reader to the complex interrelationships and biologic dependencies that occur in the forest and to show why the forester must develop a sensitivity to the existence of these relationships that goes beyond mere book knowledge. As the title of this book implies, the education required and the types of careers that forestry offers are sketched in outline.

The conflicts that occur in forest use are recognized and methods by which working agreements might be achieved to relieve these conflicts are suggested. Some forest benefits are tangible and can be measured. Others are intangible, cannot be measured, yet are nonetheless real. Where one forest use conflicts with another, as timber harvesting and intensive recreational development do, the forest manager may need the help of both polltakers and opinion formers in order to arrive at any acceptable plan of action.

Beyond this lies the philosophical question of how man is to adjust his demands on the forests to what the world's forests can sustain. Increasingly man must learn to treat the earth's ecosystems with understanding and respect. The complex human societies that man has created must adjust to the broader ecological societies of which they are a part.

Chapter 2

# Forestry Throughout the Ages

The attitudes of man toward the forest and the laws he has established to govern its use have their roots deep in his biologic needs and social makeup. Rights of use and landownership patterns are conditioned by practices dating back beyond the dawn of recorded history.

**PRIMITIVE MAN AND THE FOREST**

Prehistoric man's relation to the forest can only be surmised from the relics he has left and from the use primitive tribes today make of forests. It is possible to weave certain facts into a plausible account of this relationship. For example, the earliest known men, ground rather than tree dwellers, did use wood. Both Peking man and Java man used wood for fire at least 750,000 years ago. In addition, the Olduvai Gorge in Tanzania is the site of the earliest remains of manlike beings, known as Zinjanthropus.[5,10] Associated with these remains are specimens of petrified wood. The Gorge is in a savanna country interspersed with woodlands, which may have formed narrow ribbons of forest along the water courses in the past as they do today (Fig. 2-1). During the dry season the smaller streams would shrink to scattered water holes where animals would come to drink. Here the animals could be trapped in much the same ways as the pygmy tribes of the Ituri Forest use today.

The oldest known use of wood in human dwellings was in Africa at the Kalambo Falls site in Tanganyika. This dwelling was occupied some 60,000 years ago. Well-preserved wood was found in place in the living quarters. The wood was preserved in water-saturated soil many feet below the present surface.[3]

Early organized communities were found along watercourses that flowed through arid lands in Egypt, Mesopotamia, Pakistan, and India. Here sod huts were erected and cultivation of heavy-seeded grasses, the cereals, began. Scattered trees probably grew along the river banks as they do today. From these trees fuel, handles for tools, and timber for construction could be made. As these trees were used up, man probably sent expeditions farther and farther upstream in search of wood, which could be floated down to his villages.

Cereals were the basic element that permitted organized society to develop. They provided a concentrated food, easily cultivated, and

**Figure 2-1** Olduvai Gorge, Tanzania, site of remains of *Zinjanthropus*, the oldest manlike being.

capable of storage. The river borders that supported grasses attracted birds and animals useful for food. Man later learned to domesticate some of these animals as well as plants. Thus a civilization arose.

Separation of the forest dweller and plains dweller occurred early. The former lived a more primitive life but developed considerable skill in the use of his forest environment. At first he kept to the stream banks, where he learned to use logs, then rafts, and finally dugout canoes to float himself and his belongings from one place to another. As his skills developed in hunting, and in the use of tools and fire, he penetrated the forest, where he made crude clearings by girdling and burning trees. He also took advantage of natural clearings caused by windthrow, insect attack, or other destructive agents. Here he planted small fields and so extended his dominion over forest as well as plains.

Man learned to use shells, sinews, skins, vines, bark, twigs, grasses, clay, stones, and wood to fashion containers, clothing, simple tools, weapons, shelters, rafts, and other watercraft. Berries, fruits, nuts, and seeds added to his diet. Some plants were found to have medicinal properties that eased his pain and soothed him when ill. Others provided poisons with which he tipped his arrows to slay huge animals for which he otherwise was no match.

Primitive man learned to be a close observer of his natural surroundings. This was necessary if he was to obtain food and find his way in the forest. He learned to recognize many kinds of trees. Timber inventory parties of the United Nations Food and Agriculture Organization make use of this skill today by employing local forest-dwelling men as tree namers in the Amazon jungle and elsewhere. One tree namer was said to recognize more than 100 species and was able in some cases to place a tree unknown to him in the correct genus or family.

As the river villages prospered through use of grain and domestic animals, their need for wood increased. They desired workshops, trading centers, temples, and council chambers. By the dawn of historic times wood was extensively used for housing, furniture, sarcophagi, and carts. This is attested by the many objects skillfully fashioned from wood found in ancient Egyptian tombs. Some are still in a perfect state of preservation. The writers of Greek and Roman times as well as of the Bible make frequent reference to use of wood, including instructions on how to handle it to impede decay. Pictures in Egyptian tombs show use of the plow and other wooden tools to prepare land for sowing. They also show carpenters and shipwrights at work sawing wood and fabricating boats as early as 2700 B.C. A Roman votive stone of the first century A.D. shows, along with saws, adzes, axes, and other wood-working tools, what appears to be a hard hat worn to protect workers from flying chips, just as such hats are used by loggers and construction men today. Illustrations of these, together with a lively discussion of how trees were felled, skidded, sawed, and the lumber made into buildings is given by Makkonen.[8] He also gives information on work standards and performance requirements in effect at the beginning of the Christian era. Theophrastus, Varro, Pliny, Cato, and Virgil all wrote extensively on trees, their classification, manner of growth, description, structure, nutrition, reaction to soil and climate, and the classification of soils for various uses. They also recommended the species that were deemed best suited for various purposes.[8]

## STAGES IN USE OF FORESTS

Primitive man was apparently largely content to accept his food and shelter from the forest with little effort to modify his environment to suit his needs. The forest was, in fact, too strong for him to displace it. Only by using fire could he weaken the forest's hold on the land.

**Shifting Cultivation**  With the beginning of agriculture man found or made clearings in the forest for his crops and domestic animals. As one cleared area became infertile, he abandoned it for another. The abandoned fields were soon reclaimed by woody growth. The cycle was often repeated, giving rise to a system of shifting agriculture that persists in many places even today. It was in common use by North American Indians.

As his skill in husbandry developed, man learned how to keep land continuously productive, and so began a settled existence.

**Property Rights**  Permanent occupancy and cultivation of the land often led to the development of property rights. Forest land was at first held as communal property where any man might harvest such timber products as he required. When fuel wood became scarce, however, property rights became important in the forest. These rights took two forms: small individual holdings, and community forests, the use of which was controlled by village authorities. The growth of villages into cities and the development of agriculture itself required the use of forest products for construction and general living. Thus forests increased in value. Lumber and other timber operations solidified the rights of ownership.

Unrestricted cutting let to local wood shortages. Gradually such shortages spread. Timber shortages in China probably occurred as early as 5,000 years ago; in Egypt, 4,000 years ago. The Greeks and Romans felt timber shortages more than 2,000 years ago, and the central Europeans 1,000 years ago.[2]

**Cutting Restrictions** Timber shortages induced farsighted rulers and landowners to regulate cutting in an effort to perpetuate the timber supply. Such regulations were made by the Chinese priests, by the Roman emperors, and by the Egyptians before the time of Christ. They were carried out in France, Germany, and Great Britain before the time of the Crusades. The Chinese philosopher Mencius, who lived from 372 to 289 B.C., stated the case for conservation in his advice to King Hwuy of Leang.[1]

> If the seasons of husbandry be not interfered with, the grain will be more than can be eaten. If close nets are not allowed to enter the pools and ponds, the fishes and turtles will be more than can be consumed. If the axes and bills enter the hills and forests only at the proper time, the wood will be more than can be used. When the grain and fish and turtles are more than can be eaten, and there is more wood than can be used, this enables the people to nourish their living and bury their dead, without any feeling against any.

The stages mentioned above have not always occurred in sequence; nor have all stages appeared in every forested area. They are well exemplified in the history of the cedars of Lebanon as it is reconstructed by Lowdermilk.[6]

> King Solomon is recorded to have made a bargain with Hiram, King of Tyre, to exchange wheat and olive oil out of his Kingdom, including Palestine, for timber of fir and aromatic cedars of Lebanon for the construction of the Temple at Jerusalem. According to the record Solomon furnished 80,000 lumberjacks as hewers of wood for the wood operation in the forested mountains and 70,000 burden bearers to skid out the timbers. (Fig. 2-2)

The forests of Lebanon were incorporated into a state forest under the reign of the Emperor Hadrian, and efforts were apparently made to stop indiscriminate cutting. Throughout Roman times the forests were used to build great temples and ships that sailed in the Mediterranean. After the Roman Empire fell, conservation works were again neglected and timber cutting was unrestricted.[7]

**Forest Destruction** Grazing by goats and sheep completed the destruction of the forests of Lebanon. "Since the Arab invasion, thirteen centuries ago, the forests of Lebanon have been exploited and neglected

**Figure 2-2** Pristine stand of cedars of Lebanon in experimental forest in southwest Turkey.

until the slopes have been rendered barren, gray, and rocky, making impossible the restoration of forests equal to the original stands."[6]

What happened to the forests of Lebanon happened also throughout the forests of the Mediterranean region. Destruction of the forests by unregulated cutting, fire, and grazing of sheep and goats caused serious soil loss, silting of stream beds and harbors, and conversion of forests to an unpalatable brush cover known as "maqui."

Mute evidence of the economic loss sustained by such misuse of land is attested by the ruins of Greek and other ancient cities scattered throughout the region. Perge, Aspendos, Side, Ephesus, Pergamon, and many others had theaters seating as many as 30,000 people. Even with modern transportation few such theaters could be filled by drama-loving people now to be found in the environs. Populations of many areas are now barely 10 percent of those supported during Greek and Roman times.

Along the Indus River in West Pakistan, archeologists have uncovered three levels of settlement, each in its turn buried by flood debris following the destruction of forest cover on the mountain slopes. Such forest destruction can lead to impoverishment of man both in worldly goods and in spirit.

**Forest Restoration** Beginning in the 1940s vast land reclamation was undertaken by Spain, Israel, Italy, Greece, Turkey, and the Magreb countries of North Africa to restore forests to the slopes laid bare by past abuse. It is costly work, often involving terracing and other measures to conserve soil and moisture, and the use of potted planting stock in the field. Even so it is doubtful if wood for industrial use can currently be grown on much of the land, at least during the first forest rotation. The major objective of this tree planting is to save what remains of the soil and to protect the watersheds.

The destruction of forests in the Mediterranean during a period of some 2,000 years has been duplicated in forests of the Andes in a period of less than two generations. Here the Indians had developed terracing to a high degree, but the Spaniards found such agriculture unprofitable. Gradually one piece of terraced land after another was abandoned and more and more virgin forests were cleared for cultivation. Here the steep soils unprotected by terraces eroded rapidly under the heavy tropical rainfall so that serious gullying and complete soil exhaustion were brought about.[9]

## EARLY FORESTRY MEASURES IN ASIA, EUROPE, AND THE AMERICAS

Mainland China is believed originally to have had forests extending over about 30 percent of her land area. Centuries of overcutting, overgrazing, and fires reduced the forest to some 7 percent of the land. Two major influences were at work to augment forest destruction: first, the dividing of the land into small parcels owned by individuals; and second, the general press of the population on the land. Freehold units in China reached about 2½ to 3 acres in size. These had to be intensively cultivated and irrigated to support the Chinese families. Whole mountainsides, particularly in the Yunnan Province, were cleared for many thousands of feet above the valley floor and were then terraced for agriculture. Eventually drought brought widespread starvation to the inhabitants, the terraces were neglected, and the hillsides became badly eroded. Stunted pines occupied the depleted uncultivated soil.[2]

The People's Republic of China has taken major steps to improve land use. Through nationalization of cropland, communal farms, major irrigation works, and other agricultural improvements the efficiency of food production is being increased and labor is being released for industry and major public works. The latter works have included reservoir construction and a huge forest planting program. Some 39 million acres are reported to have been planted between 1950 and 1957.[11]

Forests fared somewhat better in India than in China. The Buddhist religion, as well as the activities of Indian monarchs and Hindu priests, encouraged tree-planting programs.[2] The unappropriated forests were held by the Indian princes, who harvested such rare woods as rosewood, ebony, teak, and sandalwood for export to other countries.

Forestry entered the Mediterranean region before the time of Christ. The Egyptians imported a number of trees for forest planting and had a chief forester as early as 3500 B.C. Reference is made to keepers of the king's forest and to actual foresters in Assyria, Palestine, and Asia Minor. A number of sacred groves were maintained in Lebanon and elsewhere. The Romans took a lively interest in trees, importing tree seedlings from the Indias, Africa, Spain, and Germany.[4] Greeks and Romans alike maintained sacred groves of trees comparable to those in Carthage, Lebanon, and elsewhere.

The Aztec king Nezahualcoyotl, who reigned in the fifteenth century,[2] mitigated the severe penalties imposed on citizens who trespassed on the chieftain's forest. He allowed the tribes to gather wood from the ground, but harvesting of standing timber was punished with death.

By the eighth century kings of England forbade trespass on royal property under fine of 60 shillings and thereby established the banned forests. By the tenth century English kings extended the royal right to chase over fields, pastures, waters, villages, and woods. Ordinances were passed by the princes and barons governing the amount of wood that might be cut for family use and the uses to which such wood might be put, including charcoal making and burning for potash. Reforestation was practiced by Henry VII of England and by the medieval city of Nuremberg in Germany.

During the Middle Ages clearing of land for agriculture had gone so far that decrees in both Austria and Germany forbade further land clearing. The purpose, in part, was to protect hunting privileges of the nobility, but it was also to ensure a fuel-wood supply for local industries and home use. The first laws to prohibit the clearing of forests in Germany appeared in 1165.[4] These were followed by ordinances in 1177, 1226, 1237, and 1291. A great number of laws were passed thereafter. The early regulation laws were simple and designed primarily to prohibit clearing of forests. Later regulations forbade pasturing newly cut areas. Some regulations directed that buildings be erected of stone to conserve wood.

Early community forests in France go back to the time of the Gauls in A.D. 500 to 510. Clovis in the sixth century took over royal and state forests. Eventually church property became royal property. After 1791 royal property was declared national land, which, with the addition of church property, brought the total up to 4.3 million acres. Forest officers were appointed as early as 1291, and by 1669 the first regulations governing timber cutting on public and private lands were established.[4]

In Switzerland, community forest ownership has dominated forest property throughout its history. The first forest ordinance regulating use dates around 1304, and the first working plan—made for the Sihlwald near Zurich—has, with revisions, remained in effect for more than 600 years. Special regulations governing cutting of the steep Alpine region were introduced early and were rigidly enforced.

In Austria the city forests were placed under a form of management as early as A.D. 1100. Various regulations governing the felling of timber were adopted from time to time. A comprehensive forest ordinance adopted in the reign of Empress Maria Theresa in 1754 revised many of the restrictions on private landowners and on the management of communal forests.

Regulations of timber-cutting operations in Scandinavia date back to 1638. A forest-planting law was established in 1647. In Norway, attempts were made to stop forest devastation in 1500, and a forest survey was authorized in 1600 by Christian IV. By 1725 a forest and sawmill commission was established, and export restrictions were imposed in 1795. Danish forest laws date back to 1557. The state forests were placed under organized forest management as early as 1781.

**European Forestry up to 1900** The rise of revolutionary movements and the downfall of feudalism in Europe caused a pronounced change in the attitude toward forest properties. Feudal holdings were seized in many countries and made a part of the public domain. Church property was confiscated. At first, some states sought to liquidate these lands by sale to increase state revenue. Germany adopted such a course but early in the nineteenth century appointed Hartig as chief of the forest administrative branch of the treasury department. Hartig curtailed further dismemberment of state land and instituted forest management. Since the Swiss had most of their lands in communal ownership from the beginning, little change occurred there. France had a difficult time in stopping the overcutting and abuse that followed the French Revolution. However, early in the nineteenth century regulations were established governing forest use.[4]

Concepts of forest management developed rapidly in Germany, France and Scandinavia during the nineteenth century. Forest properties were subdivided into compartments, and working plans established governing cutting rates and cutting areas in efforts to obtain the highest return from the soil. In France, special attention was paid to protecting watersheds against landslides and erosion, and to stabilizing sand dunes.

Regulation of cutting on both public and private land was introduced by most European states during the late nineteenth century, if not before. State forests were placed under intensive management, and efforts were made to extinguish long-established rights of users, such as collecting litter and grazing livestock that reduced the yield of the forest. Important strides were made in private forestry during this period, in part because of state regulation and in part because forest properties became sound investments for their owners. Forests constituted the base of family wealth for a number of aristocratic owners. Entailed estate was established whereby the owner contracted to operate on a sustained-yield basis and to bind his heirs to do likewise in return for certain technical aid and tax inducements offered by the state.

Formal education in forestry began about 1825 when private forestry schools were established. These were an outgrowth of the old master schools that had existed before that time. In 1816 the Cotta Master School grew into the forestry college at Tharandt, which became one of the leading forestry schools in Germany. The French school of forestry was established at Nancy in 1825. It has been in continuous existence since that time. Other colleges of forestry were established during succeeding years of the nineteenth century.

Forest research began almost as early as forestry education and has been carried forward since that time.

Europe maintained its leadership in forestry until the outbreak of World War II. The war caused a temporary setback to forestry in Europe. Wartime cutting was under less strict silvicultural control than normal. Immediately following the war, local people cut fuel wood with little restriction. Occupation forces tended to impose practices aimed at supplying immediate rather than long-term needs. The national forestry authorities were, however. soon restored and the war damage was rapidly repaired.

**Forest Practice in New Countries**  Wasteful use of forest resources is still widespread in the world today. It is especially prevalent among nations formerly under colonial administration and among those having rich forest resources that have heretofore been only lightly used. A rich timber stand represents readily harvestable and exportable wealth that can supply foreign exchange for needed industrial development. Moreover, timber cutting clears land for agricultural use. New governments are quick to realize this. An example is the Philippine nation. This country, under Spanish domination for almost three centuries, and under the United States for half a century, since 1946, has been independent. Under Spanish rule, much of the fertile, level lands along the seacoast became incorporated in large haciendas operated for sugarcane, rice, and other high-yield agricultural crops. This ownership pattern was continued under United States rule and was accepted by the new Philippine government.

When independence came, strong pressure was exerted by the Huks and other dissident elements to launch a land reform program that would enable the great mass of agricultural workers to acquire ownership of land. President Magsaysay resisted the pressure to break up the existing large land holdings but did adopt a policy, under the slogan "Land for the Landless," of opening publicly owned land to settlers. Homesteads up to 50 acres in size could be acquired by landless people on tracts declared open for settlement. Most of the publicly owned land was virgin forest. As it was cut over, certain tracts were designated for entry.

Landless people from crowded islands were soon swarming over Mindanao, Basilan, and other islands where good soils existed. They soon took up all lands opened for entry. They then began to move in behind the loggers as the superb virgin dipterocarp forests were cut, not awaiting permission from the understaffed Bureau of Forestry. Local political powers, favoring land development, encouraged the settlers by defending them against ejection by forest officers.

Soon the situation was completely out of hand. Squatters and kaingineros move in following partial cutting of virgin timber. They girdle and burn the trees left for a second cutting to plant corn, sweet potatoes, and other crops on the steep mountain slopes (Fig. 2-3). Within two years' time the fertility of the soil is exhausted through cropping, erosion, and exposure to sun. A coarse native grass, cogongrass, invades the land and holds it against both agricultural crops and reinvasion of forest. The kainginero then deserts the land and moves again behind the loggers to repeat his act of forest destruction.

Thus, by an uncontrolled land program, political forces have been unleashed that are leading to destruction of forests, soils, and social fiber. These superb forests are capable of yielding annually upward of 1,500 board feet per acre of the finest hardwood timber to be found in the world. These virgin mountain lands could support at a high level of prosperity a permanent population of at least one-third as many people as the equivalent area of the best agricultural land of the nation. Instead, they are being rendered virtually useless for decades, if not for centuries, to come.

## SUMMARY

Man was a user of wood from the earliest times of which we have evidence of his existence. His earliest organized communities were along water courses through arid or semiarid lands where he could cultivate cereals yet still find wood for use.

Once such land was fully occupied or rendered infertile man entered

**Figure 2-3** A kainginero with his wife and daughter, sowing his first crop of upland rice. *(Martin Hurwitz.)*

the forest to clear land to cultivate and to pasture his herds. Much of the land that was cleared proved to be unsuited to permanent agriculture.

After decades of cultivation or intensive pasturing soil depletion occurred due to erosion and exhaustion of mineral nutrients. This led to new clearing until the wood supply was threatened. As the need for wood became critical, concerned rulers sought to protect the remaining forest by restricting cutting rights. Some also established forest plantations.

When such rulers were followed by those less concerned with forest protection or unable to exercise control over forest use, the people resumed the forest destruction process. In parts of the Mediterranean and elsewhere in dry regions this went so far as to render much land virtually useless.

Native vegetation in moist climates occupying a stable soil mantle is usually resilient rather than fragile. This accounts for the persistence of forests, heathland, and other extensive ecosystems in lands of heavy human occupancy. Here the value of the forest encourages rulers to protect it from destruction and to extend forests by planting lands exhausted by cultivation and grazing.

Wood-using industries have become strong enough to own and manage forests for their own wood supply. A new equilibrium between man and the forest develops whereby forest growth is adequate to sustain timber cutting.

As man's activities become more and more centered in cities, he seeks to establish forest groves for water supplies, recreation, and diversion.

All stages in man's occupancy and use of forests still exist in the world today from shifting cultivation and unbridled exploitation to forest planting for timber, soil protection and restoration, and recreational use.

## LITERATURE CITED

1. Balleau, Robert O. (ed.). 1939. "The Bible of the World," pp. 429–430, The Viking Press, Inc., New York.
2. Caldwell, David H. 1949. Unpublished manuscript, New York State College of Forestry, Syracuse, N.Y.
3. Clark, J. D. 1969. "Kalambo Falls Prehistoric Site," vol. 1, "The Geology, Paleoecology and Detailed Stratigraphy of the Excavations," Cambridge University Press, New York, 250 pp. Reviewed in *Science,* **169**:44–45, 1970.
4. Fernow, Bernhard E. 1911. "A Brief History of Forestry in Europe, the United States and Other Countries," rev. ed., University of Toronto Press, Toronto; *Forestry Quarterly,* Cambridge, Mass. 506 pp.
5. Leakey, Melvin M. 1966. Finding the World's Earliest Man, *National Geographic,* **118**:420–438.
6. Lowdermilk, Walter C. 1941. The Cedars of Lebanon—Then and Now, *American Forests,* **47**(1):16–20, 34.
7. Lowdermilk, Walter C. 1941. Conquest of the Land, U.S. Department of Agriculture, Soil Conservation Service, Washington. 34 pp.
8. Makkonen, Olli, 1967, 1969. Ancient Forestry. An Historical Study. Part I: Facts and Information on Trees, *Acta Forestalia Fennica,* **82**(3):1–84; Part II: The Procurement and Trade of Forest Products, *Acta Forestalia Fennica,* **95**:1–46.
9. Meyer, H. Arthur. 1945. Conuqueros—Forest Destroyers of the Andes, *American Forests,* **51**(4):178–179, 200.
10. Payne, Melvin M. 1966. Preserving the Treasures of Olduvai Gorge, *National Geographic,* **130**:701–709.
11. Richardson, S. D. 1966. "Forestry in Communist China," Johns Hopkins Press, Baltimore, Md. 237 pp.

PART TWO

# United States and World Forestry Overview

Chapter 3

# The Development of Forestry in the United States

The story of forestry in America is inseparable from the story of the early pioneers in their struggles to make homes in the wilderness of the New World. It is also a part of the story of their indomitable spirit to create a new nation on American soil. On the one hand, it reflects the actions of a vigorous people surging eagerly forward to exploit the rich resources of the new land for personal profit. On the other, it reflects the cooperative action to build churches, schools, villages, and states, and with community-imposed restraints to equalize opportunity for achievement.

Two goals lay before the young nation: to occupy the land and use its resources, and to free the creative spirit of individualism while still guiding it in the common interest. The first goal was achieved with the passing of the frontier; the second will remain as long as we remain a free people.

The white settlers who surged into North America brought with them two major concepts of land ownership: the small owner-operated fee-simple holding that developed in the North, and the large plantation holding worked by farm laborers or slaves that came to be a dominant pattern in the South. The Dutch West India Company introduced a third concept: the patroon system. Members of the company, upon locating 50 settlers within New Netherlands, received proprietary rights to 16 miles of frontage along the Hudson River and land extending back therefrom. The settlers were obliged to make annual payments to the patroon, as the landowner was called. The rights of settlers were ill-defined and the actions of patroons' agents deemed oppressive. Friction finally led to revolt of the settlers and abolishment of the system.

The independent farmer of the North feared that competition from the slave-worked plantations would rob him of economic opportunity. This fear induced him to support the War Between the States to abolish slavery. It also led to passage of the Homestead Act that offered free land to those who would settle upon it.

The North American Indians—who numbered but some 900,000 people in 1492—occupied both open prairie and forests. Some were primarily agriculturists cultivating beans, squash, maize, tomatoes, potatoes, and tobacco. They lived along the margin of the forest or in areas that could be easily cleared. Others were mainly hunters who followed the buffalo or deer and other forest game. These made rude clearings to grow food crops and tobacco. They used the forest much as forest-dwelling primitive people of Africa and South America do today. Lacking iron tools, the Indians could make little impress on the forest except by the use of fire. Generally their clearings were small and widely scattered. Forest Indians depended mostly on game for food. Indian holdings were on a tribal basis. In intertribal struggles over land, stronger tribes naturally displaced the weaker from the best hunting and croplands.

## FORESTRY IN COLONIAL TIMES

The early colonists found on the eastern coast of North America an almost unbroken forest that stretched westward to the prairies. Extensive forests also existed in the Rocky Mountains, on the West Coast, and in Alaska. Hardwood forests occupied some 403 million acres; coniferous forests 537 million. Open woodland occupied 93 million acres, making a grand total of 1,033 million acres of forest and woodlands. This was 46 percent of the total land area of the nation.

It is small wonder that the early colonist considered the forest his enemy. The forests were dense, the trees large, and many of them of hard and durable wood that could be disposed of only by burning. The virgin forests contained windfalls and tops of trees broken by storm. They occupied the land so completely that only by heroic effort could the pioneer wrest from the wilderness a few acres of land that he might

**Figure 3-1** Replica of old French fort erected in colonial days near present site of Syracuse, N.Y.

cultivate. Fire was his natural ally in the land-clearing process. Food crops were necessarily his immediate concern, along with fuel and shelter. Roughly hewn logs were used to erect outposts and cabins (Fig. 3-1). Lumber was first cut with man-powered whipsaws, but as early as 1625 and 1631 sawmills were operating in Virginia and in Maine. Lumber was one of the first major items of export to Europe and provided in exchange weapons and other manufactured goods. Transportation by sea was important not only to gain access to European markets but also for coastal trade among the colonies and with the West Indies. Ship building became a prominent industry of New England. By the time of the Revolution output was 100 ships per year.[4]

Though getting rid of timber by burning was still a main task of the frontiersman, along the settled seacoast oak and pine timber for ship building and other construction became a matter of concern. Only 6 years after the Pilgrims landed, the Plymouth Colony enacted a law forbidding the sale or transport of timber out of the Colony without the approval of the governor and the council. Restrictions on the cutting of ship timbers were promulgated by the Massachusetts Bay Colony in 1668. In 1681 William Penn incorporated in his land purchase contracts the provision "that in the clearing of ground care be taken to leave one acre of trees for each five acres cleared, and especially to preserve oak and mulberry for silk and shipping."[3] The restrictions of 1668 were reaffirmed in the Massachusetts Bay Colony charter of 1691. The best white pines were reserved for shipmasts of the Royal Navy. Such trees were to be marked with a broad arrow. This restriction was later extended to New York and New Jersey. The broad-arrow policy was difficult to enforce. Trespass was common even though culprits were severely punished when apprehended. The punishments seemed only to encourage the colonists to go even further in defying the king's authority.[8] During the Revolution patriots in Maine, Massachusetts, and New Hampshire did their utmost to prevent the British Navy from getting control over mast and spar timbers from American forests. In the end, their efforts had a considerable effect in crippling that navy.

As timber entered more and more into the commercial life of the colonials, timber transport became of general concern. The common method was to float it down the streams. To protect this right, streams large enough for log floating were declared to be navigable waters free from restrictions by riparian owners.[4]

Laws to protect forests from fire had been adopted by nine of the 13 colonies before the Revolution. Timber trespass laws forbade cutting on communal land and private land. Fines could be up to thrice the value of the product. Other laws were enacted to establish standards for forest products for export in order to protect the reputation for high quality merchandise in the European markets.[4]

Such measures were not without purpose. Fuel for iron furnaces, glass factories, and even for such cities as Providence and Philadelphia was often in short supply during colonial times. Benjamin Franklin in his article describing the Franklin stove wrote in 1774: "Wood our common fuel, which within these hundred years might be had at any man's door, must now be fetched near 100 miles to some towns, and makes a considerable article in the expense of families."[3]

The laws to restrict timber cutting were largely evaded, since emphasis was on land clearing and timber use. Agricultural and timber products accounted for the major economic activity throughout the colonial period. Wood was the chief construction material, and the colonists used the game, fruits, and nuts of the forest as freely as the timber. Independence, brought about by removing the restrictions of the mother country, was followed by even more reckless use and abuse of forests. These actions of American colonists during the first 120 years of independence were paralleled in most forest-rich, developing nations during the period from 1945 to the 1970s. In both cases forest wealth was used with prodigality to improve the livelihood of people.

## FEDERAL FORESTRY MEASURES

**The Public Domain and Its Disposal**   An important source of conflict among the colonies even before they had won their independence was the ownership of the lands. Virginia's original charter embraced "Land, throughout from Sea to Sea, West and Northwest." Massachusetts claimed land beyond the Delaware.[9] Connecticut and Pennsylvania disputed control over the rich farming land in Wyoming Valley, Pennsylvania, where Wilkes-Barre now is situated. Actual armed conflict broke out. To remove this cause of disunity, the Continental Congress in 1780 passed a resolution urging the states to cede their western lands to the Union. From 1781 to 1802 the state lands ceded to the federal government totaled some 260 million acres.[7]

Through the various land-purchase acts, including the Louisiana Purchase, the Florida Purchase, the various purchases and cessions of land from Mexico, and the Alaska Purchase, the federal government acquired a vast public domain. It included three-fourths of the continental United States and all of Alaska, a total of 1,800 million acres.[15] The public domain became both a source of wealth and a source of embarrassment to the federal government. It was sold for public revenue, granted to soldiers in lieu of cash payments, granted to companies to promote the building of railways, and granted to states to further education. It became a source of embarrassment, because our young nation had no trained land administrators nor any police force to prevent trespass. Squatting on the public lands was forbidden by an act of 1807. The early policy for this vast estate was to preserve it for public revenue and national development.

This attitude of the Congress and federal officers in Washington was in striking conflict with the desires and actions of the frontiersmen. They occupied land wherever it was not occupied by another and used it for their own purposes. A strong demand arose to open the public lands for private settlement. The Preemption Act, passed in 1841, gave homesteaders the right to squat on land and later to purchase it. During the administration of Abraham Lincoln, the Homestead Act was passed. This act authorized the granting of 160 acres free to settlers who lived upon the land and developed it. The Timber Culture Act permitted the government to grant 160 acres of land to anyone who would plant trees on 40 acres and keep them in healthy condition over a period of 10 years. The Timber and Stone Act, the Desert Land Act, and the Carey Act all liberalized the conditions under which settlers were able to acquire title to public land through homesteading.[15]

The Swampland Grant Act permitted the federal government to grant lands that were chiefly covered by swamp to the states in which they were located to be used for state purposes.

Railway land grants in alternate sections on either side of the railway resulted in a checkerboard pattern of ownership that has created legal and administrative difficulties for land management from that day to this. The Forest Reserve Act of 1897 contained a provision that permitted railway companies and other legal claimants to lands within reserves to select at their discretion tracts of equal size anywhere within the open public lands. Many claimants were prompt to use this act to improve the quality of their holdings. Some of our nation's best timberlands passed from public ownership under this act. Administration of all these acts was weak. Little attempt was made to determine whether homesteaders lived up to the regulations.

**Stemming the Land Boom** President John Adams upon authorization of Congress spent $200,000 to acquire lands bearing live oak timber needed for naval construction. Between 1817 and 1827 Congress authorized the President to close lands to sale and trespass that supported live oak timber. Though the President acted as directed, enforcement was weak. Sales of public land bearing live oaks continued. As an effort to protect the nation's forests or even to assure a timber reserve for the Navy, the action was abortive.[8]

During the colonial period and the first one hundred years of the Republic, prominent citizens urged public action to preserve forests for sustained crops of timber and protection of watersheds. Among those to urge action were members of the New York, Philadelphia, and Massachusetts Agricultural societies; Francois-André Michaux, author of "The Sylva of North America" published in 1819; Carl Schurz, Secretary of Interior from 1877 to 1881; Reverend Frederick Starr, who in 1875 advocated research to determine how to manage forests and establish plantations; and Governor of Nebraska Jay Sterling Morton who inaugurated the first Arbor Day in 1873. None of these evoked congressional action but their arguments helped to convince many concerned citizens that action was urgently needed.

**Beginnings of Federal Forestry** The spark that started the federal government upon a permanent program of forestry grew out of a memorial by the American Association for the Advancement of Science addressed to the Congress in 1874. This memorial was prepared as a result of a talk by Franklin B. Hough of Lowville, New York. He stated:

> The preservation and growth of timber is a subject of great practical importance to the people of the United States, and is becoming every year of more and more consequence, from the increasing demand for its use; and while this rapid exhaustion is taking place, there is no effectual provision against waste or for the renewal of supply. . . . Besides the economical value

of timber for construction, fuel, and the arts . . . questions of climate . . . the drying up of rivulets . . . and the growing tendency to floods and drought . . . since the cutting off of our forests are subjects of common observation. . . . [13]

In 1876 Congress, at the Association's urging, provided for a survey of the existing forest resources and the demands that were being placed upon them by consuming industries. Inquiry also was to be made into the influence of forest on climate and was to include a study of the best methods of forest practice used in Europe. Dr. Hough was employed by the Commissioner of Agriculture at the salary of $2,000 a year to carry out this task. He prepared three voluminous reports that contained the best information then available on America's forest resources and industries. In 1883 Dr. Hough was succeeded by N. H. Egleston, who prepared a fourth report. By 1881 Congress had established a Division of Forestry under the Commissioner of Agriculture.

Two new organizations to promote forestry came to the front at this time: the American Forestry Association and the American Forestry Congress. In 1882 they merged to sponsor an American Forestry Congress. This gala affair was held in Cincinnati.[2] Hough, Egleston, and B. E. Fernow, a forester with technical training in Germany, took an active part. Governors of many states participated. The meeting was climaxed by a huge parade and a tree-planting ceremony.

**Public Forest Reserves** Carl Schurz during his term as Secretary of the Interior repeatedly urged the reserving of public-domain timberland for long-term management. Numerous bills were introduced from 1876 on but were defeated by special-interest groups. A rider attached to the act of March 3, 1891, amending the land laws authorized the President to set apart from the public domain reserves of land-bearing forests, whether commercial or not. Gifford Pinchot was to call this act the most important legislation in the history of American forestry.[10] An active program of reservation was undertaken by Presidents Harrison and Cleveland.

At the request of President Cleveland the National Academy of Sciences set up a special forestry commission in 1896 to report on the forest reserves and to recommend measures for their proper use. Charles Sprague Sargent of Harvard University was chairman. The commission recommended increasing the reserve by 21 million acres but made no recommendations for managing them, an omission that Pinchot felt was serious. A storm of protest arose throughout western states over what the local people called locking up the public lands.[10]

This protest led Congress to suspend all but two of the reserves created by Cleveland. At the same time it provided for managing existing and future reserves. This act of June 4, 1897, stated: "No public forest reservation shall be established except to improve and protect the forest, secure favorable conditions of water flow," and "furnish a continuous supply of timber for the use and necessities of citizens of the United States."[13] This act received the support of the western people and laid the groundwork for a rational policy of managing and protecting the forest reserves.

**The Roosevelt-Pinchot Era** The attitude of the people toward forestry in the nineties was shaped by many outstanding personalities, among whom Fernow, Pinchot, Hough, Schurz, and Governor Morton were conspicuous. The country was ready for a huge conservation program and awaited only effective leadership from the White House.

This came with dramatic suddenness when President McKinley was assassinated on September 14, 1901, to be followed in office by Theodore Roosevelt. The new President had become imbued with the principles of conservation through his interests in wildlife and through a friendship he had formed with Gifford Pinchot, who became Chief of Forestry in 1898. Knowing of this interest, Pinchot and F. H. Newell, who later became the first director of the Reclamation Service, urged the President to include in his first message to Congress a strong conservation statement. The President agreed. His message contained the following statement about forests:[10]

> The fundamental idea of forestry is the perpetuation of forests by use. Forest protection is not an end in itself. It is a means to increase and sustain the resources of our country and the industries which depend upon them. The preservation of our forest is an imperative business necessity. We have come to see clearly that whatever destroys the forest except to make way for agriculture threatens our well-being.
>
> The practical usefulness of the National Forest Reserves to the mining, grazing, irrigation, and other interests of the regions in which the reserves lie has led to a widespread demand by the people of the West for their protection and extension.

The President went on to advocate that administration of the reserves be united in the Bureau of Forestry.

Under Pinchot's leadership the new Bureau of Forestry had grown in activities, size, and influence. Through its service to owners of large forest properties prominent citizens learned the meaning of forestry. Meanwhile the Forest Reserves in the Department of Interior received little protection and no forest management. Moreover the Department of Interior had been engaged more in the disposal of lands than in their management in the long-term public interest.

An American Forestry Congress was convened in Washington in January 1905. Participants included the Secretaries of Interior and

Agriculture, certain other government officials, conservation-minded citizens, and association leaders of lumbermen, railways, and stockmen. The Congress' resolutions urged a dynamic federal forestry program including the immediate transfer of the forest reserves from the Department of Interior to the Bureau of Forestry (Fig. 3-2). The Transfer Act was passed February 1, 1905. and signed by President Roosevelt on that day. A month later the Department of Agriculture was given full legal authority to regulate the forest reserves.

**Regulating the Use of National Forests** The day after the forest reserves were transferred to the Bureau of Forestry, the name of which was at that time changed to the Forest Service, the Secretary of Agriculture, James Wilson, sent a letter to the Chief of the Forest Service outlining the administrative policies to guide the management and use of the forest reserves. This letter has become a classic in federal forest policy.

> In the administration of the forest reserves it must be clearly borne in mind that all land is to be devoted to its most productive use for the permanent good of the whole people, and not for the temporary benefit of individuals or companies. All the resources of forest reserves are for use, and this use must be brought about in a thoroughly prompt and businesslike manner, under such restrictions only as will insure the permanence of these resources. The vital importance of forest reserves to the great industries of the western states will be largely increased in the near future by the continued steady advance in settlement and development. The permanence of the resources of the reserves is therefore indispensable to continued prosperity, and the policy of this department for their protection and use will invariably be guided by this fact, always bearing in mind that the conservative use of these resources in no way conflicts with their permanent value.
> 
> You will see to it that the water, wood, and forage of the reserves are conserved and wisely used for the benefit of the home builder first of all, upon whom depends the best permanent use of lands and resources alike. The continued prosperity of the agricultural, lumbering, mining, and livestock interests is directly dependent upon a permanent and accessible supply of water, wood, and forage, as well as upon the present and future use of their resources under businesslike regulations, enforced with promptness, effectiveness, and common sense. In the management of each reserve local questions will be decided upon local grounds; the dominant industry will be considered first, but with as little restriction to minor industries as may be possible; sudden changes in industrial conditions will be avoided by gradual adjustment after due notice; and where conflicting interests must be reconciled, the question will always be decided from the standpoint of the greatest good to the greatest number in the long run.[10]

**Building the Forest Service** Pinchot threw his dynamic energy into welding a strong Forest Service and building up a program of national

**Figure 3-2** Gifford Pinchot, Chief of the U.S. Forest Service 1905–1910. *(U.S. Forest Service, taken Feb. 7, 1909.)*

forest administration. The task was not easy. Cattle and sheep ranchers had been grazing their animals on the public domain without federal permit and resented federal intervention. A great deal of diplomacy and sheer courage were required of the early forest officers. By fairness, scrupulous honesty, and devotion to the ideals set forth in Secretary Wilson's letter, the Forest Service established itself as a responsible public land administrative agency. Through a vigorous research program new methods and viewpoints were constantly brought to the attention of administrator and user alike. Gradually the stockmen, lumbermen, water users, and a large segment of the general public came to see their relationship with the Forest Service as one of partnership in resource management and use.

While Theodore Roosevelt was President, a total of 148 million acres was transferred from the public domain to the national forests. Opposition developed in the West and soon made itself felt in Washington. Eventually Congress withdrew from the President the right to set aside new national forests from the public domain. But before this bill was signed, President Roosevelt set aside a large additional area in national forest—enough, he thought, to establish in federal ownership all important public-domain forest land.

***Integrated Resource Management*** Gifford Pinchot's wide-ranging activities and imaginative mind led him to recognize that man's welfare was intimately related to his use of land resources. The practice on farms, rangelands, and forests, he perceived, had their impact on soil erosion, floods, inland navigation, water power generation, fish and game supply, and even on man's extraction of minerals. Instead of each being a separate activity, all were embraced in the greater task of how to use the earth for man's long-term good.[10] Promptly developing and testing this idea among his associates he sought through President Roosevelt to give it national and international recognition.

A White House Conference of Governors was called in 1908. It led to the creation of state and national commissions on conservation by some 40 of the 46 states in the Union at that time. A National Conservation Commission was created by the President following the White House Conference of Governors to consider the four great classes of resources: water, forests, land, and minerals.

In 1909 President Roosevelt called a North American Conservation Congress. Representatives assembled from Canada, Mexico, and the United States. A declaration of principles was set forth by the Congress in which the representatives recognized that natural resources are not confined by national boundaries, that no nation acting alone can properly conserve them, and that all resources should be made available for the use and welfare of man.

Pinchot left the Forest Service in 1910. The two following Chiefs, Henry S. Graves, 1910–1920, and William B. Greeley, 1920–1928, continued the program of expanding the scope of the Service and building cooperation with the states, forest owners, and forest industries. Research was greatly expanded and strengthened in both forest land management and forest products. The policy of undertaking certain broad studies on a national basis with a strong research team was inaugurated with the Forest Taxation Enquiry in 1928. This study explored in depth the effect of the general property tax on forest owners who could expect a harvest only once or twice in a lifetime. From this study eventually came a series of state approaches to forest taxation that have provided relief to many owners. Even more significant was the inauguration of periodic nationwide studies of forest resources with a view to the demands for timber, water, and other forest products and services. The nationwide forest survey was established to collect field information to provide objectivity and accuracy to such reviews (Fig. 3-3).

***Federal-state Cooperative Programs*** Beginning with the Morrill Act of 1862, which was followed by other similar acts, the federal government

**Figure 3-3** Forest officers breaking camp, White River National Forest, Colorado, 1916. *(U.S. Forest Service.)*

granted first land, then money, to the states to establish colleges of agriculture and mechanic arts. Education, research, and extension service in agriculture in the states thus became areas for national support. As the young Forest Service had an immense task to protect and manage the national forests it seemed only wise to enlist the aid of the states to protect nonfederal lands from fire and pests, and to encourage the practice of forestry on private lands. The Weeks Law, passed in 1911, authorized the federal government to contribute to the states on a matching fund basis for the protection of forests from fire. The Clark-McNary Act of 1924 extended the provisions of the Weeks Law to include cooperation in forest extension, planting, and assistance to forest owners. Other acts were to follow that also extended federal assistance to states on a matching fund basis to cover (1) control of forest pests, (2) broad help to landowners and primary wood industries, (3) research, (4) public education in fire prevention, (5) acquisition of land for recreation and the development of forests for paid recreational use. By 1966 each of the 50 states was cooperating with the federal government on forest protection, research, extension, and aid to forest owners and industries.

**National Forests in the East**   The Weeks Law authorized the federal government to acquire private land for national forests where these were necessary for the protection of navigable streams. This provided the basis for the national forests of the eastern states. The authorization was later changed to include also lands needed for growing timber, and shortly thereafter the White Mountain Forest in New Hampshire, the Pisgah in North Carolina, and the Allegheny in Pennsylvania were acquired. Later the system was extended to include 43 national forests and national-forest purchase units with lands in 23 states east of the Rocky Mountains.

**Emergency Programs**   Forestry seemed to be advancing at a gratifying rate as a result of public action until the great economic depression began in 1929. The failure of corporations and banks, the foreclosures on farms and homes, the rapid loss of jobs, and the general paralysis of industry and trade spread fear of complete economic collapse across the land. Businessmen, governors, mayors, and unemployed veterans thronged to Washington to implore the federal government to take decisive steps to reverse the drift toward economic chaos.

In 1932 President Hoover requested and Congress appropriated money for a large expansion of public works to create employment. Under President Franklin D. Roosevelt, a vast new order of public works was begun in 1933, embracing river valley development of comprehensive scope, forestry on federal and state lands, park expansion, soil conservation, flood control, along with construction of schools, post offices, highways, airports, hospitals, sewage systems, and other needed public facilities.

*Civilian Conservation Corps.*   The most imaginative and useful program to forestry was the Civilian Conservation Corps. Camps for needy young people were established on national and state forests, national parks, and soil conservation districts. The men were engaged in tree planting; care of plantations; timber-stand improvement; construction of forest roads, trails, and lookout towers; and various other forestry measures on public land. At its peak, 2,600 camps, each organized to handle 200 men, were operating. The Civilian Conservation Corps continued until 1941 when the project was closed because of war. The Corps served to acquaint many people with forestry as a major government activity. It did a great deal to improve national and state forest lands and also provided young men with a wholesome life and useful training for private jobs.

*The Lumber Code*   The National Industrial Recovery Act of 1933 set up fair trade practices among industries. Article 10 in the Lumber Code under this act specified certain forest practices to be followed by industry, which involved leaving the forest in good condition to produce another crop. The act was declared unconstitutional in 1935. Lumber associations nevertheless urged their members to continue to comply with forest practice standards of the code and many did so.

During the terms as Chief Forester of R. Y. Stuart, 1928–1933, and Ferdinand Silcox, 1933–1939, federal forestry activities were maintained at a high level with many emergency programs and a general expansion of regular activities. Land purchase for national forests was very active, for many lumbermen found it too expensive to retain their cutover lands. The prairie-plains shelterbelt program resulted in the planting of 18,510 miles of windbreaks by 1942. Flood control surveys were begun. The Forest Service urged a program of increased cooperation with private and state forestry, expansion of public forests, and public regulation of timber cutting on private lands. This three-point program was to be advocated throughout the terms of Earle Clapp as Acting Chief, 1939–1943, and Lyle Watts as Chief, 1943–1952. Vigorous opposition by the forest industries, farm organizations, and others prevented passage of laws for federal regulation of forest practices on private lands but many states took action on their own.

The Taylor Grazing Act of 1934 and subsequent acts placed 147 million acres of the remaining public domain in districts to be administered for the forage resource. The Grazing Service was established in the Department of Interior for this purpose.

Other lands under the administration of the Department of Interior were given improved management, including those in Indian reservations. The first appropriation for fire control on the public domain lands of Alaska was made in 1939.

**Forestry Activities during World Wars I and II** The complete mobilization of the nation's resources for the prosecution of the world wars made a heavy demand on forest resources and industries. Lumber, fiberboard, and plywood for construction of cantonments, overseas bases, and packaging of war materials and supplies required almost half of the national output. Amounts needed for expansion of war industries as well as for housing industry workers used most of the remainder. Special efforts were made to expand output of these commodities during a time when both equipment and labor were short in supply. The nation looked mainly to private industry to keep the goods flowing at the needed rate, and this ready market for forest products during the war years effected a recovery of prosperity in the forest industries.

With the nation involved in wars on almost a global scale it was only natural that the military should become concerned with wood supplies in other countries. Zon and Sparhawk began their classic study, "Forest Resources of the World," that was published in 1923. During World War II the Forest Service set up a special division to prepare reports on the forest resources of various lands in the Asia-Pacific area as well as elsewhere. The division has remained active since that time, expanding its concern to the general forest situation throughout the world as it affects forestry and the forest industries of the United States. Forest products research was greatly stimulated during both wars by the need to use wood, also, to substitute for other materials in short supply. Laminated beams, arches, and various other products were developed that have found a permanent market in postwar times.

**Developments Since 1945** The economy of the United States was functioning at record levels in 1945 when the Germans and later the Japanese capitulated. Most people anticipated a postwar depression, but this did not occur. The pent-up demand for consumer goods and the desire of returning soldiers to marry and establish homes caused a boom rather than a recession. Almost everything was in short supply—office and factory space, clothing, food, motor cars, and, above all, housing. Lumber, paper, and other forest products were in huge demand. The major companies, recognizing the market potential that lay ahead, began to look forward to the long-term timber supplies. As a result, forestry on a nationwide basis entered a period of the most rapid advance it had seen since the turn of the century. This time the advance was stimulated by the need for forest products and by the conviction on the part of the major timber companies that they must protect their raw-material supply.

*The Growing Need for Forest Products* Timber trend studies by the U.S. Forest Service in 1963 showed annual timber growth to exceed cut by substantial margins, yet population increases and the trend of economic growth seemed to presage a timber volume decline beginning by 1980.[14] Corporations were expanding their timber holdings and their daily consumption of wood. Forest industries and the President were urging an increase in the annual cut from the national forests, and this was allowed. Forestry and foresters shared in the booming national economy. Industrial forestry surged forward to surpass that on public lands in both intensity of mechanization and yields from the land.

The nation as a whole was enjoying unprecedented prosperity. Urban renewal subsidies stimulated slum clearance and rebuilding of central city areas. This, along with a vast migration of people from the rural South to northern cities, overtaxed urban housing. New construction of housing and office buildings was pushed. Lumber use approached the peak of earlier years, but still affluence eluded many people—the disadvantaged and many members of minority groups, such as the Blacks, Puerto Ricans, and American Indians. Young people, troubled by the Vietnam War and incensed by racial discrimination, crusaded for equal opportunity. They felt that poverty—undernourishment, poor housing, and economic want—was unforgivable in an affluent society.

**Social Justice and Environment** The nation responded. Widespread programs of regional planning, jointly sponsored by the federal and local governments, embraced transportation, urban development, and rural rehabilitation. Foresters and the forest industries were called upon to help for it was in forested regions that many of the disadvantaged people lived—Appalachia, the Ozarks, and sections of the South where industries were slow to develop. Training programs to prepare men for work in the forest and forest industries were established as well as special camps for young men of the so-called hard core unemployed. Though the training programs were short-lived, they again called attention to the forest as an underdeveloped land resource suitable for public work projects.

The Timber Resource Review completed by the U.S. Forest Service in 1958 also called attention to the growing demands for products and services from the nation's forests. The Forest Service since 1905 had practiced multiple use of national resources on a sustained-yield basis. To protect the forests from growing pressure from single-interest groups the Congress passed in 1960 the Multiple Use–Sustained Yield Act. This act authorized and directed that the national forests be managed under principles of multiple use so as to produce a sustained yield of products and services. The act declared it the policy of the Congress that the national forests were established and are to be administered for outdoor recreation, range, timber, watershed, and wildlife and fish purposes.

The report of the Outdoor Recreational Resources Review Commission issued in 1962 urged increased development of both public and private outdoor recreational resources. The Bureau of Outdoor Recreation was established soon thereafter in the Department of Interior. Edward C. Crafts, a high Forest Service administrative officer, was appointed its first director. The Bureau launched a comprehensive program of cooperation with federal, state, and private agencies in planning and developing outdoor recreational facilities. Impetus was given to the Bureau's program by establishing in 1964 the Land and Water Conservation Fund. Through this the Bureau could help to finance acquisition and development of lands for recreational use in cooperation with federal and state agencies. A requirement for state eligibility to participate was the preparation of a comprehensive statewide outdoor recreation development plan.

Interest in wilderness grew rapidly after World War II. The Wilderness Society, the Sierra Club, and many other organizations wanted to see the wilderness areas—established in national forests by executive order—given the protection of federal law. The Forest Service also favored such legislation. The Wilderness Preservation Act passed in 1964 legally established a system of such areas that includes those in the national forests. The act also provides procedures to be followed in proposing new wilderness areas to be established by congressional action.

Sponsors of the above legislation and many other citizens became deeply concerned about the environmental degradation brought about by air and water pollution, disposal of solid waste, and acts that spread ugliness over the landscape. They were not slow to point their fingers at the forest products industries for stream and air pollution and for defacing the landscape by the rapidly spreading practice of clear-cutting timber stands. The U.S. Forest Service, which had long advocated "selective cutting" but had been making increased use of even-aged silviculture for many types, came under sharp attack. Injunctions and suits were filed not only to restrict timber cutting but also to halt work on one elaborate recreational development. The competence of foresters to make important decisions on management practices was sharply challenged. A new era of forest resource management seemed to be dawning.

**State Forestry** The beginnings of state forestry in the United States go back to colonial days. The early ordinances of New Sweden (which became Delaware), of Pennsylvania (by William Penn), and of Massachusetts all recognized the importance of the forest and provided regulations for its proper use. With the coming of independence the states continued to take an interest in forestry measures. Agricultural societies in Philadelphia as early as 1791, in New York in 1795, and in Massachusetts in 1804 took notice of the importance of tree planting and urged that steps be taken to encourage it. In 1819, the Massachusetts Legislature asked its state department of agriculture to promote the growth of oak suitable for ship timbers.

In 1837, Massachusetts authorized a survey of its forest conditions with a view toward determining measures that might encourage landowners to practice forestry. In 1788, New York passed laws to regulate lumber grading to protect the good name of New York lumber in national and international trade.[17]

The surge of settlers to the prairie states after the passage of the Homestead Act in 1862 made the middle western folk acutely aware of the need for trees. Sod huts for living quarters and buffalo chips for fuel were a real hardship to people used to an abundant supply of wood. Tree planting caught the public interest.

Michigan in 1867 and Wisconsin shortly thereafter created forestry commissions to investigate timber operations, land clearing, and the importance of proper care of timberlands within the state. Consideration was given to shelterbelts and to proper management of forest areas. Similar forestry commissions were set up in New York, Connecticut, New Hampshire, Vermont, Ohio, Pennsylvania, and North Carolina.[7] These states together with Kansas and Maine had state forestry organizations prior to 1885. In that year California, Ohio, Colorado, and New York established state forest administration. With the exception of New York's, these early state forest organizations were temporary, but they laid the groundwork for permanent programs to follow.

In 1872 New York set up a commission to investigate the preservation of the Adirondack forests, which led in 1885 to the establishment of the Adirondack Forest Preserve. After 1894 citizens became seriously alarmed by the way in which the state preserves were managed. A constitutional amendment was passed prohibiting the cutting of trees in the preserve and has been reenacted at subsequent constitutional conventions.

New York's example of acquiring state land for permanent forest use was followed by Pennsylvania in 1895, Minnesota in 1899, and by many other states thereafter.

State acquisition of land for forestry came about, however, more from default of owners on taxes and from land abandonment than from deliberate state policy.[5] It was necessary that American pioneers clear forests for crops. Hill lands usually had less dense forests, were freer from malaria, and hence were often chosen for settlers' homes and fields; but soils were thin and steep slopes easily eroded. Mechanization of tillage placed the hill farmer at still greater disadvantage in competing with those on fertile level lands. The agricultural depression that followed World War I brought the problem before the country.

Land abandonment—it began in New England as early as 1850—is a painful process. As family income shrinks, savings are exhausted. Repairs to the dwelling, other farm buildings, and fences are postponed. Equipment wears out. Livestock declines in vigor, quality, and number. Crop yields decline from inadequate tillage and fertilizer. Severe family privation may occur. Eventually the farm as a whole is abandoned, or reduced to part-time operation, while the breadwinner engages in some other occupation.

In several counties in the Lake states more than 50 percent of the land became tax-delinquent. Land abandonment spread at an alarming rate, and clearing forests for farmlands virtually ceased. In New York, farms once embraced 80 percent of the land area. By 1950 this was reduced to 57 percent and by 1959 to 44 percent. The long-established trend of giving up cropland to pasture and pasture to woodland was still in evidence in 1970.

Farmers and rural bankers both recognized the economic and human waste that occurred in farming poor land. In New York they obtained passage of the Hewitt Law in 1929 to purchase submarginal farmlands for forests. By 1970 some 600,000 acres had been acquired and planted to trees. Similar programs were started in other states. In 1934 the Federal Resettlement Administration became active in purchasing lands, most of which were turned over to the states to administer.

As of 1964 the 50 states had acquired by various means a total of some 74 million acres of lands for forests, parks, game areas, and other conservation purposes. Among those with large holdings as of 1964 were the following:

| State | Million acres |
|---|---|
| Alaska | 16.0 |
| Arizona | 10.9 |
| New Mexico | 7.5 |
| Montana | 5.4 |
| Minnesota | 5.1 |
| Michigan | 4.3 |
| Utah | 3.5 |
| Idaho | 3.3 |
| New York | 3.2 |
| Pennsylvania | 3.0 |

*Source:* Ralph R. Widner (ed.), "Forests and Forestry in the American States." A reference anthology compiled by the National Association of State Foresters, 1968.

Most state forest lands, are administered for timber production, recreation, watershed protection, public hunting, and other multiple-use purposes.

The Weeks and Clarke-McNary laws proved a great boon to state forestry departments. In most states fire-control expenditures have remained the largest item in the budget of state forestry organizations. Federal aid has been a mainstay in providing financial support and in ensuring freedom from undue political pressure on forest officers employed in state organizations.

The program in Maryland might be taken as an example of state forestry. Maryland forestry first dealt primarily with protecting the lands from fire. Soon thereafter the state agreed to perform forestry services for landowners on a fee basis. Later a state extension forester was employed to stimulate forest care by farmers. Gradually the state acquired a number of forest properties which, though small in size, were managed both for timber production and for recreation. In 1943, the state passed a law regulating timber-cutting practices on private lands. It cooperates with the federal government in farm and other private forestry programs.

Similar service programs have developed in many other states. Laws pertaining to forest practice on private lands have been adopted in 17 states.

**Farm Forestry** Farm forestry has been largely a state function. The Smith-Lever Act, passed in 1914, allotted funds through the state agricultural colleges for extension work in forestry. Initial programs emphasized tree planting and field demonstrations.

It became evident by 1924 that considerable additional effort was needed. The Clarke-McNary Law provided for distributing forest planting stock to farmers and for strengthening extension forestry.

Largely as a result of the experience of the extension foresters, the Norris-Doxey Farm Forestry Act was passed in 1937 to provide advice to farmers on timber growing and marketing.

The Cooperative Forest Management Act of 1950 extended the benefits provided farmers under the Norris-Doxey Act to all woodland owners. Other federal-state programs to stimulate good forest practice on farm and other small forest holdings included those of the Soil Conservation Service, the Prairie-Plains Shelterbelt Program, the Agricultural Stabilization and Conservation Program, the Food and Agriculture Acts, and the Pitman-Robertson game management program. Most of these acts were strengthened by programs enacted by the several states. Administration varied in the several states but the state foresters came to be the chief agency through which the work was channeled in most states. By 1970 all states were involved in one or more of these programs.

## PRIVATE FORESTRY

The beginnings of private forestry in America are lost in the early history of our country. Certainly many private landowners took an interest in their forests and attempted to protect young growth from unnecessary cutting. Farmers, plantation owners, and even loggers and lumbermen set aside specific forest areas where logging was kept at a minimum or even entirely excluded. Some practiced good forest management over a long period of time with little or no aid from a technical forester.

Zachariah Allen[6] in 1820 planted some abandoned farmland in Rhode Island to oaks, chestnut, and locust. He kept accurate financial accounts. During a 57-year period he sold fuel and timber for $4,948. After allowing 6 percent interest on land, taxes, and planting costs, the venture netted $2,543—or 6.93 percent per year for the 57-year investment period.

Among the early large-scale forestry ventures in the United States were those that were begun near the turn of the century on the Vanderbilt, Webb, and Whitney estates. Gifford Pinchot was employed by George Vanderbilt as consulting forester for his 80,000-acre estate in North Carolina, later named the Biltmore Forest. Pinchot brought in a sawmill and started timber operations. The venture was successful financially and attracted the attention of other forest owners.

Ne-Ha-Sa-Ne Park, owned by Dr. Seward Webb in Hamilton County, New York, and the Whitney estate, adjoined one another. Gifford Pinchot and Henry Graves were employed to work up plans for management of these two estates. The first silvicultural rules governing cutting practices that were ever written into a logging contract in the United States probably are those prepared for the Ne-Ha-Sa-Ne forest.

The accumulation of forest lands in large family estates began in Maine as early as 1850.[12] Some of these became very large properties, of which the Coe and Pingree estate is an example. The founder, David Pingree, restricted cutting of spruce to trees that were at least 14 inches in diameter.

**Industrial Forestry** About 1912 Finch, Pruyn, and Company started a forestry program on its Adirondack holdings. Trees to be cut were marked by foresters, and the cutting budget was projected on a sustained-yield basis.

The Goodman Lumber Company in Wisconsin began its first cycle of selective cutting in 1927.[12] The Urania Lumber Company in Louisiana began operations on a sustained-yield basis about 1917.

Other companies that began forestry programs in the 1930s or earlier include Crossett Lumber Company, the Nekoosa-Edwards Company, Consolidated Water Power and Paper Company, International Paper Company, Patten Timber Company, Von Platen-Fox Company, Dead River Company, Minnesota and Ontario Paper Company, Weyerhaeuser Timber Company, St. Regis Company, Crown-Zellerbach Company, Collins Almanor Forest, Camp Manufacturing Company, St. Paul and Tacoma Lumber Company, and Champion Paper and Fiber Company.[16]

The rapid expansion of industry after World War II led paper companies to purchase forest land on a large scale in the South, the West, and to a lesser extent in the Northeast. The success of pulping of southern pines by the sulfate process brought about a general shift of the center of paper manufacture from the Northeast to the South. Three major reasons induced paper companies to engage in forestry on a sustained-yield basis: wood requirements of mills were high, as much as 600 cords per day or more; the cost of mills was in the range of $50 million each, which required a long period for amortization; and the paper companies could grow wood to pulpwood size in rotations as short as 20 years.

Once top management in such companies became thoroughly committed to forestry, they entered upon it with imaginative enterprise. Logging, tree planting, nursery operations, and other forestry operations were mechanized to a high degree to improve output per man-day. Forest inventory control was placed on a continuous basis. Cost control over operations was adopted. Pulp and paper companies were quick to recognize the benefits to be realized from Forest Service and university-financed research in such fields as tree physiology, entomology, genetics, and tree improvement. The paper companies supported appropriations for such research and some also made corporate contributions thereto. A few companies established their own experimental forests and research teams. To help assure a continuous supply of wood from small holdings, some companies further provide forestry services at their own expense to farmers and other holders of small forest properties. A few have embarked on a forest leasing plan that assures the owner a regular income from his property under a continued forest-management program.

Within a few years the leaders among forest industries have developed systems of mechanized, even-aged, forest management in the South and in the West that match in productivity of land and labor some of the best forest practices in the world. Moreover, the promptness with which they adopted research findings led to increased appropriations and spurred forest scientists to new levels of achievement.

All of these activities were placing new obligations on foresters. The demand went out for revisions of forestry school curriculums and for higher-level competence of graduates. These new demands also sharpened the distinctions between the professional forester and the forest technician. Much of the routine work was being transferred to men with technician education.

New impetus and excitement has come to the profession of forestry and it has come none too soon, for at the same time that industries are expecting new levels of performance from foresters, so also are the recreationists, wilderness exponents, and environmentalists. The latter are insisting that the forests of our land be managed by people who have full regard for all consequences of their operations to both man and his environment.

**Associations** The cause of forestry in America has been advanced by citizens' organizations throughout the land. There are well over 1,000 such organizations and probably as many as twice this number—the number grows almost yearly as people's concern for the preservation and use of open space and natural resources stimulates personal involvement. Some organizations are mainly local; others are nationwide or even international. They vary from lay and youth organizations, such as the Boy Scouts and garden clubs, to the nation's most prestigious scientific societies.

**Scientific and Professional Societies** The American Association for the Advancement of Science, as we have seen, deserves credit for stimulating Congress in 1876 to embark on a sustained federal forestry program. The Association has kept its interest in forestry up to the present writing. It publishes articles by forest research workers in its journal, *Science.* It has also printed explanations of forest policy issues, and featured at its annual meetings broad issues of environment and open-land preservation and use. Many foresters have been members of the Association and some have served on its governing council.

The National Academy of Sciences report on the forest reserves mentioned earlier was but the beginning of its long involvement in forest conservation. Its influence has been exerted mainly through the National Research Council, which the Academy organized to mobilize science during World War I; the Office of Scientific Research and Development, which played a similar role during World War II; and the National Science Foundation, which was organized after the close of that war. All three of these agencies have supported forestry and forest research.

The Society of American Foresters, founded in 1900, together with its sister societies in Canada and Mexico, represents the profession of forestry in North America. These three organizations are devoted to advancing the science and practice of forestry throughout the continent. The Society of American Foresters testifies before Congress and state legislatures on forestry issues. It also insists on public agencies employing foresters for tasks requiring their professional competence. Through the International Union of Societies of Forestry it engages in a worldwide exchange of information and in joint forestry actions.

Other scientific and professional groups have helped in significant ways. Forestry must of necessity draw heavily upon the biological, physical. and social sciences for expertise in many fields of learning. The Ecological Society almost from its organization welcomed foresters to membership and service on its council and editorial board. The Wildlife Society and the Society for Range Management were organized to enable research men and administrators to cope with the problems faced on the national forests and other public lands. These two, with the American Fisheries Society, have collaborated with the Society of American Foresters to advance their closely related interests and activities. The Forest Products Research Society seeks to further scientific knowledge of wood and its application in technology. Other societies that contribute to forestry through their science or technology include those of botanists, entomologists, pathologists, soil scientists, photogrammetrists, engineers, hydrologists, chemists, economists, sociologists, and anthropologists.

The International Union of Forest Research Organizations is supported by member institutions throughout the world. United States members include colleges of forestry and the U.S. Forest Service. Through working groups the Union promotes collaboration among forest research workers on specific projects of international concern. Its congresses, held at about 6-year intervals, are widely attended. The first congress to be held outside of Europe met at Gainesville, Florida, in 1971.

World Forestry Congresses have met periodically since 1926 to discuss various questions of forest policy, management, and technology. The meeting places have been as follows: 1926, Rome; 1936, Budapest; 1949, Helsinki; 1954, Dehra Dun, India; 1960, Seattle; 1966, Madrid; 1972, Buenos Aires.

**Lay Organizations** Citizen organizations outnumber by far all others concerned with forest conservation in the United States. Among these are sportsmen's organizations, labor unions, youth organizations, wildflower clubs, hikers' clubs, and many others. In fact, a large proportion of citizen organizations concerned with nature, outdoor recreation, parks, soil and watershed protection, wildlife, natural scenery, and environmental protection in general have shown an interest in forestry and have supported measures to protect and conserve forest productivity.

Outstanding among these citizen organizations is the American Forestry Association. Founded in 1875 and merged with the American Forestry Congress in 1882, it has consistently led in mobilizing citizen support for broad forest conservation legislation at the national level. It has also had significant influence on forestry activities in the several states. The American Forestry Congresses it has sponsored have served to highlight major forestry issues. Their resolutions have led to significant legislation by the national Congress. Year by year forestry issues are featured in annual meetings and in the Association journal, *American Forests*. The testimony that its officers and members present before congressional committees has an invaluable effect on the yearly progress of American forestry.

Many other national organizations, among them the National Audubon Society, the Friends of the Earth, the Sierra Club, the Wilderness Society, the Isaac Walton League, the National Rifle Association, and the National Parks and Conservation Association, as well as primarily agriculture-oriented organizations, such as the National Grange and the Soil Conservation Society of America, have all supported forest conservation in general as well as in their own specific area of interest.

**Forest Industry Associations** Industrial forestry has been aided and promoted by the forest industry associations. These deserve much credit for the rapid pace of advancement in private forestry. Associations date back to 1911 when the Clearwater Forest Protective Association was formed. Many others followed. At first, these were concerned with protection of forests against fire, but later they became interested in all major concerns of the forest industries. Associations were formed of lumbermen, and of men from pulp and paper, plywood, and other forest and wood processing industries. Many of these joined forces to support the American Forest Products Industries, now known as the American Forest Institute. Under its sponsorship the American Tree Farm Movement was launched that promoted forestry on industrial and other private forest holdings. It prepared and distributed nationally—to schools and to libraries—considerable educational material on forestry and its benefits to the American public. The Institute sought the cooperation of public as well as industrial foresters to promote its program and strove to dovetail its own activities with the many publicly sponsored programs mentioned above.[1, 11]

## LITERATURE CITED

1 American Forest Products Industries. 1961. "Progress in Private Forestry in the United States," Washington. 49 pp.
2 Butler, Ovid. 1946. 70 Years of Campaigning for American Forestry, *American Forests*, **52**(10):456–459, 512.
3 Chinard, Gilbert. 1945. The American Philosophical Society and the Early History of Forestry in America, *Proceedings, American Philosophical Society,* **89**(2):444–488.
4 Dana, Samuel Trask. 1956. "Forest and Range Policy. Its Development in the United States," McGraw-Hill Book Company, New York. 455 pp.
5 Fontanna, Stanley G. 1949. State Forests, *Trees,* The Yearbook of Agriculture, pp. 390–394, U.S. Department of Agriculture, Washington.
6 Holst, Monterey Leman. 1946. Zachariah Allen, Pioneer in Applied Silviculture, *Journal of Forestry,* **44**:507–508.
7 Illick, Joseph S. 1939. "An Outline of General Forestry," 3d ed., Barnes & Noble, Inc., New York. 297 pp.

8  Lillard, Richard G. 1947. "The Great Forest," Alfred A. Knopf, Inc.. New York. 399 pp.
9  Paxson, Frederic L. 1924. "History of the American Frontier, 1763–1893," Houghton Mifflin Company, Boston. 598 pp.
10  Pinchot, Gifford. 1947. "Breaking New Ground," Harcourt, Brace and Company, Inc., New York. 522 pp.
11  Sayers, W. B. 1966. To Tell the Truth. 25 Years of American Forest Products Industries, Inc., *Journal of Forestry,* **64**:657–663.
12  Shirley, Hardy L. 1949. Large Private Holdings in the North, *Trees,* The Yearbook of Agriculture, pp. 255–274, U.S. Department of Agriculture, Washington.
13  Sparhawk, William N. 1949. The History of Forestry in America, *Trees.* The Yearbook of Agriculture, pp. 702–714, U.S. Department of Agriculture, Washington.
14  U.S. Department of Agriculture, Forest Service. 1965. Timber Trends in the United States. Forest Resource Report No. 17. U.S. Government Printing Office, Washington. 235 pp.
15  U.S. Department of the Interior, Bureau of Land Management. 1950. Brief Notes on the Public Domain, Washington. 21 pp.
16  Winters, Robert K.(ed.) 1950. "Fifty Years of Forestry in the U.S.A.," Society of American Foresters, Washington. 385 pp.
17  Wolf, Robert E. 1948. "A Partial Survey of Forest Legislation in New York State," unpublished thesis, New York State College of Forestry, Syracuse, N. Y.

Chapter 4

# Major Forest Regions of the World and of the United States

Forests of some type occupy almost one-third of the earth's land surface. For simplification the U.N. Food and Agriculture Organization, Forestry Department, has grouped forests into six major formations (Fig. 4-1). Each formation is made up of a number of forest types, or ecosystems.[9]

## THE COOL CONIFEROUS FORESTS

Cool coniferous forests extend in a broad belt around the earth in high northern latitudes. They cover most of forested Alaska and Canada, occur in the northern United States, Scotland, Scandinavia, Finland, and the U.S.S.R., with a southern extension into the Himalayan region. They are also found in northern Japan. These forests provide important amounts of the coniferous wood of the world. The chief species are pines, spruces, firs, and larches.[9] The trees are generally small to moderate in size. The forests have a uniform aspect, and the timber quality also is rather uniform. The forests are excellent for pulpwood and also produce good sawlogs of moderate size. Their major drawback is the relatively

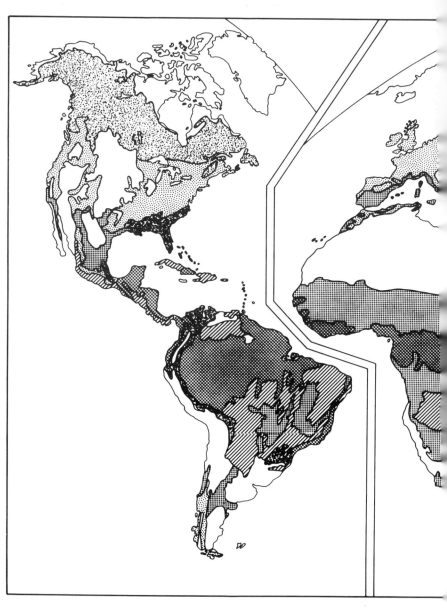

|  Cool coniferous forest |  Warm temperate moist forest |  Tropical moist deciduous forest |
|---|---|---|
|  Temperate mixed forest |  Equatorial rain forest |  Dry forest |

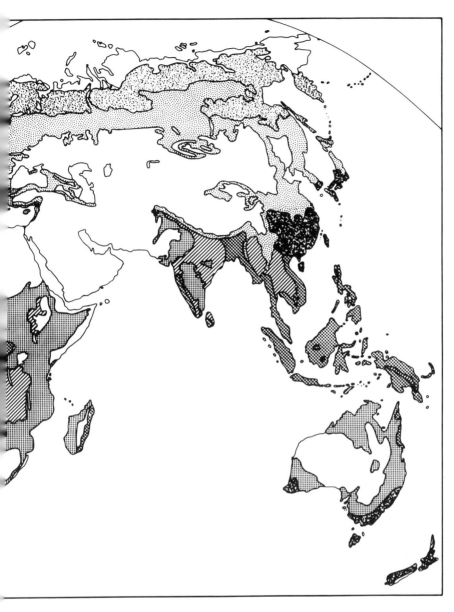

**THE WORLD'S FORESTS**

(The map does not show the actual extent of the world's forests, but only the main vegetational regions.)

**Figure 4-1** Forests of the world by major biome groups. Adapted by UN Food and Agriculture Organization, Forestry Department, from "Oxford Atlas of the World," Oxford University Press, 1962.

slow rate of growth due to the cool climate and but moderately fertile soils. In northern Russia and Siberia some of these forests are underlain by permafrost. Litter breakdown is mostly by fungi.

## THE TEMPERATE MIXED FORESTS

Temperate mixed forests occur in western and eastern North America, Europe, the U.S.S.R., northern China, Japan, the mountain regions of southern U.S.S.R.. the Himalayas, and the Andes of southern Chile and Argentina. Rich in species, they furnish such major commercial timbers as beech, oak, birch, ash, maple, walnut, yellow poplar, hemlock, Douglas-fir, and most of the pines. These forests have been the most highly used commercially of any. This is because of the high quality of the wood, the richness of species, and the fact that they occur in the well-settled and industrial regions of the world. For this last reason much of the land originally covered by these forests has been cleared for agriculture. Both the cool coniferous and temperate mixed forests are confined almost exclusively to the northern hemisphere. Together these two great forest regions furnished roughly 80 percent of the commercial timber used by the world in 1960.[9] Both will be described more fully in dealing with the forest regions of the United States.

## THE WARM MOIST TEMPERATE FORESTS

Warm moist temperate forests occur on the southern coastal plain of the United States, along the east coast of Mexico, in Japan, China, Taiwan, the Andes Cordillera, the uplands of southern Brazil, southern Australia, and throughout New Zealand. Timber growth is rapid. Many species are of high commercial value—the pines of the southern United States, the Auracarias of Brazil and Chile, the Eucalyptus of Australia, and a large number of other hardwoods. Extensive plantations of pines and other species have been made in Chile, Brazil, the United States, China, South Africa, New Zealand, and Australia. Timber yields from these plantations are high. Because of the wide distribution of their forests and the large number of species found in them no general worldwide description applies. Though conifers are found in them, they tend to revert to hardwoods in the southern United States and in Brazil. Those of the United States will be described later.

## THE EQUATORIAL RAIN FORESTS

Equatorial rain forests extend from the Yucatan Peninsula in Mexico south to embrace the Amazon Basin, the west coast of Colombia and

Ecuador, the central and east coasts of Brazil, selected areas in west and central Africa east almost to Lake Victoria, eastern Madagascar, the west coast of India and Ceylon, Bangladesh, Burma, Vietnam, Thailand, Malaysia, Indonesia, the Philippines, and New Guinea. Warm, well watered, and richest in species of all six formations, these are potentially the most highly productive forests of the world. They have the largest volume of old-growth timber, although much of it is remote. A highly complex mixture of species only a few of which are in commercial use at present handicaps exploitation. The Asian portion is well used. It furnishes the important dipterocarps of which luan, known as Philippine mahogany, is widely exported. From the American rain forests, mahogany, green heart, Spanish cedar, and balsa wood are important in commerce. From Africa, okoume, obeche, sipo, limba, and African mahoganies enter world commerce.

Because of the dense canopies of the mature tropical rain forests little light penetrates to the ground to support an undergrowth. The forests are evergreen, though some species may shed their leaves for a portion of the year. Vines of which the figs are conspicuous grow around the trunks of trees and eventually kill them by shading out the crowns of the supporting tree. By this time the fig itself virtually envelops the host trunk and is able to support its own crown. Following cutting of rain forests a luxuriant growth of both herbaceous and woody vines spreads over ground, dead trees, and young growth. It chokes out seedlings of trees so that considerable effort must be expended to get a new crop started. These forests present a great challenge to man if he is to manage them commercially on a sustained-yield basis. A few, such as the Philippine dipterocarp forests, are amenable to management and offer attractive opportunities because of the few species, rapid growth, and high value of the timber.

**THE TROPICAL MOIST DECIDUOUS FORESTS**

Tropical moist deciduous forests occur along the west coast of Mexico and Central America, the Caribbean Islands, the Orinoco Basin, most of forested South America south of the Amazon Basin, South Central Africa, India, Thailand, Laos, Cambodia, and central New Guinea. A large part of these forests is used for grazing or has been cleared for agriculture. Their best-known commercial species are the teak and sal of Asia. These forests have a dry season when the trees are leafless. This tends to discourage the dense vegetation of the rain forest and so makes these forests much easier to manage for timber crops. They are important locally and furnish valuable seed for planting as well as timber for both local use and export.

## DRY FORESTS

Dry forests are found in Mexico, east central Brazil, the Argentine, much of Africa south of the Sahara, western Madagascar, the Mediterranean coastal region, Spain, West Pakistan, India, and the sections bordering the central desert areas of Australia. These forests have been extending their range at the expense of temperate mixed forests in the Mediterranean region. Heavy grazing, fire, and steep-slope cultivation have all caused soil degradation to the point where only dry forests can now survive there. These forests produce no timber for export, although some is valuable for fuel, posts, and other local use.

## COMPLEXITY OF WORLD FORESTS

Within these six major forest formations are many different types and subtypes. These vary from the flat-crowned African thorn forests, with their uniform heights that extend but little above the reach of a giraffe, to include palm forests, tree-fern forests, mossy or elfin forests, and redwood and giant sequoia forests. The greatest variety of types is to be found in the tropical rain forests and the fewest in the cool coniferous forests. A forester from Canada may start in Alaska and move eastward across North America, Europe, and Siberia and be able not only to recognize most of the trees but also a large number of the shrubs and herbaceous plants he encounters. On the other hand, should he elect to travel from Canada southward across the United States, Mexico, Central and South America, to southern Chile, he would be confounded by the new tree species he would encounter and the complexity of unfamiliar vegetative types. Yet much of the work of foresters in the future will deal with these little-known forests, the management technologies of which are still to be worked out.

## FORESTS OF THE UNITED STATES

The forests of the United States stand intermediate in floristic complexity between those of northern Europe, which are relatively simple, and the tropical forests, which seem to be unendingly complex. To make the study of American forestry comprehensible, foresters recognize forest types or ecosystems. The Society of American Foresters lists 147 forest types, of which 97 are found in the eastern states and 50 in the western states. For further simplification, type groups known as formations and biomes may be recognized.

The major forest formations are shown in Fig. 4-2. Even formations

# MAJOR FOREST REGIONS OF THE WORLD AND OF THE UNITED STATES

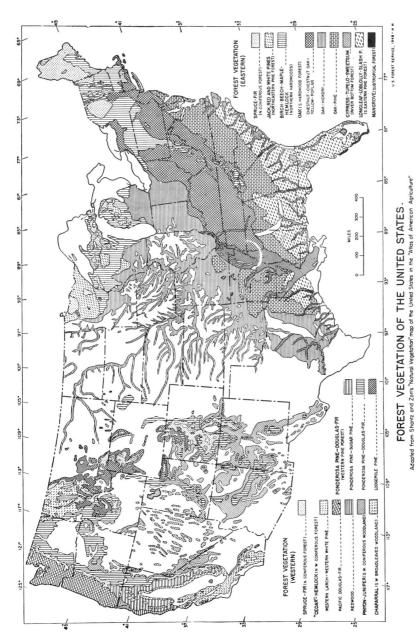

**Figure 4-2** Natural forest regions of the United States. *(U.S. Forest Service.)*

are too numerous to permit detailed discussion in a book on general forestry, especially since boundaries between formations are irregular and must be arbitrarily drawn. The major forest formations of Fig. 4-1 are much broader than those shown in Fig. 4-2.

Forest formations and ecosystems rarely follow national, state, or county boundaries for which economic statistics are compiled. The U.S. Forest Service recognizes four major geographic sections of the United States for which forest statistics are grouped. These are the North, South, Rocky Mountains, and Pacific Coast. The North includes Maryland, West Virginia, Kentucky, Missouri, and all states north of these to the Great Plains—except western South Dakota.

The South includes Virginia, Tennessee, Arkansas, Oklahoma, and states to the south of these.

The Rocky Mountain section includes western South Dakota and all states west of the Great Plains except those bordering on the Pacific Ocean.

The Pacific Coast section includes the four states that lie along the Pacific Coast and Hawaii.

Information on the above sections concerning forest area, timber volumes, annual growth increment, and annual timber harvested is given in Table 4-1. Although the area of the Rocky Mountain section is only about one-third that of the North and the South, the sawtimber volumes

Table 4-1 Forest Area, Timber Volumes, Increment, and Harvest by Sections of the United States, 1967

| Section | Area, millions of acres | Sawtimber, volume billions of bd ft international log rule | Growing stock, billions of cu ft | Average annual increment per acre, cu ft | Average annual harvest per acre, cu ft |
|---|---|---|---|---|---|
| North | 176 | 312 | 147 | 30 | 14 |
| South | 199 | 461 | 149 | 40 | 30 |
| Rocky Mountains | 66 | 403 | 101 | 19 | 13 |
| Pacific Coast | 69 | 1314 | 248 | 42 | 58 |
| All states | 510 | 2490 | 645 | 34 | 26 |

*Source:* Josephson, H. R., and Dwight Hair. 1970. The Timber Resource Situation in the United States. U.S. Department of Agriculture, Forest Service. 29 pp.

of the three sections are roughly comparable. Together these regions have less standing sawtimber than the Pacific section though they have a considerably larger volume of total growing stock. All three eastern sections have had an excess of total annual cubic foot increment over that removed by timber harvesting in recent years. The discussion that follows deals with forest formations; however, the reader will find it helpful to refer again to Table 4-1 and to Fig. 4-2 for background.

**The Northern Forest Formations** The northern forests of the United States include representatives of both the cool coniferous forests (Fig. 4-3) and the temperate mixed forests as designated by the U.N. Food and Agriculture Organization. They are characterized by spruce, fir, aspen, and birch in Alaska, the northern Lake states and Maine, and by sugar maple, beech, basswood, yellow birch, and white pine in the northeastern states. The four major type groups and approximate areas in 1969 were:[4]

| Type groups | Acres covered, in millions |
| --- | --- |
| Aspen-birch | 24 |
| White pine–red pine–jack pine | 11 |
| Maple-beech-birch | 34 |
| Spruce-fir | 20 |

**Figure 4-3** Cool coniferous forest. Old-growth Sitka spruce and western hemlock, Tongass National Forest, Alaska. (*U.S. Forest Service.*)

Five tree species have ranges that correspond more or less closely to the distribution of the northern forest in the United States. These are northern white-cedar, eastern hemlock, yellow birch, northern white pine, and balsam fir. Perhaps the last named is the best indicator of the northern forest. It is interesting that sugar maple, which more than other species gives character to the northern forest, is distributed far beyond the boundaries of the northern forest.

The northern forest is made up of a number of important forest types. The most typical are those containing beech, birch, or maple. Sugar maple tends to outnumber beech and birch on the best-quality soils, and the reverse is true on thin or light-textured soils. Intermixed in the northern hardwoods forest are hemlock and, to the north and at higher elevations, red spruce, white spruce, and balsam fir. Black spruce, tamarack, and northern white-cedar are found in the swamps that characterize the glaciated terrain of the northern states. White pine, red pine, and—in the Lake states—jack pine are found in pure and mixed stands in the northern forest.[1]

The northern forest contains many species and types of high economic value. From colonial days white pine has been preeminent for lumber and remains so today. High-quality lumber is also manufactured from red pine and from red and white spruce. Hemlock, tamarack, jack pine, and balsam fir produce intermediate-quality lumber. Northern white-cedar is usually small but produces a durable wood suitable for shingles, boatbuilding, poles, and posts. Among the hardwoods sugar maple and yellow birch are splendid woods for furniture manufacture, flooring, and interior trim. Beech likewise is suitable for flooring, handles, and other items requiring a hard, dense wood. White ash, black cherry, and basswood produce specialty woods. Ash is strong, hence useful for handles, athletic equipment, ladder rungs, and other purposes requiring high strength. Cherry is made into furniture and is extensively used for type backing because of its high dimensional stability. Basswood produces a soft, fine-grain lumber suitable for cabinetwork and is also used for veneer, cooperage, and woodenware. Paper birch that comes in after a fire is useful for dowels, stamped veneer products, and turned products such as spools. Aspen is useful for excelsior and lumber.

The most highly prized pulp species are red, white, and black spruce; and second to these is balsam fir. All may be used for both groundwood and sulfite pulp. Hemlock, jack pine, and also red and white pine may be pulped by the sulfate process. Aspen is useful for groundwood and semichemical pulp. Since 1950 many hardwood species have been widely employed for paper pulp using the sulfate and neutral sulfite processes.

The northern forests, particularly the coniferous types, grow densely and may shade out the undervegetation. Volumes in excess of 50 cords per acre may be found on lands where the largest trees are less than 18 inches in diameter. Growth rates of individual trees are not high, but volume growth per acre often equals that of the southern and Rocky Mountain forests where wide spacing of trees results from insufficient rainfall.

Under good conditions, white pine will grow at the rate of 500 to 800 board feet per acre per year, though a growth rate of 300 feet per acre is more to be expected. Mixed northern hardwoods made up of maple, birch, white ash, black cherry, and beech will grow at the rate of 300 to 500 board feet per acre. Jack pine, aspen, spruces, and balsam generally grow at $1/2$ to 1 cord per acre per year in well-stocked stands.

The northern forest region occupies the major area of forest land in New England, the Middle Atlantic states, and the Lake states. These are the most densely populated regions of the United States and are highly industrialized. Forest and forest products, nevertheless, make up a substantial part of the regional economy in each of these.

**The Central Forest Formation** The Central Forest is bounded roughly by the northern forest on the north, by the coastal plain on the east and south, and by the prairie on the west.

The characteristic forest types of the Central forest are those in which one or more species of oak are common, those in which shortleaf pine occurs, and those in which yellow-poplar, beech, silver maple, and river birch occur. The major type groups and areas in 1969 were:[4]

| Type groups | Acres covered, in millions |
|---|---|
| Oak-hickory | 61 |
| Elm-ash-cottonwood | 19 |

Associated with oaks are various hickories, shortleaf and Virginia pine, yellow-poplar, white ash, black walnut, and, formerly, chestnut. On the wetter soils silver maple, cottonwood, elm, and sycamore are found. White pine, sugar maple, and beech occur as secondary species in the Central hardwood forest, and in certain parts of it they form important constituents of the total stand.[1] Perhaps no other forest region of the United States contains a greater number of tree species having high technical uses than does the Central forest. Mast-producing species such as oaks, hickories, walnut, and beech also afford the Central hardwood forests great potentialities for wildlife.

The forests of the Central states furnish a wide variety of products. Black walnut, shagbark and shellbark hickories, and beech produce edible nuts prized by many people (Fig. 4-4). The chestnut oak produces bark of high tannin content. Hickory, ash, and oak are used for tool and implement handles. The various oaks, of which white oak is the most valuable, are used for railway ties. tight cooperage, furniture, veneers, and many other products. White oak is particularly sought for ship timbers.

The two most valuable woods for cabinetwork, furniture, and a variety of special uses are black walnut and yellow-poplar. Both are straight grained, uniform in texture, easily worked, strong, and durable. Because of its high resistance to shock, decay, and warping, black walnut is the preferred wood for gunstocks. Highly figured veneer is cut from black walnut stumps. Yellow-poplar produces a veneer suitable for core stock and for highly technical uses.

Useful lumber is also cut from gum, silver maple, beech, sugar maple, and basswood. Black locust, cottonwood, American elm, sycamore, river

**Figure 4-4** Central hardwood forest. Shellbark hickory, Turkey Run State Park, Indiana.

birch, hackberry, buckeye, and a host of other minor species have a variety of uses. Locust is valuable for outdoor posts. Elm makes a tough veneer suitable for baskets and crating.

The conifers of the Central forest include Virginia pine, eastern red-cedar, and shortleaf pine. Virginia pine seldom reaches sufficient size to be valuable for use as lumber, but it is used extensively for sulfate pulp. Eastern red-cedar produces a fine-grained, aromatic wood valuable for manufacturing cedar-lined chests and closets as well as pencils. Shortleaf pine produces lumber comparable in quality to that of the other southern pines.

The growth rate in the Central forest is somewhat more rapid than that in the Northern forest, and the stand density is about the same or somewhat less. Growth rates of 300 to 500 board feet per acre can be attained annually on well-stocked stands. Even higher growth rates could be expected were it not that agriculture has crowded the forest off the most productive soils. The Central region supports a great number of wood-using industries and therefore provides a good market for the better-quality trees that are produced.

**The Southern Forest Formations** The Southern forest region begins with southern Delaware and extends along the coastal plain, including all but the tip of southern Florida, and then extends west to eastern Texas and Oklahoma. The region sends a narrow tongue up the Mississippi Valley to southern Missouri and Illinois. Primarily, this is a pine region with swamps filled with baldcypress, water tupelo, red and black gums, swamp oaks, and various other species, including evergreen magnolia, cottonwood, willows, and bays.[10] From the standpoint of timber production the Southern region is potentially the most important in the United States.

The southern coastal plain is a relatively level, sandy plain. The inland piedmont, up which the Southern forest advances, is rolling to hilly in topography and has loam and clay soils. It is characterized by the longleaf, loblolly, slash pine type groups and the cypress-water tupelo-sweetgum groups. Upland hardwoods are highly valuable, including oaks, hickories, cucumber tree, yellow-poplar, and many others useful for furniture and construction.

| Type groups | Acres covered, in millions |
|---|---|
| Loblolly-shortleaf pine | 54 |
| Longleaf–slash pine | 26 |
| Oak-hickory | 56 |
| Oak-gum-cypress | 35 |
| Oak-pine | 24 |

***The Southern Pine Forest*** The Southern Pine forest is made up predominantly of four species: loblolly, shortleaf, longleaf, and slash pines; but pitch pine and pond pine occur to a limited extent.

Loblolly pine, which is rapid growing and is the most aggressive in invading old fields, grows in moderately dense stands and attains a height of approximately 100 feet and a diameter of up to 30 inches. The timber is of medium quality and, when properly manufactured and seasoned, is widely used for general construction.

Shortleaf pine produces lumber similar in quality to that of loblolly pine, but the tree grows somewhat more slowly and attains slightly smaller size.

Longleaf pine, which grows all along the coastal region, extending into western North Carolina and down along the coast as far west as Texas, is one of the most interesting pines. The needles attain lengths up to 12 inches. For the first 4 or 5 years of the tree's life it may remain in the grass stage, but during this time it develops a deep taproot and may attain a diameter of almost 1 inch at the ground surface. The huge terminal bud remains close to the ground and well protected by thick layers of bud scales. While in this stage, the plant resembles grass. After it becomes well established, the tree begins to grow in height. Shooting up quickly it continues to grow rapidly until it reaches maturity. Longleaf pine may attain a diameter of 24 inches or more and a height of 90 feet. The trees tend to grow rather widely separated, as considerable soil space is needed to obtain ample moisture.

Longleaf pine is one of the most fire-resistant of all conifers. While it is in the grass stage, fires may consume the needles yet fail to kill the terminal bud. After the tree has attained a height of 20 feet or more, fires may burn through at frequent intervals with only limited damage to this tree.

Longleaf pine produces a strong, heavy, dense lumber with a high resin content. It is highly prized for general construction. Longleaf is the principal producer of gum for naval stores.

Slash pine, the most rapid growing of all southern pines, resembles longleaf pine in many respects. Slash pine has a restricted natural distribution, but it has been successfully planted outside its natural range. It is almost equal to longleaf pine as a producer of naval stores. Both of these pines reproduce poorly after the stands reach maturity. Undergrowth of palmetto, various grasses, and resinous shrubs may exclude pine seedlings unless removed by controlled burning (Fig. 4-5).

Since 1935 pulp production in the southern states has expanded at a rapid rate. High-grade kraft paper may be made from southern pines by the sulfate process. The fibers are long and the paper is very strong. It is

**Figure 4-5** A selectively cut slash pine forest. (*South Carolina State Commission of Forestry and American Forest Institute.*)

used for bags and wrapping. Highly intensive timber management is practiced by pulp and paper corporations in the South. Slash and loblolly pines are the favored species. These are often grown in plantations, some from specially developed genetic stock. Fertilizers may be applied. Stands may be harvested at 20 years of age for pulpwood and at 40 for sawtimber. Operations may be highly mechanized. Mechanical timber harvesters have been developed for limbing, felling, and bucking such trees. Even without using mechanical harvesters, operations are often mechanized. Following felling, the land is cleared of slash by burning or windrowing. It is then gone over with bulldozers or heavy disk plows to eliminate brush and palmetto. After the soil settles it is planted by machine to a second crop of pine. Productivity per acre and per man-day are both high.

Open Southern forests support grass useful for grazing livestock. Cattle can be grown along with pine trees in the South, though tree growth is generally better where grazing is excluded. Where pasturing is moderate, the damage done is less than the value of the meat produced.

**Southern Hardwood Forests**  The river bottoms and adjacent uplands of the South support extensive hardwood forests. Among commercial species are ash, cottonwood, blackgum, maple, various oaks, walnut, and yellow-poplar. Associated with these in swamplands are baldcypress and white-cedar. Much of the land is subject to periodic flooding, which tends to control species composition. These forests produce some high-quality timber suitable for veneer, tight cooperage, lumber, specialty stock, piling, and other products. Growth rates are relatively rapid, 500 to 700 board feet per acre per year for good stands. Cutting cycles can be as short as 8 years.

**Western Forest Areas**  The Rocky Mountain and Pacific Coast sections share many forest type groups. The combined area of the major groups as of 1969 was as follows:[4]

| Type group | Acres covered, in millions |
| --- | --- |
| Douglas-fir | 37 |
| Ponderosa pine | 36 |
| Lodgepole pine | 16 |
| Fir-spruce | 15 |
| Hemlock–Sitka spruce | 10 |
| Other softwoods | 10 |
| Hardwood types | 11 |
| Total | 135 |

**The Rocky Mountain Forest Formations**  The Rocky Mountain forest is bounded by the plains region on the east, the Sierras and Cascades on the west, Canada on the north, and Mexico on the south. The forests in the Rockies are located almost exclusively in the mountains. In the foothills may be found the Rocky Mountain juniper, the bur oak, and western live oak, and in the south mesquite and the piñon pine. These trees, growing in open groves or as scattered individuals, carry out the struggle between the true forest vegetation found at higher elevations and the desert scrub, sagebrush, and shortgrass plains at the lower elevation. Moisture and protection from too severe exposure to sun and wind favor the trees. Droughts, fire, and open exposures favor desert scrub and grasses.

Proceeding up the mountain slopes, one encounters ponderosa pine, blue spruce, more aspen, and often scrub oaks and maples. At middle elevations lodgepole pine, western white pine, grand fir, western larch, Douglas-fir, and white fir occur. In Idaho and western Montana are some of the best forests to be found anywhere in the Rockies. Western white pine, western hemlock, and ponderosa pine form splendid stands with volumes per acre in excess of 20,000 board feet. The trees attain large size, with diameters of 3 feet or more and heights in excess of 140 feet. They produce excellent lumber. From Colorado southward the trees are generally much smaller and the lumber value of the timber is modest. These forests do contain many trees of pulpwood size, but water supply is generally inadequate to support pulp mills.[1]

At higher elevations lodgepole pine, Douglas-fir, Engelmann spruce, and limber pine appear (Fig. 4-6). At high elevations mountain hemlock, alpine fir, Engelmann spruce, whitebark pine, and bristlecone pine are the common species. Just as the forest thins out at low elevations, so also does it thin out at high elevations. Here the trees must find shelter against the high winds that sweep across the mountain crests and against the driving snow that tends to flatten them down to the ground. They must be able to perform their life functions in the short summer period when the mountain peaks are free from snow. Ravines and protected slopes where the snow does not drift too deeply are forested at higher elevations than exposed ridges. Timberline occurs at about 11,000 to 12,000 feet in the central Rockies, at 8,000 to 10,000 feet near the Canadian border.

In the intermountain or great-basin region the forests are found only on the isolated mountain ranges that occur in this area. Their distribution

**Figure 4-6** Strip cutting of lodgepole pine on the Gunnison National Forest, Colorado. (*U.S. Forest Service.*)

is limited to a relatively narrow belt varying from about 7,000 to 10,000 feet in elevation.

The chief products of the central and southern Rocky Mountain forests are water, forage for domestic livestock, and recreation. Timber operations for increasing water yield have been given serious consideration. The value of these mountain forests as range for domestic livestock is often of considerable importance to local ranchers. Forage yields are low because of short growing seasons, but the grass becomes available when plains and foothill ranges have passed their productive peak.

The southern Rocky Mountain region has a total of 73 million acres of forest land, of which only 20 million acres are commercial. Most of the forest land is in public ownership, and forest industries consist primarily of sawmills and industries preparing forest products for local use.

The northern Rocky Mountain states—Wyoming, Idaho, Montana, and the Black Hills region of South Dakota—have a total of 53 million acres of forest land, of which 34 million acres are commercial. Much of this forest land produces valuable timber with rapid growth rate, and potentialities for sustained-yield forestry are good. Of the commercial forest land, the greater part, 22 million acres, is in national forests. An additional 4 million acres are found in other forms of public ownership. Private ownership totals 8 million acres, of which half is in large holdings. Manufacture of forest products is primarily for lumber, poles, and ties. The use of high mountain country by recreationists is almost certain to grow. Range management is a major activity of the region and will undoubtedly remain so in the years to come.

**The Pacific Coast Forest Formations** The forests of the Pacific Coast are the grandest and most awe-inspiring to be found anywhere in the world. These are exceptionally heavy timber producers. Here are found the nation's most valuable forest trees (Fig. 4-2): the Douglas-fir, ponderosa pine, sugar pine, western white pine, redwood, Sitka spruce, western red-cedar, Port-Orford-cedar, and, in the Sierras, white fir, red fir, and the bigtree.

The most valuable commercial stands occur in the North Pacific region, where Douglas-fir and western hemlock are in mixture with Sitka spruce, silver fir, and western red-cedar. Here, stands of great density occur, up to 100,000 board feet an acre or even higher. Douglas-fir trees 6 to 8 feet in diameter and containing seven to twelve 16-foot logs are not uncommon. The forests of this region attain their density and impressive development because of the high rainfall and favorable temperature. Winter snows are rare along the Coast, and dry periods are infrequent and relatively brief. Beyond the Coast range are dry interior valleys that tend to be devoid of trees. As the Sierras and Cascades are approached, the forests again resume sway and hold it up to timberline, which occurs at about 12,000 feet in the south and 8,000 feet in the north.[1]

The Pacific Coast states together contain more than half the reserve sawtimber volume of the United States. From these uncut forests of fir, pine, and redwood comes the major volume of high grade softwood lumber.

The forests of the North Pacific region require very heavy logging equipment. Logging and lumber manufacture represent a major item in the economy and receive important attention from legislatures and citizens. Three of the Pacific Coast states have adopted forest-practice laws. The pulp and paper industry is also developed on the Pacific Coast. Douglas-fir plywood manufacture is a unique feature of West Coast industry.[3]

Along the Pacific Coast from the Oregon border down almost to San Francisco lies the redwood region, the world's most impressive forest. Here, magnificent red-barked trees from 6 to 12 feet or more in diameter lift their crowns 200, 300, and even 350 feet above the ground. The forest floor is covered with a deep bed of needles that cushion the step. An individual tree contains enough wood to build a modern house, and the stands on a single acre would build a village.

Redwood logging is a difficult and slow operation. The huge trees must be very carefully felled if their stems are not to be shattered when they strike the ground. Often it is necessary to prepare a cradle of branches and small trees to break the fall of the giants. The bark on these massive trees is 5 to 6 inches thick. The logs must, therefore, be handled with sling equipment rather than grabhooks. After logging, a huge amount of debris is left on the ground to be disposed of, ordinarily by burning, before a second crop can be started.

Redwood trees attain ages in excess of 1,000 years. Fortunately, redwood can be used in small sizes and grows very rapidly. The growth of redwood and Douglas-fir forests may exceed 1,000 board feet per acre annually under good growing conditions.

The Sierra forests contain giant Sequoia, ponderosa pine, sugar pine, western white pine, whitebark pine, Jeffrey pine, lodgepole pine, Douglas-fir, white fir, red fir, and incense-cedar. These also are impressive forests. The sugar pine may attain a diameter of 6 feet or more and contain five to seven logs 16 feet long. The Douglas-fir and ponderosa pine attain large size and have high value for timber use. The stands are considerably less dense than those along the coast. The Sierra forests are drier than those of the northern Pacific Coast. Hence, protection against fire requires the greatest vigilance. These high slopes, once burned over, are often difficult to reproduce to trees, especially if they become overgrown with mountain mahogany, manzanita, Ceanothus, and other shrubby species. A high degree of skill is required of the forester to maintain the Sierra forests in high productivity and to protect the watersheds from erosion and flash floods.

**Alaska** The forests of Alaska are worthy of special attention. They cover a huge area, 118 million of the state's total area of 365 million acres. They are almost wholly publicly owned. At present all is noncommercial except for some 5.8 million acres along the coast. As much as 40 million acres of interior lands may ultimately have commercial value of at least local importance. Almost all of the commercial forest lies within the Tongass and Chugach National Forests where management is under the direction of the U.S. Forest Service.[2] An additional 400,000 acres of commercial forest are in federal and state ownership. Under the terms of the Statehood Act of July 7, 1958, the state of Alaska was granted and entitled to select 400,000 acres of public domain as well as 400,000 acres of National Forest lands for community and recreational development. In addition, it could select for state use an additional 102,550,000 acres from the open public domain—an area slightly larger than the state of California. The state of Alaska thereby acquired some fabulously rich oil lands.[11]

Together, the noncommercial forest and tundra make up most of Alaska's land area. This is a huge land area that seems destined to remain in essentially a wild condition for decades to come. As such it does have its appeal and challenge. Free of auto roads, it can be penetrated only by plane, boat, or sled. On the whole, wildlife is abundant. Aside from huge herds of caribou with their trailing wolf packs, the state has Sitka deer, brown bear, Dall sheep, and mountain goats to attract the sportsman and amateur photographer. Game fish are plentiful in many of the streams. For the less adventurous, Alaska presents much of North America's most spectacular mountain and seacoast scenery along the Inland Passage where huge glaciers come down to the water's edge.

The coastal forests are made up largely of hemlock and spruce. The forests extend along the coast for almost 1,000 miles, yet occupy less than 10 percent of the area of the state. Much of the commercial timber is within $2^1/_2$ miles of tidewater. Timber growth is favored by a relatively warm and humid climate tempered by the Japan Current.

The coastal timber is impressive. Stand volumes may run as high as 100,000 board feet per acre, and over extensive areas they average 35,000 board feet per acre.[2] Over a rotation of 110 years, second-growth stands are estimated to be capable of producing some 70,000 board feet per acre. The Tongass Forest has some 92 billion board feet of gross volume on land that is considered to be commercially operable. Western hemlock makes up 64 percent of the volume, Sitka spruce 29 percent, western redcedar and Alaskan cedar 7 percent. Similar timber stands are found on the Chugach Forest, though individual trees and stand volumes are lower than on the Tongass. The timber quality on both forests is well suited for pulpwood.

The interior of Alaska is cold and dry. Much of the land is underlain by permafrost. The forests are extensive, made up largely of black and white spruce, with mixtures of birch and aspen. At present, these forests are little used for timber except locally.

Growth on interior forests is estimated at 2 billion board feet annually.[11]

Much of the land has been ravaged by fires. Lightning fires are common. Because of lack of transportation other than air and because of limited resources for fire fighting, the problem of fire control on interior forests is a staggering one (Fig. 4-7).[8] Although fires cause terrific damage, it is recognized that throughout the ages they have shaped the character of Alaskan forests.[6] Burning off a coniferous forest with its heavy layer of organic matter exposes mineral soil with its favorable seedbed for aspen, birch, and spruce. Also, exposure to sunlight increases the depth of unfrozen soil.[6]

**Tropical Forests** Tropical forests in the continental United States are confined entirely to the southern tip of Florida below Lake Okeechobee. These are subtropical rather than tropical. The land is mostly

**Figure 4-7** Fire crew preparing to board a plane bound for a fire in Alaska's interior forests. Spruce forest borders the landing strip. (*U.S. Bureau of Land Management.*)

swampy and covered with thickets of mangrove and other saltwater shrubs. The coconut palm is found along the coast. The tropical forest in Florida has little value for commercial timber production, though it does protect the soil from erosion when heavy seas occur. Within the swampy lands of southern Florida live many large birds and animals, including alligators, that require the solitude of remote areas for their homes.

The state of Hawaii, in the subtropic latitudes of 19 to 22°N, consists of eight main islands and several islets along a 390-mile chain. The total area is 4,111,000 acres. Kauai, the westernmost of the major islands, is the oldest and has the deepest soils. Originally, the better soils on all islands were well forested with sandalwoods and koa, known commercially as Hawaiian mahogany. The level, deep soils on all islands have been taken over for cultivation of sugarcane and pineapple. Forests remain on the steep mountain slopes and the thin-soiled lava flows. A large part of the forest area is covered with desert scrub, of which mesquite is a typical species at the lower elevations. Most of the forest is found between 2,000- and 8,000-foot elevations. The forests have been badly abused through overcutting and grazing by wild goats, pigs. and cattle. Hawaiian forests are now managed mainly for watershed protection and for recreational use. A small amount of timber is harvested locally. The rainfall on many of the Hawaiian Islands is excessive. One mountain on the island of Kauai has more than 800 inches of rainfall a year. The forest officers in Hawaii are attempting to plant burned-over areas and denuded forest lands with tree species that will be particularly valuable in maintaining the soil in condition to absorb moisture rapidly.[5,7]

The United States has real tropical forests in Puerto Rico and the Virgin Islands. Puerto Rico, an island about the size of Connecticut, has a great diversity of vegetation. At its eastern end rainfall is in excess of 200 inches annually, and dense tropical rain forests can be found. Here are giant tabanucos, guaraguaos, ausubos, and the valuable *Magnolia splendens*. The forest is highly diversified, supports large trees, and has a rapid growth rate (Fig. 4-8). Many of the woods are hard, heavy, durable, have beautiful figured grains, and, once properly seasoned, are suitable for cabinetwork.

Most of Puerto Rico's forests are found in a range of mountains extending along the island from east to west. On the rest of the island the demands of agriculture and grazing are so heavy that space does not exist for growing timber. Coffee is cultivated at moderately high elevations under the shade of trees. Agricultural research has demonstrated that coffee can be grown without shade in Puerto Rico if the soil is properly fertilized. The yield of sun-grown coffee can be tenfold that of shade-grown.

**Figure 4-8** A 48-inch tabanuco tree, Luquillo Experimental Forest, Puerto Rico. (*U.S. Forest Service.*)

The moderately dry sections of Puerto Rico grow mahogany, silk cottonwood, Spanish cedar, teak, and other valuable trees. The tabanuco will also grow on moderately dry lands. Only cactus and lignum vitae can survive on the driest sections of the island. Lignum vitae is one of the heaviest and densest woods anywhere. It is richly impregnated with waxes and is therefore valuable for making stuffing boxes for the propeller shafts on ships.

Even though the population is still rapidly growing, the pressure of people on the land seems to have lessened considerably since 1935. Emigration, industrialization, and the growth of tourism have been largely responsible for the change. Local oil refineries have practically done away with the need to cut wood for domestic fuel. Tobacco growing that depleted large areas of mountain soil has declined. Even lands once devoted to cane and citrus crops have been diverted to the grazing of dairy cattle to supply milk for a population that used but little of it in 1935. Cities, on the other hand, have grown, and use of rural lands for residences has expanded. Recreational use of forests has increased greatly. The forests do seem to be holding their own if not in fact being expanded somewhat.

The three Virgin Islands have but limited commercial forests.

## LITERATURE CITED

1. Barrett, John W. (ed.) 1962. "Regional Silviculture of the United States," The Ronald Press Company, New York. 670 pp.
2. Bruce, Mason B. 1960. National Forests in Alaska, *Journal of Forestry,* **58**:437–442.
3. Guthrie, John A., and George R. Armstrong. 1961. "Western Forest Industry: An Economic Outlook," The Johns Hopkins Press, Baltimore. 324 pp.
4. Josephson, H. R., and Dwight Hair. 1970. The Timber Resource Situation in the United States. U.S. Department of Agriculture Forest Service, Washington. 29 pp.
5. Judd, C. S. 1935. Forestry in Hawaii, *Journal of Forestry,* **33**:1005–1006.
6. Lutz, H. J. 1960. Fire as an Ecological Factor in the Boreal Forests of Alaska, *Journal of Forestry,* **58**:454–460.
7. Rhett, Beverly. 1947. Sugarwater Forests, *American Forests,* **53**(1):8–10, 47–48.
8. Robinson, R. R. 1960. Forest and Range Fire Control in Alaska, *Journal of Forestry,* **58**:448–453.
9. U.N. Food and Agriculture Organization Secretariat. 1966. Wood: World Trends and Prospects, *Unasylva,* **20**:1–134.
10. U.S. Department of Agriculture Forest Service. 1969. A Forest Atlas of the South, Southern Forest Experiment Station, New Orleans and Southeastern Forest Experiment Station, Asheville, N.C.
11. Zumwalt, Eugene V. 1960. The Alaska Public Domain, *Journal of Forestry,* **58**:443–447.

Part Three

# Forest Biology

Chapter 5

# Forest Ecology

The concept of system in the life processes that take place on the earth was introduced in Chap. 1. In this chapter a closer inquiry will be made into this system and its subsystems in order to learn what man can and cannot do to make earth's forest ecosystems serve him better than they now do. This is important, for forests cover almost one-half of the land surface that can be permanently inhabited by man.

Forestry deals with life and the elements that go to make up or to influence life. It may be thought of as ecology applied to the management of forest lands. Ecology deals with the interrelationships among plants, animals, and the environment. Its scope will become clearer if we consider as its basic unit the ecological system—or ecosystem.

**THE ECOSYSTEM**

All life on earth—together with those environmental elements that support it—constitutes an earthwide system. But as such it is too vast and complex for a penetrating study of how it performs as a system.

Ecologists therefore study subsystems of relatively uniform composition and life processes. It is these that they refer to as ecosystems.

An *ecosystem* consists of the plant and animal life that inhabits a given area of land or water together with the elements of the physical environment that support, influence, and interact with the living forms present. It embraces the green plants that manufacture their own food, the herbivores that feed on the green plants, the carnivores that feed basically on the herbivores, and the fungi and bacteria that decompose plant and animal remains into simple chemical substances that plants can reuse. The system is able to function by using solar radiation to supply the energy by means of which the green plants photosynthesize sugars, starch, cellulose, and fats from carbon dioxide and water, releasing oxygen in the process. It requires, in addition to carbon, hydrogen, and oxygen, the several other chemical elements that are essential to living plants and animals. From these, using the energy of the sun, all of the complex syntheses, hydrolyses, reductions, oxidations, and polymerizations that living organisms perform in their life processes take place.[3]

The two basic constituents that flow through the ecosystem and unify it are energy from the sun and basic chemical nutrients. Chemical energy, the result of photosynthesis, is at its highest in the green plants. It is progressively dissipated as it flows through the herbivores, carnivores, and decomposers, all of which release it as heat. Much of this heat leaves the system by radiation of the earth to outer space. Where oxygen supply is low, as under water, a small amount of plant and animal remains may accumulate, ultimately to be converted to fossil fuels. Of the chemical end products, carbon dioxide and water are released to the air. The mineral nutrients go to the soil where they become available for recycling in the system. Small amounts are carried away by water runoff or are transported from the system by animals or by man as harvested products.[6]

Ecosystems are both real and abstract. They are real in that they exist in specific areas of land and water. They are abstract in that the same type of ecosystem may be found in several different land areas. We may think of a jack pine ecosystem on a particular section of land or of the jack pine ecosystem as it occurs generally throughout the Lake states.

Ecology deals with ecosystems, their composition, structure, and dynamics.[3] Forestry deals with the establishment, management, and use of forest ecosystems for human benefit. It is appropriate to inquire more deeply into the operation of forest ecosystems to explore what their productive potentials may be. Much can be learned about ecosystems by considering the flow of energy, water, carbon, nitrogen, and other mineral elements through the system.

## THE FLOW OF ENERGY THROUGH ECOSYSTEMS

The sun as a thermonuclear reactor supplies the energy for the earth. Every minute the earth as a whole receives from the sun as much energy as mankind uses in a year.[8] About half of that which reaches the earth's outer atmosphere is depleted by reflection from clouds and dust or is absorbed by gases of the atmosphere and reradiated to space. Of that which reaches the ground level, only about half lies within the visible region of the spectrum that green plants can use for photosynthesis. Even of the energy that falls on a green leaf, much is reflected, some is transmitted, much is used to transpire water, and some is reradiated to the environment. Theoretically only 14 percent of total radiation is available for photosynthesis. At most, 2 to 3 percent is actually converted into sugars and other plant substances.[8] The net amount of solar radiation so converted or fixed by the average forest ecosystem is in the order of 0.3 percent in sunny climates and up to 1.0 percent in cloudy regions.

As a converter of solar energy to chemical energy, therefore, the forest ecosystem appears to be inefficient. However, because of their large area of leaf surface, some six to 20 times that of the vertical projection of the tree crown on the forest floor, forests tend to surpass other forms of vegetation in the amount of solar radiation fixed by photosynthesis.[8] The actual amount of carbon taken from the air and converted into plant tissue by forests can be as high as 4 tons per acre per year. Considered on a global basis, therefore, the total amount of solar energy converted by photosynthesis is enormous.[6, 7]

Forest canopies tend to reduce the light intensity at the forest floor to but 1 to 20 percent of that in the open. Few green plants can photosynthesize enough food to support their life in light intensities of 1 percent or less, hence they are sparse beneath dense tree canopies.

Estimates of the actual amount of energy fixed per year by three types of forest vegetation are given below in kilogram calories per square meter of forest surface:

| Forest type | In K Cal/m² |
|---|---|
| Temperate deciduous forest | 4,800 |
| Temperate coniferous forest | 11,200 |
| Tropical rain forest | 20,000 |

These estimates are indicative of the general range for the three forests and should not be considered as actual averages.[3]

Of the total product of photosynthesis by a green plant, some 23 percent is used directly by the plant to carry on its own life processes and is released as respiration of the living cells in stem, leaves, flowers, fruit, and roots. The remainder theoretically is available to herbivores, but under normal (as opposed to starvation) conditions they consume as little as 5 percent and no more than 30 percent. Some of the product goes into temporary storage, some may be removed from the system, and some goes directly to decomposition.

Of the energy consumed by herbivores, again a large amount is used in respiration and dissipated as heat, some is decomposed, and perhaps 3 percent may be used by carnivores. Eventually the remainder is decomposed. Whether we think of total energy fixed or mass of living substance, biomass, the same principle holds (Fig. 5-1). The energy or mass that remains unused in each step of the food chain, as green plants are fed upon by herbivores, herbivores by primary carnivores, and primary carnivores by secondary carnivores (carnivores that feed on other carnivores), is around 70 to 97 percent.[3] This accounts for the fact that travelers to East African game parks will see considerable grass, shrubs, and small trees for every gazelle, zebra, or wildebeeste, and 100 or more of such animals for every lion, leopard, hyena, or cheetah.[1]

So at each step in the energy flow through the ecosystem some energy is lost to respiration, some to decomposers, some to temporary storage or export. Also the chemical substances formed tend to grow in

**Figure 5-1** Nature's densest biomass occurs in redwood forests, Del Norte, Calif. (*U.S. Forest Service.*)

complexity from the primary producer, the green plant, to the herbivore and the carnivore. All substances of living organisms are eventually decomposed to simple compounds that the plants can absorb to recycle through the system. Thus the system obeys the laws of thermodynamics in that it neither creates nor destroys energy or mass, and that energy is transformed ultimately into heat that cannot reenter the system. Without the continuous supply of solar energy for photosynthesis the ecosystem would cease to function.

## THE WATER CYCLE

Water performs two indispensable roles in the ecosystem: it is one of the basic raw materials for photosynthesis, and it is a solvent and transporting agent for moving essential chemicals into and through the living elements of the ecosystem. Higher plants and most animals take up water in liquid form. Its availability determines to a large degree the type of ecosystem that can be supported. The actual amount of precipitation is less important than the ratio of rainfall to evaporation from a free water surface. In general where this ratio is less than 0.16 deserts prevail, where it is 0.20 to 0.60 grasslands take over, and where it is 0.80 to 1.60 forests occur.

Water enters the forest ecosystem from the atmosphere mostly as rain, snow, or hail. Some may be condensed on plant and soil surfaces. The leafy canopy may intercept up to 11 percent of that from heavy storms and up to 80 percent of that from light showers. The upper leaf litter also prevents some from reaching the soil surface. Intercepted water is evaporated to the atmosphere. Some water flows directly into streams whence it is transported out of the forest ecosystem. The most useful to plants and animals is the water that enters the soil. Here it dissolves basic plant nutrients that enter through the roots and are transported in the sap stream to the leaves. The leaves give off this water through transpiration to the atmosphere. A small amount, about 1.6 percent of that which falls, may be converted by photosynthesis into sugars and other plant products. The actual amount of water converted to plant tissue by a forest during a year is impressive—up to 5 tons per acre. During the cycle this water may have become a constituent of cellulose, starch, fats, lignin, proteins, or ribonucleic acid as it moves from tissue to tissue through plant and animal bodies before being finally released as simple $H_2O$.

Forests affect the hydrologic cycle of the earth because they transpire large amounts of water—perhaps as much as 10 to 20 tons per acre on a warm sunny day. The amount forests use can be temporarily reduced by thinning and sometimes reduced as much as 25 percent by complete removal of the forest. Worldwide, the total energy used by transpiration is estimated to be 10 percent of that reaching the land. Vegetation thus serves as an air conditioner of global proportions.[9]

## THE CARBON CYCLE

Carbon is present in the atmosphere in the form of gaseous carbon dioxide at a concentration of 0.03 percent. Through the open stomata, closeable pores in the leaves, carbon dioxide comes in contact with moist leaf cells into which it passes. Here, by the action of chloroplasts energized by light, it is combined with water to form carbohydrate substances. Oxygen is released that passes through the cell walls and out of the intercellular spaces through the stomata. The carbon, as the water, enters into many chemical compounds as it cycles through the system. Should a wildfire burn over the forest, great amounts of carbon are directly released to the air in the smoke. Wood that is used by man in buildings also returns eventually to the air as carbon dioxide, either slowly through the process of weathering and decay, or rapidly through combustion.

The oceans are believed to serve as a great reservoir of carbon dioxide, interchanging it freely with the atmosphere to keep the relative concentrations in equilibrium.[3] The forests serve a like role by increasing the rate of photosynthesis when the concentration of carbon dioxide rises and decreasing it when it falls.[2,5] The oceans play another role in that certain water plants, such as *Elodeas,* are capable of precipitating from sea water calcium carbonate, and this compound sinks to the bottom to become limestone. Such limestone, if uplifted by geologic action, may be decomposed by organic and other acids to release carbon. Thus an upland ecosystem may release carbon that had been bound up in limestone many geologic ages ago. This exemplifies how ecosystems may be interrelated in both time and space.

## THE NITROGEN CYCLE

Nitrogen in molecular form is present in the atmosphere at a concentration of about 79 percent. In this form it is not available for use by tree species. It must first be fixed in the form of nitrates. The fixation process is carried on by a number of species of bacteria and by blue-green algae. Small amounts may be fixed by lightning discharges. Bacteria, the chief agents in fixing nitrogen in forests, are of two types: symbiotic with other plants, and free living. Symbiotic bacteria are associated with leguminous plants of which locust and Kentucky coffee tree are examples. Alders and other forest plants also have associated nitrogen-fixing bacteria. The bacteria cause the formation of nodules on the roots in which nitrogen is fixed. It moves directly from the nodules up the stem to leaves and other growing regions.

Free-living bacteria also fix nitrogen in the soil. Several species are involved. Nitrogen-fixing bacteria use carbon compounds as a source of energy to perform the fixation process. The total amount of nitrogen fixed by both types can be as much as 500 pounds of nitrate per acre per year in clover fields.[3] Plants convert nitrates into amino acids and proteins. These in turn can be used by herbivores and carnivores. Eventually all plant and animal protein is decomposed by bacteria and fungi. These convert proteins into ammonium salts that may be reconverted into nitrates in one or more steps by nitrification bacteria. Other bacteria are capable of denitrification whereby fixed nitrogen is converted back to molecular nitrogen, which prevents accumulation of nitrogen in soil at levels toxic to plants. Of all mineral nutrients in soil, nitrate is the one most likely to be in deficient supply.

## CYCLE OF OTHER PLANT NUTRIENTS

Sulphur, potassium, phosphorus, calcium, magnesium, iron, and the other elements that are essential for the growth of plants, as well as those additional ones essential to animals and that must be obtained from plants, also go through cycles in the ecosystems analogous to those for water, carbon, and nitrogen. Many of these elements, especially magnesium and calcium, tend to concentrate in the leaves of trees. Others are concentrated in the fruits. The concentration in the main stem of trees is low. Hence wood may be removed from the forest with little or no reduction in those elements that result in forest soil fertility. It has been found for instance, that the amount of nutrient elements removed by timber harvesting is less in some areas than that returned each year by accumulation of dust.[6]

## SITE FACTORS

The above discussion has concentrated on the movement of energy and nutrient elements through the ecosystem. The rate of movement and the total biomass attained, as well as the annual accretion thereto, are determined by the capacity of the environment to supply the needed growth elements in balanced amounts and by the inherent capacity of the several species of organisms to make full use of these elements. The environmental growth factors, or site factors, may be grouped under three headings: climate, soils, and biotic.[10]

**Climate** *Solar radiation* not only provides the energy for photosynthesis that fuels the ecosystem, but is responsible for climate as it is expressed through temperature, precipitation, air movement, seasons, and length of day. All of these profoundly affect ecosystems. As the vertical component of solar radiation is highest at the equator and lowest at the poles, ecosystems tend to attain their highest biomass, complexity, and abundance of species in the tropics, and their lowest in the polar regions. The extremes of broad ecosystems are represented by the tropical rain forest and by the arctic-alpine tundra.

*Temperature* controls the rate of most chemical reactions. Photosynthesis of green plants is very low at temperatures below freezing. Only a few plants, among them some coniferous trees, are able to augment their net chemical energy reserve at temperatures below zero degrees centigrade. During January and February loblolly and white pine respiration exceeds photosynthesis.[4] Even at 10°C photosynthesis is low. As temperature rises to about 38°C, photosynthesis increases to a maximum and returns to zero at about 50°C. Few plants can endure a temperature of 50°C for more than a few minutes. The life processes of plants and animals respond to temperature much as does photosynthesis. Warm-blooded animals, however, operate within a very narrow body-temperature range.

Low-incident solar radiation, therefore, becomes one of the major factors in accounting for slow growth, small sizes of trees, and restricted species composition of the northern forests. It has its concomitant effect on the type of ecosystems to be found in arctic and subarctic regions. This is particularly true of the populations of soil fauna that play such a large role in the decomposition of litter. In the northern forests fungi are chiefly responsible for this function, whereas in temperate regions earthworms, mites, and many other creatures are active in litter decomposition.

*Precipitation.* next to temperature, is the major climatic determinant of the lushness of ecosystems. The minimum amount of rainfall required to produce a closed forest canopy is roughly 30 to 40 inches in the tropics, 20 to 30 inches in the temperate zone, and 10 to 20 inches in the arctic regions. Rainfall may be excessive in the tropical regions, especially at higher elevations where the soil may remain continuously saturated. This gives rise to the dwarf mossy or elfin forests with their many orchids, bromeliads, and other epiphytes growing on the tree branches and trunks. Difficulty in getting an adequate supply of nutrients limits plant growth where continuous saturation and water movement leaches away the mineral elements.

At the opposite extreme are the deserts, where little runoff occurs. The excessive evaporation causes mineral salt accumulation in the surface layers, and the so-called alkali soils result. The rain that does fall in many arid regions occurs in heavy downpours, producing high runoff that may flush away the accumulated alkali. To exist in deserts, plants must either perform their life functions during the short period in which the soil is moist, or develop structures that enable them to endure long periods of drought. Trees are generally restricted to water courses. Some develop very deeply penetrating root systems that keep them supplied with water. Trees tend to use water copiously when soil moisture is abundant, and much more slowly when the soil becomes dry. As soil becomes dry, photosynthesis in trees declines markedly.[4]

*Length of day* as well as daily temperature regulates seasonal activity of many plants and animals. Examples in plants are dates of bud opening, flowering, fruit maturation, and tissue hardening. In animals migration dates, breeding and nesting times, and hibernation may also be induced by day length. This response minimizes injury and mortality that would occur if temperature alone controlled behavior. For example, prolonged unseasonable cold in spring may delay bud opening and bird migration but unseasonable warmth in late winter does not induce them. Similarly, trees will shed their leaves and harden buds in the fall even though frost may be delayed for several weeks. Further, within a single species, such as the ponderosa pine that grows over a wide geographic range, climatic races exist. For this reason foresters are careful not to move tree seed more than 1 degree in latitude or 1,000 feet in elevation lest the progeny prove nonhardy or slow growing.

*Wind* influences vegetation along seacoasts, on mountain ridges, and on the western plains of the United States. Trees at timberline on mountains tend to be misshapen. Their live crown may extend at a sharp angle away from the prevailing wind. Some may even become prostrate from wind-driven snow, ice, and soil particles. Winds of hurricane force may fell trees over wide areas. Tornadoes are even more destructive, but the size of the area affected is usually restricted. Damage by hurricanes and tornadoes is generally infrequent. Even though they may destroy many trees or almost an entire forest they seldom cause a permanent change in the nature of the ecosystem.[10]

Forests greatly impede wind. In temperate forests wind velocity may be but 5 to 30 percent of that over open land. Tropical rain forests, with their dense growth of vines and thick leaf cover, may show but 0.4 percent as much wind velocity at the ground level as they show above their crowns. Winds have a beneficial effect in bringing moisture-laden clouds that precipitate water, in bringing fresh supplies of carbon dioxide, and in carrying in mineral particles that collect on the leaves ultimately falling to the ground to augment plant mineral supplies.

**Forest Soils** Trees must obtain their water and mineral nutrients mainly from the soil. Shallow soils cannot supply adequate water and nutrients to support luxurious forest stands, any more than those composed of gravel and coarse sands and having a low water table can support a dense, high-forest cover.

Soils are formed by the action of climate and ecosystems on the parent rock material. Climate, through seasonal, diurnal, and other temperature changes, and through the dissolving power of the water it supplies, breaks down rock fragments into coarse and fine particles classified ultimately as sand, silt, and clay. Climate is also responsible for the leaching of soluble nutrients from the soil into stream channels.

Vegetation adds annual layers of litter to the soil surface, which supports soil organisms that fragment the litter and carry the particles below the surface where they come into intimate contact with mineral particles. Humic acids tend to break down complex minerals, releasing chemicals useful to plants.

Glaciers, water, and wind transport and deposit soil particles. They also abrade rock, grinding it into coarse and fine particles. In general, soils formed of transported material tend to be better for plant growth than those formed by weathering of rock in place. During the transporting stage rock fragments from many sources may be mixed together, thereby generally assuring a more complete supply of the various needed minerals than might occur in residual soils. The process of deposition tends to assort the sizes of fragments into layers, often with a good mixture of fine and very fine material superimposed. Thus good water-holding capacity and adequate drainage are afforded. Windlain soil is known as loess. It and good river bottom lands are favored for agriculture. Sandy outwash plains from the glaciers of the ice age often tend to be low in mineral nutrients as compared with the well mixed material in moraines.

Soils of arctic and subarctic regions tend to be shallow and to have undergone little weathering compared with those of temperate regions. They also may have an underlying frozen layer known as permafrost. Permafrost impedes soil drainage and also prevents the roots from penetrating to the deeper layers to replenish the plant nutrients carried away by stream flow.

Tropical soils are subjected to intensive leaching. Soluble silica and other minerals leach downward leaving the insoluble iron and aluminum compounds that impart a red or yellow color. Many of the plant nutrients that are available are concentrated in the thin layer of rapidly decomposing organic material from which they are promptly taken up by roots and recycled through the ecosystem.

The action of water and to some extent of soil organisms in moving organic matter and solutes downward in the soil tends to cause definite layers or horizons to form in forest soils. Immediately below the organic layer is the leached layer or *A* horizon from which soluble minerals are carried downward to be deposited in the *B* horizon. This is the zone of accumulation and enrichment. Below this is the *C* horizon that is made up of unmodified parent material. The character of the horizons is determined to a large extent by the type of forest supported. And in turn the type of soil, its acidity, nutrient- and water-supplying capacity determines the type of forest that will grow on it. In general hardwoods favor and are favored by soils of granular structure, neutral reaction, and generous calcium content. Conifers are able to grow on acid soils and those relatively low in available nutrients. They tend to impart strong differentiation to the soil horizons.

**Life in the Soil** The writer started his forestry career as a plant physiologist. While attending the Botanic Congress in 1935 he met the famous soil scientist Sir John Russell. Sir John chided plant physiologists for using water cultures with highly purified chemicals rather than soil. "Soil," Sir John said, "is not sterile sand and water solutions. Soil is alive." And so it is. It is filled with rotifers, bacteria, fungi, nematodes, algae, round and flat worms, insects and their larvae, other arthropods, and even small mammals scurrying about to make their living from the leaf litter and from each other. The type of forest profoundly influences the soil fauna and flora supported. Broadleaf trees tend to support a rich fauna, while conifers, especially in cold regions, tend to support fungi as the main form of decomposers. Generally the soils that support the richest diversity of species tend to be the most fertile. It is these creatures burrowing through the soil, mixing it and decomposing organic material and minerals that cultivate soil for the forester. They keep the soil porous, making channels for root and water penetration. They aerate the soil. And most important of all they decompose organic material, releasing essential minerals for recycling through the ecosystem.

**Biotic Factors** The living members of an ecosystem form a third important category of site factors, the biotic factors. The various forms are highly interdependent. Populations may fluctuate widely with food and water supply. The various species feed on one another. One species may compete with another, and individuals of the same species may compete with one another. They also perform services for one another. When catastrophe strikes, such as a forest fire, numbers of individuals may be decimated. The survivors quickly multiply to restore the normal population levels. Among plants and animals that produce many progeny in a year, restoration is very rapid. The capacity of a species to multiply under favorable conditions is spoken of as its biologic potential. The site factors that hold the population in check are referred to as environmental resistance. Environmental resistance may be brought about by unfavorable climate, inadequate water and soil nutrients, or action of parasites, predators, or competitors.

**Limiting Factors** The total substance of living forms of any ecosystem, the biomass, is the product of the site factors. An individual site factor has a minimum, an optimum, and a maximum. For the effect of temperature on photosynthesis we have seen that these three are, respectively, about 0, 38, and 50°C. As temperature rises above 0°, photosynthesis increases steadily up to about 38°. But should water supply or light be inadequate the increase stops before the 38 degree optimum is reached. Liebig formulated this relationship in his law of the minimum: When a reaction depends upon a number of factors, the rate at which the reaction proceeds depends upon the factor that is in minimal supply. Agriculturists use this principle to determine the kinds and amounts of fertilizer to apply to crops. Wildlife managers use it to ameliorate the factor that is limiting the population of a game species. Foresters use it when they open an overhead canopy to promote establishment and growth of tree seedlings.

## ECOLOGICAL SUCCESSION

Ecosystems have been described as self-contained and self-perpetuating systems of high stability. Actually but few are stable over long periods. It is characteristic of an ecosystem that as it develops it changes the soil and other factors of its environment. It may thereby prepare the way for its own replacement. It may take millennia for a lichen ecosystem on exposed rock surface to be replaced. Only a few years are required for an aspen–gray birch cover to crowd grass and weeds from an abandoned field in the northeast. Within some 30 years the gray birch itself dies out and later the aspen dies also. Seedlings of red maple, ash, and elm slowly take over. Eventually the seedlings of sugar maple, basswood, beech and hemlock invade and so reestablish the climax forest that the farmer's forebears cleared from the land (Fig. 5-2).

Originally it was supposed that once the climax stage was reached it

# FOREST ECOLOGY

**Figure 5-2** A climax ecosystem of western red-cedar and Alaska cedar, Revillagigado Island, Tongass National Forest, Alaska. (*U.S. Forest Service.*)

would maintain itself indefinitely if undisturbed by man. As one tree died another would take its place so that a continuous cover of high forest of large trees would be maintained. Unfortunately few climax forests seem to behave this way. If man does not cut the forest, some other destructive agent seems to take over. Competition among the large trees becomes acute. A dry climatic cycle, an insect outbreak, a windstorm, or some other destructive agent may cause the large trees to die in groups. As openings form, border trees are exposed to increased stress from drying winds. So a magnificent stand of climax forest eventually breaks down, in the process replacing itself by seedlings that grow up in the openings. The prevalence of root mounds from overturned trees testifies to the importance of windstorm as a factor in forest breakdown and replacement in the northeastern states. When such a catastrophe occurs, wood-destroying insects and fungi multiply at an exponential rate to feed on the dead or overturned tree trunks and branches. With the aid of bacteria they quickly decompose the woody material, returning its basic substance to the soil and air. After so doing. even more quickly their populations subside to normal numbers. A new forest having much the composition of the old thus takes over, but it is made up of young and thrifty individuals.

Change and breakdown are an inevitable stage in the life of every forest stand. It may be affected by timber harvest for human use, but if it is not, then nature in her own way will eventually bring about a rejuvenation of the forest. Such natural rejuvenation is often disturbing to man when it occurs in an old forest he sought to preserve for posterity to enjoy (Fig. 5-3).

In the breakdown stage of any forest ecosystem most of the energy that is released contributes nothing to soil building. The small portion that is so used is mostly that from the leaves, not from the main trunk or heavy branches. Man may therefore remove timber from the ecosystem without harming it, impoverishing the soil, or even necessarily interfering with the effect of the forest on stream regimen. The removal of plant nutrients by a timber harvest seems to be about one to 10 times that annually replaced by weathering and rainfall.[7] Man may in the process of harvesting

**Figure 5-3** Breakdown of old-growth forest, Admiralty Island National Forest, Alaska, 1938. (*U.S. Forest Service.*)

improve conditions for birds, game, and other useful wildlife, and may even augment water yield from the land until the forest cover is reestablished. This is fortunate for man as otherwise he could not supply his needs for timber and paper. Careful harvesting of timber crops need cause no more lasting damage to the forest than harvesting field crops does to a well managed farm.

Forest ecosystems tend to be resilient in their response to disturbances. Were this not so they long ago would have disappeared from the earth. Aside from major climatic changes, fire, overgrazing, or land-clearing by man, forests tend to hold the land once they become established. They are subject to destruction by various soil, water, and air pollutants, but these are usually local. Only when some highly important local element in the system is fragile, such as soils on steep slopes, is the system as a whole fragile. Another example is the tropical rain forests of the mid-Amazon basin where leaching of nutrients from the light sandy soil is very rapid. Once the forest with its leaf litter and humus layers is destroyed, nothing remains to hold the mineral nutrients from washing away within a year or so. It is then no longer possible within a reasonable time to restore a high forest of the vigor of the one removed.

## COMPLEXITY OF FOREST ECOSYSTEMS

In a highly simplified generalization, the philosopher Korzybsky characterized plants, animals, and man as respectively binders of energy, space, and time because of their capacities to photosynthesize carbon compounds, move about, and record thought. The use of the terms green plants, herbivores, predators, and decomposers in the treatment above is also a highly simplified generalization.

In reality exceedingly complex interrelationships exist both between the major categories of life and within them, as individuals compete with one another and species compete with species for continued existence. A variety of complex protective devices have developed. Thorns tend to protect certain trees from overbrowsing by herbivores. The unpalatable taste of certain plants and insects provides a degree of protection from some that would feed upon them. A host of symbiotic relationships also develop, such as the various plant and insect specializations that assure pollination as well as provide food for the insect. Each herbivore tends to have its favorite food plants and will shun others if favored ones are plentiful. Each carnivore also has its preference for a rather limited number of prey species. Even decomposers specialize. For example it has been found that as many as six separate species or instars within a species of mites successively participate in the breakdown of leaf litter.

## MAN IN THE ECOLOGIC SYSTEM

Wherever man lives he has an impact on the ecosystem. It may be light, as in the case of primitive people, or profound, as in the case of high-consuming affluent societies. The impact is exerted chiefly by how man obtains his food and raw materials, how he processes them, and how he disposes of his wastes.

It has already been mentioned that in the Mediterranean region former forests have been converted to semiarid brush lands by timber cutting and heavy grazing of sheep and goats. Overgrazing alone has been responsible for serious erosion and flood damage in Pakistan and India. In China erosion of mountain slopes followed intensive cultivation. In Brazil steep land has gone through a complete cycle that started with forest clearing for coffee. Sugar cane growing followed when the land had eroded so that it would no longer support coffee. Cattle grazing followed when cane would no longer grow, and finally tree planting for mine timbers and pulpwood is returning the land to forest. Many decades will elapse, however, before the new forest will restore the soil productivity to a level that again will support coffee or even a high quality forest. Analogous cycles of land use change have occurred in the United States on the hill lands and stony fields of the Northeastern states and the piedmont of the South.

Steep lands are by no means the only ones to be impoverished by agricultural use for which the land was unsuited. Extensive fields on the southern coastal plain that had been laboriously drained by slave labor also quickly became infertile under cultivation. Most have since returned to forest. Plowing the western prairie eventually created the dust bowl of the 1930s that is now largely controlled by shelterbelts and improved cultivation practices. Nevertheless it remains vulnerable should a new dry cycle return.

In the Lake states fires raged through the slashing left from harvesting heavy stands of white and red pine. These consumed the pine litter and humus, leaving an ash cover over the sandy soils beneath. Plant nutrients in the ash rapidly leached through the light soil making it unfit for agriculture and even unfit to support a stand of pine such as was logged. Low value jack pine seeded in to restore a forest cover over much of the land but some areas were taken over for decades by ceanothus, various heath plants, and other low shrubs. Logged hardwood stands in the Lake states, if not taken over by aspen following wildfires, were not infrequently occupied by a dense grass cover of little value for grazing but still forming an ecosystem capable of keeping out tree seedlings for an indefinite period.

Under the care of foresters, forests are gradually reclaiming much of the land they formerly occupied in the Lake states. It has been necessary over much of the land to start with such pioneer species as jack pine and aspen forests, rather than with the more valuable white pine and hardwoods.

Ecosystems vulnerable to logging are more likely to be encountered where moisture supply is limited, as in Douglas-fir stands on the eastern slope of the Cascades, than where moisture is abundant as it is on the western Cascade slopes.

The damage to land caused by steep-slope cultivation, overgrazing, and careless logging have long been recognized by people in many parts of the world. Newer forms of ecosystem abuse have resulted from oil spills on land and sea; air and water pollution by industrial plants, municipalities, and motor exhausts; misuse of persistent pesticides and solid waste disposal. Where pollution is concentrated forest ecosystems may be weakened or destroyed completely, as in the neighborhood of smelters. This topic will be dealt with further in later chapters.

Fortunately for man many ecosystems withstand human use remarkably well. Those on moist sites such as the western Cascade slope and the grasslands of New Zealand are good examples. Man has converted former forests to field crops that have been cultivated successfully for 4,000 years in some regions, notably parts of the Middle East and the Ile de France country north of Paris. Man-established forest ecosystems have been under management for some 650 years without any sign of decline in yield of forest products.

## LITERATURE CITED

1 Brown, Leslie. 1965. "Africa: A Natural History," Random House, Inc., New York. 300 pp.
2 Hardh, J. E. 1966. Trials with Carbon Dioxide, Light and Growth Substances on Forest Tree Plants, *Acta Forestalia Fennica,* **81**:1–10.
3 Kormondy, Edward J. 1969. "Concepts of Ecology," Prentice-Hall, Inc., Englewood Cliffs, N. J. 209 pp.
4 Kozlowski, Theodore T. 1962. "Tree Growth," The Ronald Press Company, New York. 442 pp.
5 Mikola, Peitsa. 1969. Comparative Observations on the Nursery Technique in Different Parts of the World, *Acta Forestalia Fennica,* **98**:1–24.
6 Ovington, J. D. 1962. "Quantitative Ecology and the Woodland Ecosystem Concept," in J. B. Cragg (ed.), "Advances in Ecological Research," 1:103–192, Academic Press, Inc., New York.
7 Ovington, J. D. 1965. "Woodlands," English Universities Press, Ltd., London. 144 pp.
8 Reifsnyder, William E., and Howard W. Lull. 1965. Radiant Energy in

Relation to Forests. U.S. Department of Agriculture, Forest Service, Washington. Technical Bulletin No. 1344. 111 pp.
9 Smith, Frederick E. 1970. Ecological Demand and Environmental Response, *Journal of Forestry,* **68**:752–755.
10 Spurr, Stephen H. 1964. "Forest Ecology," The Ronald Press Company, New York. 352 pp.

Chapter 6

# Forest Care and Use

If people are to enjoy to the fullest the benefits that forests make possible, then men must care for the forests and use them well. Use of the forest involves collecting and processing products, taking advantage of forest services, and enjoying their amenities. Caring for forests includes protecting them from destructive agents, which will be covered in the following chapter, and husbandry of forests. Husbandry and long-term use are inseparable functions, for the use to be made of a forest determines the type of care and husbandry it should receive. If timber production is the dominant use, then husbandry involves planting or selecting from natural stands the kinds of trees most useful for timber and providing these with the root and crown space they need for their rapid development. If municipal water supply is the dominant use, then the forest should be managed so as to protect the quality of the water and such steps as may be feasible should be taken to augment the annual flow. If the forest is dedicated to intensive recreational use, such as occurs in large city, county, state, and national parks, then forest care involves avoiding littering, undue compaction of soil around areas of concentrated use, protecting birds, animals, trees, flowers, and shrubs from human molestation, as well as providing trails, picnic areas, camp grounds, lookout points, and other improvements that permit intensive use without destruction. This requires also selecting trees and other vegetation that are resistant to heavy use where heavy use occurs, and assuring light use where rare and particularly attractive plants and animals are found. If the dominant use is preservation of wilderness values, then forest care involves keeping human impact at a minimum. Analogous care should be provided if the dominant use is for game production, soil protection, or environmental amenities.

Multiple use is the most common in commercial forests. Here a careful balance in husbandry must be applied so as to achieve high-level sustained yield of the several benefits in proportion to their overall contribution to human welfare.

Natural forest ecosystems tend to endure for long periods of time yet are continually changing. This change may be minor, or it may result in one ecosystem being replaced by another. Some of the most impressive forest ecosystems, and certainly those most valuable for timber production, are subclimax systems. To be perpetuated they require periodic fire, insect outbreaks, windthrow, or the intervention of man.

This chapter deals with how man may use and modify forest ecosystems to make them best serve his needs. As machines and new technologies are introduced and perfected, as general affluence increases, and as man's priority of use changes, so does the extent to which he is able, both purposely and inadvertently, to modify the ecosystem. Because of his imperfect knowledge, man's efforts to modify the forest sometimes lead to outcomes he neither anticipated nor desired. Such unpleasant surprises are just as likely to occur in the future as they have in the past.

The basic fact must nonetheless be faced that to live in organized societies man must modify and even replace natural ecosystems with those better suited to supplying his needs. He is obliged to operate with incomplete knowledge and this probably will always be the case. Corresponding incompleteness of knowledge exists in medicine, agriculture, engineering, and public affairs.

*Silvics* is the term used for the science of biology, including ecology, that relates to the management of forest ecosystems. It deals with the genetics, physiology, structure, and interrelationships of trees with the other components of the forest ecosystem. It also covers the responses to be expected from trees to the conscious effort by man to manipulate the ecosystem to supply his needs.[11]

*Silviculture,* literally culture of the forest, is the art of using the sciences embraced under silvics to produce high yields of valuable crops and services from the forest. Silviculture is limited by the nature of the forest ecosystem involved, by the forester's understanding of it, and by the expenditures he can afford to make to achieve a desired result. It is also limited by something else—the sensitivity of the forester for the forest. A French forester, upon seeing a particularly perceptive example of good silviculture, expressed it this way: "If I were a tree this is the kind of forest in which I would wish to live." Such sensitivity to the harmony of a well-ordered forest comes about only from intimate acquaintance with and respect for the forest.

Practice of silviculture is an art in that it must harmonize the desires of the owner with what he can afford to invest, and with what it is possible to accomplish within the scope of the ecological processes with which he must work. It is an art also in that it requires the forester to combine knowledge of many sciences with economics through exercise of well-informed judgment.[10]

It follows from the above that silviculture must be varied with type of forest, species makeup, climate, and soils as well as with the owner's wishes. To help the forester do this intelligently, publications on the silvics of forest trees, regional silviculture, and silvicultural guides for major forest types have been prepared.[2,3,7,8,13,14]

Silviculture applied to the growing of timber crops is on a relatively firm foundation since costs of many operations have been worked out and anticipated returns can be predicted within reasonable limits.

Less progress has been made in quantifying responses of forests to silvicultural treatments applied to increase water yields, wildlife-carrying capacity, recreational benefits, and environmental amenities, though all have been among objectives of foresters and public and private forest owners. Uncertainties of response and imprecise measures of recreation and amenity values have complicated the evaluation process.

What can a forester do to improve on the natural forest or the one left to him from past misuse by former owners? He can make the forest more productive by:

    1  Controlling species composition
    2  Controlling stand density
    3  Restocking unproductive areas
    4  Protecting against destructive agents and salvaging damaged trees
    5  Controlling the length of cutting cycles and rotations
    6  Organizing the forest so as to facilitate harvesting, management, and use

7 Protecting the site and the nonmonetary benefits of the forest[10]

8 Increasing the water yield of a forest by manipulating the cover so as to favor snow accumulation and to reduce evaporation

9 Increasing its wildlife-carrying capacity by favoring food- and cover-producing trees and plants, by creating openings and border strips, and by avoiding large areas dominated by a single species or age-class

10 Cutting dead and decadent trees to enhance the appearance of a forest and make it a safer place for campers and other visitors

11 Providing access, creating vistas, and developing a range of stand ages and composition to bring about variety that makes the forest more inviting to man

Forest vegetation tends to spread until it occupies the available growing space. When this occurs one or more of the growth factors—water, light, carbon dioxide, or mineral nutrients—become limiting. Trees then begin to die from competition. The forester accomplishes the improvements cited above largely by harvesting or killing surplus and unwanted trees so that the growth potential of the forest may be concentrated on its most desirable elements.

The care man may apply to enable the forest to serve him well must be based on knowledge of what happens during the life of a tree and of a forest.

**Life of a Tree in the Forest** Most forest trees start out as seedlings. The seed of a forest tree may be very small and light. Tiny seed such as that of the trembling aspen may be carried miles by the wind and must germinate within four or five days or the seed dies. Other seeds are large, such as those of the oak, chestnut, walnut, and, largest of all, the coconut. These fall directly to the ground. They may roll a short distance, but unless transported by man, animals, or water, they germinate near the mother tree. Large seeds may also be short-lived. They can be killed by excessive drying; consequently, if they are to germinate, they must find a moist place on the forest floor. Rarely do these seeds live through the first summer unless they germinate. Seed of hawthorn, basswood, and junipers are long-lived, rarely germinating before the second summer following ripening.[12]

Each seed has stored within it reserve food to support the life of the seedling—for 3 or 4 days in the case of aspen, as long as 1 year in the case of oak. As soon as the reserve food in the seed is exhausted, the seedling is on its own. The organic food that is converted into stem, root, branch, and leaf tissue must come from the photosynthetic activity of the leaves. The seedling must have access to water, carbon dioxide, and light to provide its food substance. Its roots must take up a number of mineral elements that are essential to the life of the tree. These include nitrogen for protein synthesis; phosphorus for nucleoproteins and energy transfer; potassium apparently for enzyme action and translocation of carbohydrates; sulfur for amino acids and growth substances; calcium for cell walls; magnesium, a constituent of chlorophyll; iron to promote chloroplast synthesis and respiration; manganese for chlorophyll synthesis; zinc for synthesis of growth regulators; copper for enzymes; boron for sugar translocation; molybdenum for nitrogen metabolism. Other elements in trees include aluminum, sodium, and silicon found in relatively large amounts and others found in smaller amounts. What role, if any, these play in the tree's living processes is still to be determined.[6] Favorable temperature is also essential for photosynthesis and growth.

The rate at which a tree grows depends largely on the rate at which it can be supplied with its basic growth needs—mineral nutrients, water, carbon dioxide, and light. If any one is absent, growth slows down or ceases entirely; and if it remains absent for long, death will ensue. Fortunately, throughout the moist tropical and temperate zones most elements are available in adequate amounts for satisfactory tree growth. The one that is most likely to be limiting is moisture supply. Occasionally, a soil is found that is deficient in calcium, potassium, sulfur, or some other one element. An interesting case of deficient supply of potassium occurred on formerly heavily farmed sand soils on the Pack Demonstration Forest at Warrensburg, New York. A red pine plantation badly stunted quickly resumed normal growth where pine or hardwood slash from healthy trees was distributed and allowed to decay. This led to a series of experiments applying various individual and combinations of mineral salts until the deficient element, potassium, was revealed.

Experiments have shown that trees may make substantial response where nitrogen is applied to poor soils. Carbon dioxide is always present in the air in sufficiently large amounts to support tree growth. Sunlight is always ample in the temperate and tropic zones except beneath the shade of a dense forest overstory.

Many tree seedlings survive best if protected from the hot rays of the sun that heat and dry the soil. Usually, seeds that germinate best in the shade produce seedlings that are capable of surviving for a considerable time in low light intensities. These are said to be shade-tolerant. Those that germinate best in the open may require abundant sunlight if they are not to weaken and die. These are said to be shade-intolerant seedlings.[2]

**Hazards to Survival** From the very beginning the young tree seedling faces conditions that tend to destroy it. A dry hardwood leaf that is blown accidentally over a tender pine seedling may, if packed down by rainfall, be sufficient to cause its death. A small insect may clip the stem or destroy the growing point before it can form leaf tissue to support life. An animal may tread on the seedling or a bird may eat it. A hailstone may break it off, and even rain may beat it into mineral soil. Literally hundreds of seedlings must start for each tree that grows to maturity.

When the seedling gets larger, a rabbit may eat it, a disease may strike it, or insects may consume it.

When the seedling reaches sapling size, it must compete with its fellow trees and sometimes with older trees that shade it and send their roots into the same soil areas. Wind, lightning, fungi, or insects cause death of the older trees, and so the sapling gets its chance to get ahead. When it attains pole size or larger, wind, sleet, snow, and ice all present hazards to survival.[11]

**Maturity and Death** The growth of leaves, stems, and roots is regulated by plant-growth substances called plant hormones, or auxins. These auxins are complicated chemical substances that tend to inhibit growth in one region and stimulate it in another. Growth substances are largely responsible for the expression of dominance in terminal buds and in the main root system.

As the tree grows to maturity, it reaches the period when it begins to produce flowers and fruit-containing seeds to perpetuate the species. Seed production is generally favored by an abundance of sunlight and moisture. Consequently, after partial cuttings the remaining trees tend to develop a larger crown and to produce abundant seed. Good seed years may occur annually in some species, biennially in others, and at irregular intervals of 3 to 9 years in many species.[6, 12]

Seed production becomes more bountiful as the tree approaches maturity. When abundant seed is formed, less food reserves remain to support the growth of the tree. When growth rate declines, the tree is unable to extend its root or crown system fast enough to meet its needs. So, gradually, the overmature tree weakens. Heart-rotting fungi grow in wounds caused by broken limbs, fire scars at the base, or injuries caused by man or animals. Eventually, the weakened tree is either thrown by wind or dies where it stands.

This is the fate of a tree in an undisturbed forest. If the trees are of uniform age all may tend to weaken and die over a relatively short period of time. Such a forest is said to have become overmature and to break down. It is then replaced by a new forest.

Throughout the life of a forest stand, individual trees are dying due to competition and other causes or combination of causes. Harvesting such trees before death occurs is a means of salvaging wood already grown before deterioration sets in, and of helping to keep the remaining trees vigorously growing.

## CLIMAX FORESTS, VIRGIN FORESTS, AND WILDERNESS

To apply effectively the concepts of climax and subclimax forests, introduced in Chap. 5, some further understanding is needed.

The climax forest should be distinguished from the virgin forest. A virgin forest is one that has not been disturbed by man's activities or heavily grazed by domestic livestock. It may represent almost any stage in the succession from the pioneer trees to the climax forest.

The climax forest, on the other hand, is composed only of those species that can maintain possession of the land indefinitely without the aid of man or fire. Neither the climax forest nor the virgin forest is necessarily ideal for man's use. Examples will make this clear. On Star Island in Cass Lake, Minnesota, no commercial timber cutting has been permitted. The soil is basically uniform throughout. On the island can be found every major stage in forest succession for the region. The pioneer jack pine forest occupies the land most recently burned, but even though this forest is mature, few seedlings or poles of jack pine can be found. Instead, the longer-lived red pine, where mixed with the jack pine, is taking over. The red pine also reproduces only moderately well, so it is being replaced by the still longer-lived white pine. The white pine forest is the grandest of all. The trees attain heights of up to 120 feet and ages of over 400 years, producing a forest of high esthetic value and utility. But white pine also is giving way to the maple-basswood climax. Ill-adapted to the light sandy soil, the trees attain heights of only 40 to 50 feet and ages of 150 years. Unimpressive to visitors and useless to the logger, it nevertheless can hold the soil indefinitely against the pines and other species and therefore represents the climax.

It stands to reason that something had changed to bring about the replacement of jack pine by maple-basswood. Soil analyses showed a slight increase in the calcium content of the upper soil layers. Also, there was an increase in the under vegetation and its litter that made it more difficult for pine seedlings than for those of maple and basswood to become established. A key factor in the change, though, is *tolerance*. This is generally thought of as the capacity of a plant to grow in the shade of others. It involves also the capacity of a species to endure in the shade even though it may grow very slowly. In this capacity red pine exceeds jack pine, white pine exceeds red, and maple and basswood exceed white pine. Tolerance usually plays a key role in forest succession.

In the virgin forests heavy stands were interspersed with open areas, burned areas, windfall areas, and areas where the timber died from attack by insects, tree diseases, or other causes. Droughts undoubtedly had a great deal to do with the dying of timber over substantial areas. Practically all our pine forests of the southern coastal plain, the Douglas-fir forests of the West Coast, and the jack pine and the red pine forests of the Lake states, owe their origin to fires.

Wilderness areas are set aside by the federal government on national forests, national parks, wildlife refuges, and other federal lands to conserve their primeval character for scientific use and public enjoyment. They may embrace tundra, rugged mountains, and any stage in forest succession from pioneer to climax. They do pose a question as to how far man should go in protecting them from fire and outbreaks of insects that can destroy completely the current character of these wilderness areas. Some intervention by man will be required if a wilderness is to be perpetuated in the condition that prevailed at the time the white man first came to America.

## SILVICULTURE OF NATURAL FOREST ECOSYSTEMS

The forester exerts conscious effort to manipulate and improve both natural and planted forests. In the one case the forester is dealing with long established and more or less stable ecosystems; in the other with newly developed and rapidly changing systems.

**Climax Types** If the forester were fortunate enough to start with virgin forests of climax types that are ideally suited for human purposes, assuming that such forests exist, his task would be to perpetuate these forests. The northern hardwood forest of the eastern United States came close to meeting this specification except that now it is no longer virgin. Most northern hardwood forests have been cut at least two times in the past. In each cutting the best trees were removed for sawlogs and the small, crooked, and defective ones were left standing. As good timber became scarce, loggers were obliged to cut smaller sized and defective trees. Such repeated "high grading" has depleted most northern hardwood forests of valuable growing stock. The forester's task is to repair it. If timber growing is the owner's objective he has two choices. One is to cut the remaining merchantable trees and deaden the standing culls. A new forest of sprouts and seedlings will usually spring up. The owner has a long wait, however—30 to 40 years—before he can again expect to harvest merchantable timber (Fig. 6-1).

The second choice is to select the small trees of good species and form for a future cutting. The forester will then plan to remove the others

**Figure 6-1** Examining defect in mature sugar maple, northern hardwood type, Upper Peninsula Experimental Forest. (*U.S. Forest Service.*)

in one or two cuts. He will also seek to get rid of the cull trees, thus creating small openings in which seedlings may become established. Thereby he reduces the waiting period between successive harvests.

The forest owner may be lucky enough to have acquired a northern hardwood stand clear-cut some 50 years ago for fuelwood or chemical wood. In such a case he may have a thrifty stand reaching merchantable size for sawlogs and pulpwood. If this is so, the forester will wish to remove trees by periodic thinning to promote continued rapid growth. The less thrifty trees, unwanted species, and poor quality stems are progressively removed in early cuttings to concentrate the volume increment on the trees of highest value. Each year the annual value added to the stand is larger than the year before. The prudent owner would want to prolong this period of rapid value increase as long as possible. But the time eventually comes when tree vigor declines and growth falls off. A new crop must then be started. Actually it may already be on the ground, at least where small openings exist. These openings can be enlarged and more can be made to encourage the establishment and growth of seedlings. Gradually, over a period of years, all the veterans are removed, but by this time new trees will have reached harvestable size. Following such a pattern the stand is converted from even-aged to uneven-aged. This is known as the selection system of silviculture.

This method of stand management maintains the forest intact. It is especially recommended for use on steep slopes and strategic watersheds where it is necessary to keep a continuous forest cover on the land. The practice should be varied to fit conditions on the ground.

The above discussion presents the case from the viewpoint of the owner and the forester. There is another viewpoint to consider—that of the logger and the equipment he has available. He prefers a management system that enables him to make a profit, or at least high wages. With modern equipment this can be achieved most readily if he is permitted to take what he wishes and pay little heed to what remains. This is the way most loggers operated in the past and many are slow to change. It is to the advantage of the owner and forester that the logger make a good living so that he will stay in business. It is to the advantage of the logger that the forest owner stay in the business of growing timber and permitting its harvest. It is thus up to the forester to work out the best compromise he can that is fair to both owner and logger and that will, above all, perpetuate a healthy forest.

Many possibilities exist between clear-cutting large areas and following a single tree selection system. Clear-cut areas can be reduced in size even down to one acre or less. So reduced, the cutting method is called group selection. This allows the maneuvering of logging equipment and keeps disturbance of the forest within bounds. At subsequent cuttings, groups are enlarged and new ones are made as young trees grow up in the openings. A market has developed for hardwoods of pulpwood size that admits flexibility to operations. Selection hence becomes a feasible as well as desirable method for perpetuating northern hardwoods for timber use. For maximum value yield they should be managed on relatively long rotations with intermediate thinnings and improvement cuttings to concentrate the growth on the trees of high quality. It has already been stated that a forest ecosystem is limited to converting some 0.3 percent of solar energy to wood. This can be done by many small, slow-growing trees of poor quality or by a few rapidly growing trees of high quality. It is usually possible to reduce the wood volume of dense stands of young trees by up to 40 percent without causing a significant decline in annual increment per acre.

The above general principles can be applied to the management of climax forests of hemlock and true firs of the West Coast, luan forests of the Philippines, and the spruce-fir forests of Siberia, though local economic conditions may dictate other practices.

**Subclimax Types** Subclimax forest types of the United States include the pine forests of the Northeast, Lake states, South, and Western states, the Douglas-fir and ponderosa pine of the West, and forests of various subclimax hardwood species. These forests generally produce much more valuable timber crops than the hemlock, fir, or climax hardwood forests that tend to replace them. Where timber production is a major objective, maintaining stands of subclimax types is therefore desirable. This generally requires making substantial openings in the forest and applying some type of treatment to the undergrowth that otherwise might crowd out seedlings of desired species.

Clear-cutting in strips or patches followed by slash disposal and either natural reseeding or planting has become common practice in Europe and is used extensively in the United States. The results have generally been good from the timber grower's standpoint. The new young stands do best if they are thinned periodically. This can sometimes be done satisfactorily with machines while the trees are still too small to be merchantable. Thereafter commercial thinnings can be done periodically up to the time of final harvest and reproducing the stand (Fig. 6-2).

Let us take as an example the relatively open-grown stands of pine on the southeastern coastal plain of the United States. These usually have a ground cover of grass, palmetto, and other shrubby plants that choke out young pine seedlings. Prescribed burning may be all that is required to prepare an adequate seedbed for pine regeneration. If not, more drastic means, such as burning, heavy disk plowing, and other soil treatments may be needed. The pines are usually established by planting nursery-grown stock. A thinning regime such as mentioned above is desirable to hasten the development of the new forest stand. However, by wide spacing of trees capable of rapid growth a pulpwood crop can often be grown without the need of thinning.

**Pioneer Types** The pioneer types, such as aspen–paper birch in the northeastern United States, lodgepole pine in the Rockies, jack pine in the Lake states, and loblolly pine in the southeast, seem best suited to short rotation management in even-aged stands. Following cutting these will reproduce naturally to their own species or gradually revert to more valuable types, provided a dense undergrowth of shrubs and grass is not present. Where such undergrowth occurs it may be necessary to remove it by controlled burning, disking, or other means. Commercial thinnings are often possible in loblolly pine and may be possible in the others, thereby increasing the total yield over the rotation. Both jack pine and lodgepole pine hold their cones on the tree unopened for many years. These supply an abundance of seed following clear-cutting to reseed the forest. They

**Figure 6-2** Students thinning a white pine stand, Pack Forest, Warrensburg, N.Y. (*C. Wesley Brewster.*)

need only a favorable seedbed and freedom from severe competition to restock the area. Details on the care and management of the several species and types are given in books on silvics, regional silviculture, and general silviculture.[3,10,14]

For all three ecologic stages certain broad principles and practices are applicable. Periodic thinnings planned to favor final crop trees will bring these to harvestable size more quickly than will occur in unthinned stands. Removal of unwanted species and poor quality individuals will tend to maintain health and vigor of those desired for final harvest. Pruning crop trees, if followed by periodic thinning to favor their rapid development, makes possible early production of veneer and quality sawlogs. Application of fertilizers has been found to stimulate growth in soils deficient in mineral nutrients. Control of natural succession toward a climax forest is especially important on soils that are adequate to sustain good forests of subclimax or pioneer species but are too infertile for high production by climax species.

## SILVICULTURE OF MAN-MADE FORESTS

Man-made forests may be thought of as of two kinds, those that make use of seeding or planting to reproduce a natural forest of the same kind that has been harvested, and those that involve replacing the original forest with an entirely different type of forest or that result from the planting of bare land to forest trees. The first kind need not result in destroying the natural ecosystem that existed before the mature trees were harvested. It is generally the one most likely to prove successful inasmuch as all components for sustaining the forest are already present. This practice is being widely followed in the southern pine forests of the United States and is the standard procedure in European and Japanese forests. The care and management of the stands so established differ perhaps in intensity but not in overall methods from the management of natural stands of the same species.

Planting grassland or abandoned cropland to trees as well as converting tropical hardwoods to Eucalyptus means transforming the existing ecosystem into an entirely different one. As man usually brings to the planting site only the young trees with the small amount of nursery soil that adheres to their roots, he is doing little to create an ecosystem favorable for his planted trees. This may or may not develop naturally in time. If it does the plantation succeeds, otherwise it fails.

It is sometimes amazing how rapidly an ecosystem becomes established in a forest plantation on abandoned agricultural land. In the northeastern states, for example, many old fields have been planted to the exotic Scotch pine. By the time these plantations had reached the age of 20 years, seedlings of white ash, red and sugar maple were to be found growing under the pine canopy. In fact one experienced forester replied to the question, "How can we establish more white ash forests?" with the statement, "Plant Scotch pine." By the time the Scotch pine reaches merchantable size enough hardwoods of pole size are generally present to make a good crop of their own after the pine is removed. Shrubs and herbs of the northern hardwood forest also gradually become established, thus forming an ecosystem approaching that in natural forests. A similar natural reseeding tends to occur in plantations of native red pine, especially on loamy soils.

**Establishing Forests on Open Land**  Extensive fire-swept lands of the Lake states and the South, worn-out pastures, badly eroded fields, and other cleared and abandoned lands are destined to remain unproductive of timber for decades unless they are planted to trees. Unclothed with woody growth, they may become a source of flood runoff, silt, and debris that menaces other lands. Protective tree plantings are needed on denuded mountain slopes and on the prairie-plains region to minimize wind erosion and exposure of farm homesteads. In 1970 the total area in need of forest planting was estimated to exceed 70 million acres.

Extensive tree planting has been carried out by the United States Forest Service and by Michigan, New York, and other states. Counties and municipalities, and even school districts, have engaged in tree planting. Stimulated by extension foresters and various crop-reduction programs, farmers have planted considerable areas to trees. Since 1945 Christmas-tree growers have greatly expanded their activities. Of special significance have been the large areas of private planting established by many pulp, paper, and other timber companies since 1945. These plantings are strictly for commercial wood production. Many have shown gratifying returns on the capital invested.

**Tree Seed** Tree seed may differ radically from agricultural seed. It is borne high in the trees where it is difficult to observe and collect. It must be collected as soon as it is ripe, before wind dispersion occurs. Tree seed is adjusted to germinating in the natural forest habitat, and to many means of dispersal. Some of it is well protected against destruction. Basswood, for example, must lie for 1 to 2 years on the forest floor while the seed covering is softened by action of microorganisms before sufficient water will penetrate to enable the embryo to swell and emerge. Cottonwood, willow, elm, maple, and most seed that ripens in early spring is short-lived. Collecting cones or tree fruits, extracting, cleaning, storing and pretreating seed, testing it for viability, and getting good establishment of seedlings in a forest nursery are all exacting tasks that require technical knowledge.[12]

Seed may be sown directly in the forest where the seedlings are to grow. This is a desirable practice for heavy-seeded species such as oak, walnut, and hickory. Care must be taken to protect such seed against rodents, however, lest they dig up and devour every seed. If a proper seedbed can be prepared, the seedling will spring up in the place where it is to grow permanently. This is a decided advantage in that it avoids root disturbances that occur during planting.

**Nursery Practice** The majority of tree seeds are so small and the seedlings so delicate that if they are sown directly on the planting site, survival is exceedingly low. To plant successfully such species as northern white-cedar, white spruce, hemlock, many of the pines, larch, and fir, it is desirable to grow the seedlings for one to four years in a nursery.

Seedbeds are carefully prepared, and the seed, after appropriate pretreatment, is sown in rows or broadcast in the beds, covered to protect it against soil drying, and watered daily until it germinates. The nursery beds must be weeded, watered, and properly fertilized to produce a satisfactory crop of seedlings suitable for forest planting. The nursery must also be protected against diseases and insects, birds, mice, and other destructive agents. The job of a nurseryman is an exacting one. He must be on duty daily to discover early and correct promptly any unfavorable factor that may be operative in his nursery. The task of the nurseryman is not to grow fine-looking seedlings, but to grow seedlings that are vigorous and rugged enough to stand up under field conditions (Fig. 6-3).

Seed and nursery practice has been mechanized and rationalized considerably since 1930. Seed extractories have been built to maintain precise drying conditions. Seed storage houses have accurate artificial temperature control. Seeding, weeding, and lifting of nursery stock has been mechanized. Chemical weeding is now used widely in place of hand weeding.[1] In Finland tree seedlings are grown in plastic greenhouses on fertilized peat with carbon dioxide enriched air. Seedlings so grown attain in one year a size equal to those produced in two years in exposed beds. No weeding is required in the greenhouse beds.[9]

**Figure 6-3** Operating a root pruner, Chittenden Nursery, Michigan. (*U.S. Forest Service.*)

**Field Planting** Field planting of tree seedlings often requires proper preparation of the planting site to receive the seedlings. This can be done by furrowing, scalping, burning, or completely cultivating the soil. This last practice has proved the best way of converting lands invaded by brush to forest.

Each seedling must be carefully set in the ground at the proper time of the year if it is to have a good opportunity for survival. Spring is normally the best time to plant, but in many parts of the country trees can be successfully planted in the fall. The modern tree-planting machines have greatly facilitated field-planting operations on sites on which such machines can be used efficiently.

**Plantation Care** For the first 2 to 5 years of their lives, plantations must be protected against both fire and browsing animals.

Competition from herbs, shrubs, and weed trees must not be excessive. Protection of newly planted trees by an overhead canopy tends to enhance first-year survival, but if older trees are left standing after the seedlings once become established, they will stunt the seedlings' growth.

Plantations may require thinning at least once before the trees are of sufficient size to bring in enough income to cover the cost of the thinning. Thereafter the plantation may be managed up to harvest time in much the same way as is described for natural stands of similar composition. This is assuming that a more or less natural ecosystem develops to supply the needs of the growing plantation. If it does not, special study may be necessary by experts in forest planting to determine the cause of unhealthy development and how the condition may be corrected. In the prairie-plains region, lack of mycorrhizae in the soil and other soil deficiencies may be encountered. Nutrient deficiency can be encountered in sandy outwash soils. Where tree growth has been absent for several decades other types of deficiencies may be encountered.

At the time of final harvest the decision must be reached whether to clear-cut and replant or to remove the planted trees and allow those that may have seeded in to form the new stand.

**Forest Planting in the Tropics** Because of the great complexity and diversity of tropical forest ecosystems and the low numbers of commercial species per acre, foresters have established plantations as the surest means of growing commercially valuable timber in the tropics. In this they have enlisted the help of the shifting cultivator under the so-called taunga or parcelero system. Instead of fighting the shifting cultivator, they assign him a specific area of land to clear and to grow his crops on for one or two years, but they also require him to plant and care for trees interspersed with his food plants. At the end of the 2-year period he is assigned a new area to clear and plant, thus extending the forest plantation. In this way the actions of the shifting cultivator are legalized, he clears the land as he would do in any case, and for slightly more effort he establishes a valuable forest plantation as well.[5] As the plantation grows he may be afforded work in clearing it of vines, in performing thinnings, and in other forest operations. Thousands of acres of tropical forest plantations have been established by this method at modest cost to governments.

Growth of forest plantations in the tropical regions tends to be rapid. It is possible to grow trees to pulpwood size in 7 to 10 years and to saw-log size in 20 to 30 years. Annual yields per acre may be as high as 3- to 10-fold those of temperate zone plantations. Where good markets for plantation timber exist, such plantings pay off handsomely. Teak is one of the most widely planted species. Under ideal conditions a teak plantation may produce wood worth as much as $500 per acre per year in European or American markets.

Where tropical forests are made up of species 80 percent of which have high commercial value, silvicultural practices similar to those outlined above for temperate zone forests seem to work out well. For the very complex forests the taunga system seems to be the only one feasible until sufficient knowledge is accumulated to use the many different species that occur, and to perpetuate useful stands for the future. The current practice of harvesting only the few choice species seems destined to degrade the forest rather than to perpetuate its utility.[5]

**Planting Trees in Arid and Semiarid Zones** Arid zone planting is receiving much attention throughout the Mediterranean region, parts of Africa, the Indian subcontinent, China, Australia, the U.S.S.R., and the semiarid regions of the United States. In most cases the primary aim of such planting is for soil and watershed protection, recreation, amenities, and other forms of environmental improvement rather than for the production of commercial timber. In some areas wood may be of such high value that returns from its sale alone justify such planting. The planting sites are often badly degraded from past cultivation or from grazing by cattle and goats. Rainfall is infrequent and often occurs in heavy downpours. Long periods of hot dry weather are the rule and put heavy stress on the planted trees. To establish forest plantations under such extreme conditions requires rigorous attention to technical detail throughout every step in the operation. The Spanish foresters began a program of tree planting on their degraded semiarid soils in about 1945. Their efforts have been highly successful. Their first and most difficult step was to win the confidence and support of the local people so that grazing animals would be kept off the area planted. This task was approached with great understanding and respect for the wishes of the people. Assurance of adequate income from improvements to cropland and pasture was a feature of the program. Great flexibility was permitted the local forest officer in working out such cooperation. This, together with the careful attention to species selection, site improvement, and planting practices, has made the program highly successful and popular. Similar practices are being used by other Mediterranean nations.

Forest planting on semiarid lands degraded by past steep-slope cultivation and heavy grazing requires special techniques for success. Planting stock, grown in degradable plastic containers that may be set out on the planting site, is often used. Soil may actually be transported to each hole in which a seedling is to be planted. On steep slopes, trees may be planted on terraces that catch any rain and thereby minimize soil washing. Stones may be placed around the tree to shade it. Even irrigation may be used to get the seedlings started. These are but a few of the techniques used in the Mediterranean region, Pakistan, and in the western United States where it is important to establish forest cover on unfavorable planting sites.

The U.N. Food and Agriculture Organization has helped many nations to establish successful forest plantings in arid regions. A thorough and authoritative book was published on this subject in 1968.[4]

## FORESTS FOR ENVIRONMENTAL IMPROVEMENT

The places where forests are most needed for environmental improvement are generally those where growing conditions are least favorable. Examples are unstable mountain slopes, avalanche areas, rocky slopes eroded under mountain agriculture and overgrazing, windswept areas on the western plains and along seacoasts, and, above all, areas of concentrated residential, commercial, and industrial use. Many of the principles outlined for planting forests on difficult sites and for growing timber for commercial use may be applied directly or with modification to forests for environmental improvement. One such principle is the provision for constant renewal through replacement of declining trees with vigorous young ones. Though the layman's concept of environmental forestry may be the preservation of existing trees, a program of forest renewal is actually the key to success. Where public use is heavy, as along streets and highways, in parks, cemeteries, campgrounds, and picnic areas, trees that present a hazard to life or property should be removed, but their replacements should already be on the ground.

The selection system of silviculture has its place in parks, urban tree corridors, and even with street trees after they become so large as to form a more or less continuous canopy. The natural forest has many trees that have died from overcrowding, insect or fungal attack, or other causes. Also, forest trees usually have dead branches. The public generally appreciates a forest most if it has a healthy appearance, as is brought about by frequent thinnings and by pruning dead branches from stems. Large dead branches may also present a hazard to forest visitors. They sometimes fall without warning even on days with little wind. Opening vistas and glades also adds to the amenities of forests.

Special problems encountered in cities are paved streets that lead to soil dryness and entrapped toxic gases from leaky mains and sewers, air polluted by smoke and motor cars, salt spray from winter roads, dust and oil films that may clog leaf pores, and general lack of a rich soil fauna and flora to keep the soil aerated and porous for root penetration. Planting and caring for trees in such difficult environments should be planned on a case-by-case basis using the techniques, if required, of both forester and arborist.

## MECHANIZATION OF SILVICULTURE

Forest operations in terms of the weight and bulk of material handled and the extensive land area involved can be classed as heavy industry. Generally speaking, forestry as such has not been a leader in modern industrial developments. Logging large-sized West Coast timber has long been mechanized, but practices in the small-sized timber of the East and South lagged seriously until the 1960s. The advent of the timber-harvesting machine has ushered in an entirely new order of mechanization of forest operations. A variety of machines have been constructed. It is general practice to develop a system of operations with consecutive steps that can be mechanized in a single machine or in two or more machines. The ultimate aim is to eliminate the man working unprotected on the ground. Harvesting machines will be dealt with further in Chap. 12. Here, only their impact on silviculture will be considered.

Harvesting machines work best in timber of uniform size and utility, growing closely spaced on land suitable for easy maneuvering of the large, heavy machines. Such conditions can be found in the spruce-fir forests of Canada, Maine, and the Lake states, and in pine forests of the Rocky Mountains, Lake states, and southern coastal plain. They can also be provided in man-made forests.

Foresters formerly looked askance at mechanized woods operations because of the damage they cause to trees left for a second cutting. Often machines also left the forest covered with debris, making it difficult to establish the new timber crop. But the trend toward mechanization is in full swing and seems unlikely to be reversed in North America, Europe, or in the Soviet Union where harvesting machines have long been under development. Can the forester devise a type of silviculture to meet the impact of the machine age?

It appears that he has already done this in the South, but at a risk. He has developed genetically superior strains of pines. Seed is produced in seed orchards. Nursery-grown stock from such seed is machine planted on open land and land recently clear-cut. Spacing is wide enough for the trees to reach pulpwood size within 18 to 20 years, and without thinning. Where thinning might be needed, small-sized tractor-mounted tree shears and rubber-mounted skidders can perform row thinning and even cut trees from intervening rows. Mechanized harvesters reap the final crop leaving the land covered with logging debris 1 to 3 feet deep. This is disposed of by controlled burning. Heavy disk-type plows, brush cutters, and rippers drawn by powerful tractors break up the remaining debris, uproot the palmetto and other shrubby growth, and mix their parts with mineral soil so that rapid decay results. Two such soil cultivations may be needed spaced 3 or more months apart. After allowing the soil to settle, machine planters are brought in to set out the superior seedlings that are to grow into the next crop.

Yields per acre are high. Costs of operation are within acceptable limits. Quality of the final product is uniform. Output per man-day is high, hence few laborers are required and those that are can be well paid.

What then is the risk? The risk is in developing a monoculture not only of a single species, but also one of essentially uniform genetic makeup. The question is, can such a forest be maintained indefinitely on short rotations without serious risk of loss from tree pests or unfavorable environment? The hope in the South must be placed on the skill of experts in pathology, entomology, genetics, and soils to detect such hazards early and to act promptly and effectively to minimize the damage (Fig. 6-4).

Neither the terrain nor the ecosystems of the mountain forests of the East and West seem to be suitable for efficient use of harvesting machines. Nor do such machines seem to fit in well with long rotation silviculture with intermediate cuttings, such as is necessary for production of high quality sawtimber and veneer logs. Some approach that is less damaging to residual trees and is highly flexible will be needed.

Until 1972 effort has been concentrated mainly on mechanizing but one stage in the forest operation—timber harvesting. What is needed is equal development effort given to all stages of timber growing without

**Figure 6-4** Protecting controlled pollinated cones against squirrels. Institute of Forest Genetics, Placerville, Calif. (*U.S. Forest Service.*)

permitting one to upset the others. There is good reason for believing that this can be brought about.

## LITERATURE CITED

1. Abbott, H. G., and E. J. Eliason. 1968. Forestry Tree Nursery Practices in the United States, *Journal of Forestry,* **66**:704–711.
2. Baker, F. S. 1934. "The Theory and Practice of Silviculture," McGraw-Hill Book Company, New York. 502 pp.
3. Barrett, John W. (ed.). 1962. "Regional Silviculture of the United States," The Ronald Press Company, New York. 670 pp.
4. Goor, A. Y., and C. W. Barney. 1968. "Forest Tree Planting in Arid Zones," The Ronald Press Company, New York. 409 pp.
5. Haig, I. T., M. A. Huberman, and U Aung Din. 1958. "Tropical Silviculture," Food and Agriculture Organization of the United Nations, Rome. 190 pp.
6. Kramer, Paul J., and Theodore T. Kozlowski. 1960. "Physiology of Trees," McGraw-Hill Book Company, New York. 642 pp.

7   Leak, William B., Dale S. Solomon, and Stanley Filip. 1969. A Silvicultural Guide for Northern Hardwoods in the Northeast, U.S. Department of Agriculture, Forest Service Research Paper NE-143, Northeastern Forest Experiment Station, Upper Darby, Pa. 34 pp.
8   Marquis, D. A., D. S. Solomon, and J. C. Bjorkbom. 1969. A Silvicultural Guide for Paper Birch in the Northeast, U.S. Northeastern Forest Experiment Station, Upper Darby, Pa., Paper NE-130. 47 pp.
9   Mikola, Peitsa. 1969. Comparative Observations on the Nursery Technique in Different Parts of the World. *Acta Forestalia Fennica,* **98**:1–24.
10  Smith, David Martyn. 1962. "The Practice of Silviculture," 7th ed., John Wiley & Sons, Inc., New York. 578 pp.
11  Toumey, James W., and Clarence F. Korstian. 1947. "Foundations of Silviculture upon an Ecological Basis," 2d ed., (rev. by C. F. Korstian). John Wiley & Sons, Inc., New York. 456 pp.
12  U.S. Department of Agriculture, Forest Service. 1948. Woody-Plant Seed Manual, Miscellaneous Publication 654, Washington. 416 pp.
13  U.S. Department of Agriculture, Forest Service, Central States Forest Experiment Station and North Central Region. 1962. Timber Management Guide for Upland Central Hardwoods, Columbus, Ohio. 33 pp.
14  U.S. Department of Agriculture, Forest Service. 1965. Silvics of Forest Trees of the United States, Agricultural Handbook No. 271, U.S. Government Printing Office, Washington. 762 pp.

Chapter 7

# Protecting Forests

Throughout many geologic ages forests have been subjected to fire, windstorm, land uplift and subsidence, inundation, and desiccation; to attack by insects, diseases, animals, and man. They have advanced to colonize new lands when conditions were favorable and retreated when unable to persist. Why then should forest protection be a concern of man?

Man protects forests for two reasons: first, because he needs their products and services, and second, to prevent wildfires that start in the forest from spreading to his homes, villages, and workshops. Man is learning that complete protection against all agents that cause loss of timber or other damage to the forest is neither possible nor desirable. Forests provide a home for forest dwelling animals and plants, and these in turn make their contribution to the health and perpetuation of forests. Even fire and windstorm play their role by reducing the accumulated organic matter when it reaches a level at which it interferes directly with the renewal of the forest by young seedlings. Over the long run such agents, though they kill trees, tend to perpetuate the forest. The task of the forester is not to attempt to eliminate tree-destroying pests from the forest; it is to control their populations and behavior so that he may supply his own needs for forest products and services. It is equally important to control man's own wanton or misguided actions that imperil the sustained yield of forest products, services, and amenities.

Forest trees, in common with other long-lived plants and animals, have no fixed life span. Some, such as aspen, rarely exceed an age of 70 to 90 years, but eastern hemlock may live for 700 years and the giant Sequoia for 4,000 years. Even the most massive of all trees may be surpassed in age by the giant cypress of Mexico and the gnarled and weather-beaten ancient bristlecone pines of Arizona. Specimens of the latter have been found with over 4,600 annual-growth rings. Few trees die because of age alone. Instead they become broken by storm, are struck by lightning, are weakened by competition for soil, water, and nutrients or by seed bearing, are attacked by insects, or break off from the onslaughts of heart-rotting fungi.

Groups of trees growing together in forest stands also have no fixed age. But even-aged stands, like individual trees, reach a stage of maturity in which they contain the maximum of sound-wood volume. Thereafter, the stand thins out. Individual trees become defective. Fallen stems and broken tops indicate decadence. In such stands many trees may weaken at about the same time, thus presenting insects and disease organisms an opportunity to build up to outbreak proportions. Such stand decadence may occur in the ecological climax types, such as the beech-birch-maple type, as well as in the aspen–paper birch pioneer type. Open conditions favor windthrow and allow the sun to dry the surface litter so that fires, if started by lightning or other causes, spread rapidly.

Forest fires occurred in North American forests long before the white man arrived. Some were deliberately set by Indians, others were started by lightning. Should they occur today, fires of the magnitude of many of those of the past would do inestimable damage.

The objective of forest protection is to control fires, insects, and other destructive agents by managing the forest so that conditions conducive to damaging outbreaks are minimized.

## PROTECTION FROM FIRE

Man has always feared forest fires, and with good reason. The Peshtigo fire in Wisconsin in 1871 burned $1^{1}/_{4}$ million acres and took 1,500 lives.[13] The same year a 2-million-acre fire in Michigan took 138 lives. In 1894 the Hinckley fire in Minnesota burned over 160,000 acres, taking 418 lives. The great Idaho fire in 1910 burned 2 million acres and cost 85 lives. The 1918 Cloquet fire in Minnesota took 432 lives. The Tillamook fire of 1933 in Oregon burned 270,000 acres. Fires in the same area burned again in 1939 and in 1945, bringing the total burned-over area to 310,000 acres.[35] In 1947 a huge fire in Maine burned over 220,000 acres, destroying 1,000 homes and taking 16 lives.[34] Total damage was estimated at $30 million. Even small fires can take life. On August 5, 1949, Montana's 5,000-acre Mann Gulch fire killed 13 people, 12 of them experienced smoke jumpers. The crew was caught despite their taking every reasonable precaution against a disaster.[10] The same year a district ranger was killed fighting a small fire on the St. Jo National Forest.

Man should not delude himself that such fires are only an occurrence of the past. In 1969 forest fires burned almost 6.7 million acres. In 1970, 525,000 acres were burned in southern California during a week-long fire. This conflagration destroyed 900 homes and caused the death of 14 persons.[1]

**Man's Attitude toward Forest Fires** In early days fires were generally looked upon as a benefit in that they burned off woody growth obstructing agriculture. A vast difference in magnitude exists between the slowly creeping surface fires that spread but 150 feet per hour and the "blow-up" fires such as those mentioned above. Blow-up fires are an awesome sight. They may travel at a speed of 4 miles per hour. Their flames leap 150 feet into the air, releasing energy that sends a convection column of smoke thousands of feet upward. Burning embers may be carried one-half mile or more in front of the fire to start new fires that feed the blow-up character of the conflagration. They are essentially three-dimensional rather than merely a surface phenomenon.[6] Such fires cannot be stopped by means currently at man's disposal. Only a change in burning conditions or fuel supply can reduce their energy output sufficiently for man to resume direct control measures. Such fires account for the major areas burned, the greatest damage caused, and the greatest expenditures for control. It is such fires that have changed man's attitude from tolerance toward fires to respect and fear.

During roughly the first half century of federal, state, and private forestry in the United States, expenditures for protecting forests from fires exceeded all other forestry expenditures. Since 1945 fire control activities have been integrated with other forest management work. Progress has been gratifying. Even in the bad fire year of 1960 less than one-fourth as much forest land was burned as in 1945.

**Effects of Forest Fires** Forest fires destroy ground vegetation, including young tree seedlings, small and even mature timber, wildlife, and surface soil organisms that play a key role in the decay of litter and in maintaining the tilth of forest soil. They strip the soil of its protective vegetative cover giving rise to avalanches, accelerated soil erosion, and augmented floods. A highly destructive flood occurred in Salt Lake City in 1945 because of a watershed of about 600 acres that had been burned more than 11 months before the flood came. The fire itself did little direct damage, but the flood caused damage of $347,000. A 225-acre fire near San Antonio was responsible for a flood causing damage of $47,000.[14]

**Controlled Burning** Notwithstanding the disastrous effects of uncontrolled fires, the forester in America has been obliged to use fire in the forest. Its first use was for the disposal of coniferous logging slash. This was usually piled or windrowed and burned after snow cover came. In the pine belt of the South, grass and pine needles accumulate on the forest floor, forming a heavy "rough." Burning this rough under careful control during the winter eliminates fuel that might give rise to disastrous fires in spring and fall. Such controlled burning tends to discourage hardwoods and favors the reproduction of longleaf pine, a very fire-resistant species. It also lessens damage by brown-spot needle disease of longleaf pine. Similar problems of disposal of accumulated dry vegetation and litter occur in the pine lands of southern New Jersey and the eucalyptus forests of Australia.[19] Prescribed broadcast burning of slash is accepted practice in the West Coast Douglas-fir logging areas.[7] Prescribed burning has also been used as a silvicultural tool for thinning overdense ponderosa pine and at the same time reducing fire hazard.

**Causes and Prevention of Forest Fires** The objective of fire-control planning is to prevent as many wildfires as possible and to suppress those that do start before they get large. Nationwide, only one forest fire in ten is due to lightning; nine are due to man's carelessness. Smokers, incendiaries, and debris burners cause two-thirds of all fires. Railroads, campers, lumbering operations, and miscellaneous causes account for the remainder. Man-caused fires have been reduced by shutting down logging operations during hazardous periods, installing spark arresters on locomotives, requiring permits for debris burners, enforcing state laws against incendiaries, closing forests to sportsmen in severe seasons, and constantly calling the attention of the public to the necessity for caution with fire in the woods.

*Presuppression Activities* Presuppression activities are closely related to fire prevention. The causes of fire are analyzed, and a major plan is prepared to provide for prompt fire detection and suppression. Fire lines and roads are constructed to furnish access and lines from which fires may be effectively fought. Equipment is located at strategic points. Forest fuel-types are mapped and efforts are made to reduce fuel on areas where fires are most likely to start. Fire-danger rating stations are established, at which weather readings are systematically taken and measurements made of the moisture content of forest fuel. Fire-danger meters are devised to integrate the hazards resulting from temperature, humidity, fuel moisture content, days since last rain, wind velocity, season, and time of day.

Fire-danger meters and burning-index meters permit calculating the rate of spread of fire. The fire dispatcher, with other information available to him, can thus determine the size of the crew to dispatch to bring the fire under control. Seasoned judgment must, of course, be used with these aids.[15]

The key to quick action is prompt detection. Trained watchers in lookout towers are on constant alert during hazardous periods. They spot the fire, determine its location and probable size, and telephone the information to the district ranger. Ground and air patrols may be used when visibility from lookouts is poor.

**Fire Suppression** A small fire can often be controlled by one man or by a very small crew. Small fires are fought directly with shovels, rakes, fire swatters, water from backpack pumps, and other simple tools. The crew first knocks down the hot spots, cuts off the head of the fire, and then proceeds to build a line around its flanks and rear.

On large fires camps must be established with a supply system, workshop, communication center, and fire-camp headquarters. A relatively large organization may be involved, with a thousand or more men working on the fire line. Coordination becomes a major item, and communication must be established between the fire boss and the foremen who are responsible for the several sectors. Heavy equipment may include bulldozers, heavy fire plows, and tank trucks with hose for spraying water on hot spots (Fig. 7-1).

**Kinds of Forest Fires** *Ground Fire* The ground fire occurs where the fire burns in a heavy layer of duff or peat such as accumulates in a swamp. Such fires may burn to a depth of 6 feet or more and are extremely difficult to put out without digging out the burning material. They burn slowly and are not a serious menace in themselves, but they may smolder for weeks and then suddenly set surface material afire.[6]

*Surface Fires* These fires, which are the commonest of all, burn the dry vegetation and forest litter. Where tree growth is sparse, such fires may travel rapidly. Wind accelerates rate of spread.

Surface fires can be fought with all types of equipment and by many stratagems. Small fires can be fought on the advancing front, quickly knocked down, and the flanks then extinguished. Fire fighters usually

**Figure 7-1** A Mathis fire plow is used to suppress wildfire on the Ocala National Forest, Florida. (*U.S. Forest Service.*)

attack large fires on the flanks, working toward the head. On very large fires, lines are prepared in advance of the fire. As the surface fire approaches, backfires are set and are drawn by the draft of the head fire down toward it. The two fires then meet some distance back of the fire line and burn themselves out, but constant vigilance must be exercised to extinguish any spot fires that occur from sparks thrown over the fire line.

**Crown Fires** These burn in the crowns of coniferous trees, are exceedingly hot, travel rapidly, and can run many rods ahead of the accompanying surface fire. Usually the heat from the surface fire dries out the needles and heats them to the ignition point. Crown fires can best be fought by putting out the surface fire during the night and early morning, when the crown fire generally dies out. The only other available method may be to fight such fires by means of backfires that burn off the surface material that would provide the heat necessary to keep the crown fire going.

After any fire has been brought under control, mop-up operations begin and continue until the last spark is out. Mop-up involves digging out burning stumps, burying logs, felling burning snags, and digging out fire in peat or heavy litter.

**Use of Aircraft**  Aircraft are used for patrol, observation, and transport of men, equipment, and supplies. Airplanes equipped with tanks holding up to 4,000 gallons of fire-retardant liquids have been used in squadrons to attack forest fires directly. These are effective in stopping or retarding small fires until ground crews arrive. Helicopters are also used to spread fire retardants. Sodium calcium borate and swelling bentonite clay are used in slurry form as retardants.[26] Monoammonium and diammonium phosphates with and without thickeners are also reported to have given good control on going fires. It now appears possible to dissipate electric charges causing lightning by seeding clouds with silver iodide or dry ice[31] (Fig. 7-2). Helicopters have been used for scouting, liaison, and landing men near the fire line. They can be used by the fire boss as a base of control. They can transport equipment and evacuate injured men. They are also used to transport special crews that jump from the craft to the ground to attack the fire directly or to prepare spots where the craft may land with men and supplies.[26]

**Figure 7-2**  Skyfire radar unit on a mountain top. Munitalp Foundation officers Fuquay and Schaefer record data on a charged cloud, using a cloud theodolite, while U.S. forest officer Barrows talks with cloud-seeding aircraft. (*U.S. Forest Service.*)

**Fire-control Organizations** A fire-control organization is of necessity large and far-flung, and a maximum of responsibility must rest on the men on the ground. On national forests the district forest ranger is usually the fire boss. In state forest organizations the district foresters have this responsibility. The state forester backs them up with extra equipment and manpower if these are required but does not attempt to take over the job unless the district organization breaks down.

The northeastern states have enacted a fire-control compact with central coordination that permits one state to furnish men and equipment to a neighbor state if a fire threatens both states or is too large for one state organization to handle. The Province of Quebec joined the compact in 1969, thus making it international.

**Review** After a large forest fire has been extinguished, there remains the job of reviewing it. This includes determining where and how it started, and reviewing initial and all subsequent actions taken to judge their efficacy in hastening the time when final control was achieved.

## PROTECTING FORESTS FROM INSECT OUTBREAKS

Insects are believed to kill more trees than any other destructive agent. The western pine beetle is estimated to have killed 25 billion board feet of timber from 1921 to 1946; the Black Hills beetle, 2.5 billion feet from 1895 to 1946; and the mountain pine beetle, working in California, has caused damage to sugar pine, western white pine, and lodgepole pine totaling 20 billion board feet.[5, 27] The Englemann spruce beetle killed 4 billion board feet of spruce timber during the period from 1942 to 1948. In an outbreak that began in 1909 and continued for almost a decade, the spruce budworm destroyed 250 million cords of spruce and fir pulpwood in Quebec, Maine, New Brunswick, and Minnesota. A second outbreak began in Ontario in 1935. By 1947 most of the mature fir and a considerable part of the white spruce stands had been killed on an area of over 20,000 square miles. The amount of timber killed was enough to supply the pulp mills of Canada for more than 3 years.

The gypsy moth has defoliated thousands of acres of hardwood forests in the northeast. Whole mountainsides of trees have been stripped bare of leaves converting a pleasant landscape into one reflecting only desolation and death.[33] The insect continued to spread even though extensive efforts were made to confine it. Acceptable control measures were still being sought in 1971. The brown-tailed moth defoliated extensive areas in the west. Both of these moths were introduced from Europe. The forest tent caterpillar has defoliated extensive areas of aspen forests

**Figure 7-3** White pine tree badly deformed by repeated weevil attack.

in the Lake states. Other insects have been the cause of less spectacular losses yet the total damage caused has been high.

Insects also slow down the growth rate of trees by feeding on the foliage, the developing fruit, the roots, or the bark.[32] By selective attack they may virtually eliminate a species from the forest. Wood-boring insects, together with the decay that often follows their attacks, may so weaken the stem that it breaks off in windstorms. By feeding on tree seedlings insects may prevent a cutover stand from reproducing to desirable species. They render much standing timber virtually worthless by boring in wood and permitting access to decay fungi. Insect attack may upset cutting cycles and budgets, disrupt plans for recreational use, modify unfavorably the habitat for wildlife, and affect the water regimen of forest streams.

Not all trees killed by insects represent a forest loss. Some may be thought of as thinning or deaths that would have been inevitable in any case.

Insects causing damage to forest trees may be grouped under the following types: sucking insects such as aphids and scales, defoliators, bark beetles, and insects that destroy wood. A few insects such as the pine-shoot moth and the white pine weevil destroy the terminal buds or terminal shoots and hence deform the trees. (Fig. 7-3.)

**Aphids and Scale Insects** The sucking insects, aphids and scales, do not kill forest trees directly, but they may so weaken the trees that they succumb to attacks by other insects or to disease. An example is the beech scale, which, in conjunction with the Nectria fungus disease, has caused widespread dying of beech in Maine and New Hampshire. The aphids and scale insects suck plant juices from the stems, leaves, and other succulent tissue of the tree, and a few attack roots. Aphids and scale insects are subject to predation by the ladybird beetle, by other insects, and to a limited extent by birds. Reliance for control in the forest has been mainly on cultural measures, natural agents, and introduced insect predators and diseases. Various insecticidal sprays have also been effective. Aphids that depend upon ants to carry them about may be controlled by destroying the ant colonies.

**Defoliators** The serious defoliators are the larvae of certain moths and sawflies and, to some extent, the adults and larvae of some beetles.

Examples are the tussock moth, forest tent caterpillar, larch sawfly, Leconte sawfly, gypsy moth, brown-tailed moth, and spruce budworm. Most are native insects. All have caused serious damage when outbreaks occurred.

The 17-year locusts. or cicadas, may sweep over considerable areas of forest, feeding heavily on leaves and damaging branches in which they deposit eggs. May beetle adults cause some defoliation, though this is rarely serious. The larvae, white grubs, feed on roots and cause heavy losses in nurseries and young plantations.

**Bark Beetles** These beetles are the most destructive of all insects in terms of the quantity of timber destroyed. They burrow beneath the bark, feeding on the inner bark and cambium, where they find considerable quantities of sugars and starches. Although the healthy trees often produce a vigorous flow of sap that kills the insects, trees infested in late summer, and those that have a limited moisture supply or are unable to resist the attack for other reasons, are girdled and killed. These beetles provide avenues for the entrance of fungi and bacteria and may carry spores from infected trees to healthy trees.

Defoliators and bark beetles have a tendency to fluctuate widely in population. They may be present in the forest for a long time and cause no appreciable damage, and then within 2 or 3 years develop a huge population that causes a devastating outbreak.

Most bark beetle outbreaks have occurred in overmature timber. They represent one type of rapid stand breakdown that was mentioned in Chaps. 5 and 6.

**Wood-Destroying Insects** Wood in service is attacked by various wood-destroying insects, the most serious of which are termites, carpenter ants, and powder-post beetles. The powder-post beetles will work in tool handles, furniture, flooring, and barn timbers where the moisture content is considerably less than that required by termites. No insect can work in wood dried to 6 percent moisture content.

Insect attacks on green logs result in degrade of the lumber and other products manufactured there-from.

**Control of Insect Damage** Insect feeding on trees cannot be prevented without drastically upsetting the natural ecological processes. Nor can insect outbreaks be prevented from occurring where overmature forests cover large areas, as in established wilderness areas. Man can, however, use a number of measures in managed forests to keep timber losses due to insects within acceptable limits. These include good silvicultural practices, reduction of favorite foods or habitat conditions, introduction of parasites and predators, and direct control.

Plantations of a single species over large areas should generally be avoided. They tend to be an open invitation for insect or disease outbreaks. Severity of attack by white pine weevil, which deposits its eggs in the developing terminal shoot of sun-exposed trees, can be reduced by planting pine beneath hardwoods. But unless they are kept in check, the hardwoods often prove to be a more serious menace to the pine than does the weevil. A more successful practice is to plant the pines close together. Spruce budworm damage can be minimized by removing the mature balsam fir before the insect reaches outbreak proportions. Reduction of western bark beetle damage has been possible by learning to recognize the types of trees that the beetles will attack and removing these in early logging operations.[23]

Damage to wood in service can be prevented largely by keeping the wood dry and by preventing direct contact between the wood and the soil. Certain timbers that must be located closer than 18 inches to the soil can be protected by impregnating them with various preservatives.

**Direct Control of Insect Outbreaks** Attempts to control insects by direct methods proved costly in the past. Shortly after World War II, DDT (dichloro-diphenyl-trichloro-ethane) became available for use in controlling forest insect pests. It was followed by other chlorinated hydrocarbons and various other highly toxic, broad spectrum insecticides. Field tests with ground and aerial spraying stopped outbreaks of such defoliators as tussock moth and gypsy moth within minutes after application. It seemed that man had the answer to controlling his insect enemies. But thoughtful biologists questioned the possible side effects. It was soon found that birds, fish, and reptiles were also killed. Forest entomologists felt impelled to seek other approaches to insect control.

Recent effort has been directed toward finding various biological and biochemical measures that can be directed against each specific insect pest. Of some 520 predatory insects introduced over an 80-year period, 20 have been effective. Viruses and bacterial diseases with specific prey and host requirements have been tried with some notable successes. The bacterial insecticide *Bacillus thuringiensis* has shown promise for controlling certain defoliators. A strain of this bacillus under test in 1970 was about 100 times more effective than that first in use.[22] Nuclear polyhedral viruses attack the Douglas-fir tussock moth and the European sawfly. The rearing, sterilization, and release of large numbers of males has proven an effective control measure against species, such as the screw worm fly in Florida, that are concentrated in a relatively small area and that have females that mate but once. Sterilized males may be used after populations have been reduced to low levels. Pheromones are complex chemicals that animals and insects emit to convey messages to others of the species. Sex attractants are an example. These are being studied intensively to learn if their use might upset behavior of harmful insects and thus interfere with population growth. Chemicals that prevent metamorphosis have been discovered and methods of using these on individual species have been devised. Other developments in the rapidly breaking field of chemical ecology offer exciting new avenues for approaching insect control on an individual species basis. These in conjunction with other control measures offer promise of stemming outbreaks of both native and introduced insect pests of the forest. This subject will be dealt with further in Chap. 8.

Another approach that offers some promise of reducing damage from both insect and fungal pests is to breed resistant varieties of trees. This is an involved problem with many possible avenues of approach including those of breeding trees that themselves produce chemicals that repel or at least do not attract insect pests.

Major reliance must be on keeping timber stands vigorous through proper management practices. Short rotations, harvesting before senescense begins, and cleaning up debris that serves as a breeding ground for pests can reduce population growth thereby keeping losses within bounds.

## TREE DISEASES

Tree diseases seldom cause spectacular losses such as those from wildfire or insect outbreaks. Much may be hidden from sight, such as root, butt, and stem rot. Striking exceptions have been the rapid spread of introduced fungal diseases. The chestnut blight, introduced in 1904, has eliminated chestnut as a commercial tree throughout its range. No other disease or insect has surpassed its record of destruction in America. The Dutch Elm disease introduced about 1930 seems to be well on its way to eliminating the American elm unless effective measures can be found to exterminate its chief carrier, the European elm bark beetle.

Diseases kill or damage seedlings, young trees, and mature trees. These diseases are of two main types: nonparasitic and parasitic. Nonparasitic diseases are those caused by drought, sunscald, winter injury of various types, improper nutrition, smoke or gaseous pollution of air, flooding, soil pollutants, salt spray from winter highways and from sea spray blown inland by hurricanes; ozone, sulfur dioxide and photochemical reactions on automobile exhaust gases. Studies in the San Bernardino Mountains in southern California have led investigators to conclude that if smog from automobile gases continues to pollute the air at 1969 rates, ponderosa pine will be virtually eliminated from the forests of the area.[4] Some nonparasitic diseases, such as sunscald and smoke injury, occur only locally. Others, such as drought damage and wind burn, may be widespread.

Few soils of the United States are so deficient in mineral nutrients that trees cannot survive. Many are infertile, however, and support only small trees, as do the sand plains of New Jersey. Glacial outwashed sands of sections of the Lake states and New York, following cultivation, have been seriously reduced in their capacity to support vigorous tree growth. The poorest soils are often found on rocky ridge tops and on lands recently elevated from the sea.

Locally, trees may be damaged by chemical wastes, such as the acid water from mines, smokes, and smelter fumes, and by other toxic chemicals in heavy concentration. The outstanding example is the copper basin of Tennessee. Here smelter fumes, mainly sulfur dioxide, have resulted in the death of all trees for a distance of about 8 miles from the smelter.[20] Trees are also readily killed by flooding caused by dams. Unfavorable weather will be discussed later.

**Parasitic Diseases** Parasitic diseases of trees are caused by bacteria, fungi, viruses, nematodes, and mistletoes.[2,11,16,21,25,29] Fungi cause more losses than any other type of disease, particularly in the larger trees. Nematodes are destructive of seedlings and of newly germinated seed. The phloem necrosis disease of elm and brooming of black locust are caused by virus infections. The parasitic organism causing a disease is called a pathogen.

**Fungal Diseases** Many tree diseases, such as the chestnut blight, are spread by spores carried by the wind.[8]

The rusts are highly specialized fungi. A classical example is the white pine blister rust fungus, which was first found in North America at Geneva, New York, in 1906. It occurred on cultivated gooseberries and currants that serve as its obligatory alternate host. These were promptly destroyed. By 1909 a new infestation was found on eastern white pine seedlings that were imported from Europe and shipped widely over the Northeast and the Lake states. Attempts to destroy these were unsuccessful. By 1915 all hope of eradicating the fungus from the United States was abandoned, since it was learned that new infections had been introduced on the West Coast. These have spread on the western white pine and sugar pine.

The fungus attacks all five needle pines. It enters the trees through the needles, then spreads down the twigs and branches to the main stem, which it kills by girdling the bark.[18] The rust cannot spread directly from pine to pine but must spread from pine to currants or gooseberries. On these it causes only moderate damage and may spread from one currant bush to another. In the fall of the year spores are produced that can infect only pine. Vast sums have been spent to eradicate wild currants and gooseberries throughout the commercial range of eastern and western white pine. Only partial success has been achieved. The U.S. Forest Service has consequently ceased planting western white pine and no longer favors it in cleaning and weeding operations.[24] Eastern white pine still holds its own as a forest tree even though many trees become infected with the rust and eventually die.

The Dutch elm disease, an excellent example of a disease in which the causal agent is carried by an insect, broke out in Europe and was found in the United States in 1930. The fungus is parasitic on living trees and saprophytic on dead trees. Its yeastlike spores produce toxins that are carried by the sap stream throughout the tree and so cause its death. The disease organism is carried from diseased to healthy trees by the elm bark beetle. Beetles were introduced at the same time as the disease in various importations of elm logs used for producing high-grade veneer. The Dutch elm disease had become so widespread by 1971 that it threatened the survival of the species. Methoxychlor is a biologically degradable insecticide with low toxicity to birds and wildlife. Though expensive, it is effective against the European bark beetle, the chief carrier of the Dutch elm disease. A control technique through use of sex pheromones was also being sought in 1971.

Year in and year out heart rots are responsible for more damage to timber than any other diseases. The fungi are spread by windborne spores that infect the tree through injuries caused by logging, ice breakage, fire scars, insect attack, or other injuries that expose the heartwood of the tree. The fungi causing heart rot are of several types, but the group known as the polypores are the most common and widespread. They produce bracket-shaped, spore-bearing conks on the tree trunk (Fig. 7-4). Heart rot has infected many mature trees in western conifer stands, causing heavy degrade or complete loss of much timber that was formerly of commercial value. Many of the trees left standing after logging in eastern forests were those already infected. Others have since been attacked by heart-rotting fungi. These occupy space that could otherwise be used for growing commercial timber. Some of them may serve as den trees for wildlife.

No feasible method has been found for preventing heart rot in forest trees. The only effective control is the practice of intensive forestry. Care should be taken to prevent injury by fire, since much heart rot gets its start through fire scars. Where selective logging is carried out, care should also be taken to avoid injuring the base of trees. The annual loss from heart rot is estimated at 1.5 billion board feet.[16]

The fungi that cause heart rot in trees also cause decay of slash and fallen trees, a necessary function in the forest.

**Figure 7-4** Heart-rotting fungus on sugar maple. (*C. Wesley Brewster.*)

Heart rot, in common with other actions of decomposing organisms, appears to be far from a simple process. According to Shigo[28] the decay proceeds by stages with each stage dominated by a single decay organism or a group of organisms. Only after one group has completed its work does another take over to continue the decay process. Acquaintance with these stages and the external evidence of each enables an experienced forester to judge the extent of decay that has occurred and how rapidly it may be expected to progress.

Wood-rotting fungi cause considerable loss to wood in service as well. It has been estimated that approximately 10 percent of the annual production of lumber in the country is used each year merely to replace wood that has decayed in service.[2] Decay in wood can be controlled by keeping the wood dry. Yet many uses of wood, such as for railroad ties, telephone poles, and piling, necessitate its being located in places where it becomes wet. In these cases, decay is impeded by treating the wood with toxic chemicals such as creosote, zinc chloride, or pentachlorophenol.

The causal agents in deterioration of wood in service, the damage they do, and measures for prevention and treatment, are dealt with in a well-illustrated book by Findlay.[9]

**Mistletoes** The eastern mistletoe, and to some extent the western mistletoe, are carried by birds. These are parasitic seed plants that grow on the branches of pines and hardwoods, sending their rootlike haustoria down into the conductive tissue of branches. The eastern mistletoe becomes quite large. It produces seed covered with a sticky, gelatinous material that adheres to birds' feet, thereby spreading the infection from one tree to another. The dwarf mistletoe causes witches'-brooms and other damage to conifers, particularly to ponderosa pine. The seeds of this plant are forcibly ejected for a distance of 20 to 40 feet horizontally.

No effective control has been found for the dwarf mistletoe except to clear-cut the stand, burn the slash, and start a new forest.

**Pest Control and Research** The federal government takes many steps to prevent introduction of tree pests and to control introduced and native pests that damage forests. The maintenance of a plant quarantine and inspection service at all ports of entry has been helpful in keeping to a minimum the number of serious diseases that have been introduced.[12] Plants to be imported should be grown under quarantine until it is clear that they are free from disease.

The individual states also have active programs of inspection of nursery stock and various control measures to minimize the spread of diseases and insect pests. Together with the federal government the states finance research aimed at the following objectives.

    1  To develop reliable methods of keeping abreast of insect populations and disease incidence, and of the factors that cause rapid changes, thereby to enable prediction of future outbreaks and the probable damage that may result.

    2  To control pests in tree nurseries and to prevent their spread to the field plantings.

    3  To suppress outbreaks that do occur so as to keep losses to an acceptable minimum.

    4  To develop management techniques that minimize losses.

    5  To improve the quality of standing timber and to lengthen the life of wood in service.[17, 33]

In 1969, the National Academy of Sciences recognized the quality of research in forest diseases by electing to membership Dr. George H. Hepting, an outstanding forest pathologist. He was the first forest officer to be so honored.

## PROTECTING FORESTS AGAINST ANIMALS

**Domestic Animals** No measure is available of the total damage caused to American forests by domestic animals. Grazing is practiced the least in the northeast and in the heavy forests of the Pacific Coast since these regions produce little forage in the forest. Formerly woodlands were commonly grazed in the Central and Middle Atlantic states but the practice has declined in recent years. Grazing was widespread in 1970 in southern and in most western forests except for the heavy stands along the Pacific Coast.

Grazing is defensible in the West where timber stands are of low density and a considerable amount of grass can be found between the trees. This is true to a certain extent of the pine forests of the South also. Throughout the rest of the United States little grazing value would exist in the woods, provided the trees were growing as densely as they are capable of growing.

Cattle, sheep, horses, hogs, and goats compact the soil, reducing infiltration of water, inhibiting root growth, and reducing aeration. Erosion is a particularly serious menace in western forests that are subject to heavy grazing by sheep and cattle.

It is an anomaly that grazing by domestic animals continues to reduce output of forests and even to destroy them, while at the same time in the same country planting up depleted grazing lands to forests is being practiced. This points to the need to consider the total economic output of the land in terms of the uses that may be made of it.

**Beneficial Effects of Grazing** In pine forests of the South and open conifer forests of the West properly regulated grazing may have some beneficial effects. The animals can consume grass that, when dried, would become a serious fire hazard. Browsing by sheep and goats may help to eliminate unwanted hardwoods in an area being reproduced to conifers, but control to achieve the desired effect is difficult.

**Protection against Wildlife** It is a maxim that any animal that lives in the forest must get its food from the forest, and this often comes in part from the woody vegetation—either directly, in the form of leaves, twigs, buds, and bark, or indirectly, in the form of nuts, fruits, and fungi that grow on the live or dead trees.

Where squirrel populations are dense, they have been known to destroy an entire seed crop. Mice, chipmunks, and birds eat tree seed in large quantities. Grosbeaks are reported to have destroyed entire seed crops of certain southern pines. Birds also eat the young seedlings. In coniferous nurseries seedbeds must be screened lest the birds destroy every emergent seedling.

Rabbits, mice, deer, and other browsing animals will eat young tree seedlings. Deer and rabbits both tend to browse more heavily on hardwoods than on softwoods, thereby favoring the reproduction of conifers. White-tailed deer tend to feed selectively on hardwoods, preferring the valuable timber trees, sugar maple, yellow birch, and white ash to the low value beech.[30] Deer have often been known to degrade hardwood seedling stands by selective feeding.

The bark of trees is eaten in winter by rabbits, mice, squirrels, porcupines, deer, and elk. Mice may destroy sugar maple by gnawing bark beneath the snow during winter. Porcupines, where abundant, may cause noticeable damage to both hardwood and softwood trees by gnawing bark. They may girdle trees at almost any height, and they also cut off twigs. Elk gnaw the bark from young aspens and alders during the winter.

Ruffed grouse, squirrels, porcupines, deer, rabbits, and many other animals feed on the buds of trees. The actual damage caused by this feeding is difficult to appraise but is considered of minor significance.

Beaver cutting is normally limited to trees of 2 to 4 inches in diameter, but they have been known to cut down trees of 18 inches and even larger. Beaver eat the bark from felled trees and store branches underwater for winter food. Dams erected at the outlet of a low or swampy area may back up water, flooding several acres of land. Most trees thus inundated die. Yet beaver damming universally promotes water storage, thus serving to maintain water tables and to regulate stream flow.

The most spectacular damage to forests by wildlife is that caused by elephants in Africa. They push over, break, and uproot trees to get at the foliage and twigs. They may debark trees they cannot push over. They have been credited with changing large areas of forest to grasslands in the Murchison Falls National Park in Uganda and elsewhere. Other game herds, especially gerenuk and giraffes, feed extensively on tree foliage. Browsing herds can cause considerable damage to tree growth in many parts of East Africa, but often the trees are of little value compared with the game animals they support.[3]

**Control of Wildlife Damage** Wildlife-damage control can best be effected through good wildlife management combined with good forest management. The wildlife population is likely to be most destructive during those periods of the year when food is least plentiful. If timber-cutting operations are carried out at this time, deer, rabbits, and other animals that inhabit the forest may collect to browse on the branches of felled trees.

From time to time wildlife populations in local areas may become excessive and result in severe damage to the forest. This will require direct action. Hunting and trapping are usually the most effective for game and fur animals. Live trapping and transport to other areas has rarely succeeded in reducing population. Fencing is costly. Nature eventually will bring about an adjustment through predators, disease, and starvation, but such adjustments are often delayed until severe damage to the wildlife range occurs. Furthermore, natural adjustments may severely deplete the seed stock.

**Unfavorable Weather** Native forests, though not immune to injury, withstand normal vicissitudes of weather. They may be seriously injured by abnormal weather. Plantations of exotics are likely to be particularly susceptible. Drought injury can be minimized by planting drought-resistant species, by eliminating vegetation that competes for soil moisture, and by preserving moderate overhead shade. Tree seedlings hardened in the nursery by exposure to soil dryness are likely to survive better on dry sites than unhardened stock.

High temperature sometimes causes loss to tree seedlings, particularly in nurseries. This can be prevented by watering the seedlings during the hottest part of the day. Heat loss is rare in forest plantations but has been known to occur. Usually it is spotty and confined chiefly to places where air drainage is poor. Air temperatures that kill seedlings must be as high as 120 to 130°F.

Frost action on heavy soil may heave fall-planted trees from the soil or lift them so far out that they die. Frosts may also stunt growth and actually exclude trees from "frost pockets." A frost pocket results from radiation of heat from the soil to the sky. On clear still nights air thus cooled, if impeded from flowing away, accumulates. Frost that extends over wide areas may also cause severe damage to trees, especially late spring frosts after the buds have opened and vegetative growth has started.

Winter needle burning may occur on warm clear days while the ground is still frozen. Under such conditions the roots are unable to supply moisture to replace that lost by rapid transpiration. Sunscald in winter appears to be caused by rapid freezing of tissue newly activated by solar heating. Summer sunscald may occur in the South due to overheating of the cambium by direct solar radiation.

**Wind** High winds will uproot trees or break them off. Moderate winds will break the tops. Cyclonic storms of tornado type will sweep through forests, leveling trees in their paths. Hurricanes cause severe damage to forests (Fig. 7-5).

Although forests cannot be completely protected against damage by windstorm, damage can be minimized by proper density and by maintaining good borders at the edge of the forest to prevent the wind from sweeping under the crowns.

**Rain** Heavy rainfall sometimes causes land slips, gullying, flooding, and over-land flow that may wash trees out by the roots. Most forests,

PROTECTING FORESTS

**Figure 7-5** Single survivor of a virgin red pine stand, Chippewa National Forest. (*U.S. Forest Service.*)

however, keep the soil sufficiently open so that the water soaks in and causes little erosion or damage. Snowslides or avalanches may harm forests in high mountains. Ordinarily only limited steps can be taken to prevent these, such as keeping vegetation on the upper slopes and promoting the growth of woody vegetation wherever possible. Hail may strip the leaves from trees, but such storms are generally local and recovery from such damage usually is rapid.

**Snow and Ice Damage** Wet snow, and particularly rains in freezing weather, may overload tree crowns and cause extensive breakage. A single ice storm may damage trees over thousands of acres. Trees planted out of their natural range are especially susceptible. Hardwoods are usually damaged most by ice, conifers most by wet snow. Trees with narrow crowns usually resist damage better than those with broad crowns. Trees in stands that are neither too dense nor too thin usually have sturdy stems that enable them to resist snow breakage.

**Diastrophism** Locally forests may undergo damage resulting from earthquakes and volcanic action. In 1959, a severe earthquake occurred in the upper Madison River Valley of Idaho. Extensive faulting resulted that tilted Hegben Lake, thus creating some new land and inundating other land. Minor damage occurred to the forest because of diversion of streams and falling boulders. The most spectacular event, however, was a huge slide that sent trees, boulders, and soil cascading into Madison River Canyon, creating a natural dam almost 500 feet high. A lake formed behind the dam, killing the trees that were flooded. Tidal waves resulting from earthquakes may damage low-lying forests along the seacoasts but seldom penetrate far inland.

Lava flows and deposits of volcanic ash have destroyed considerable areas of forest in Hawaii and in Alaska. The Mt. Katmai eruption in Alaska in 1912 buried thousands of acres, killing the tree growth. Fine ash, in fact, drifted all the way around the earth from this eruption.

**LITERATURE CITED**

1  Barrows, J. S. 1971. Forest Fire Research for Environmental Protection, *Journal of Forestry,* **69**:17–20.
2  Boyce, John Shaw. 1961. "Forest Pathology," 3d ed., McGraw-Hill Book Company, Inc., New York. 572 pp.
3  Brown, Leslie. 1965. "Africa: A Natural History," Random House, Inc., New York. 300 pp.
4  Cobb, Fields W., Jr., and R. W. Stark. 1970. Decline and Mortality of Smog-Injured Ponderosa Pine, *Journal of Forestry,* **68**:147–149.
5  Craighead, F. C., and John M. Miller. 1949. Insects in the Forest: A Survey, *Trees,* The Yearbook of Agriculture, pp. 407–413, U.S. Department of Agriculture, Washington.
6  Davis, Kenneth P. 1959. "Forest Fire: Control and Use," McGraw-Hill Book Company, Inc., New York. 584 pp.
7  Dell, John D., and Lisle R. Green. 1968. Slash Treatment in the Douglas-fir Region—Trends in the Pacific Northwest, *Journal of Forestry,* **66**:610–614.
8  Diller, Jesse D., and Russell B. Clapper. 1965. A Progress Report on Attempts to Bring Back the Chestnut Tree in the Eastern United States 1954–1964, *Journal of Forestry,* **63**:186–188.
9  Findlay, W. P. K. 1967. "Timber Pests and Diseases," Pergamon Press Series of Monographs on Furniture and Timber, vol. 5, New York.
10  Forbes, Robert H. 1949. The Smoke Jumpers Carry On, *American Forests,* **55**(10):18–19.

11  Gill, Lake S., and Jess L. Bedwell. 1949. Dwarf Mistletoes, *Trees,* The Yearbook of Agriculture, pp. 458–461, U.S. Department of Agriculture, Washington.
12  Gravatt, G. F., and D. E. Parker. 1949. Introduced Tree Diseases and Insects, *Trees,* The Yearbook of Agriculture, pp. 446–451, U.S. Department of Agriculture, Washington.
13  Guthrie, John D. 1936. Great Forest Fires of America, U.S. Forest Service, Washington. 10 pp.
14  Hammatt, R. F. 1949. Bad Business; Your Business, *Trees,* The Yearbook of Agriculture, pp. 479–485, U.S. Department of Agriculture, Washington.
15  Hardy, Charles, and Arthur P. Brackebusch. 1960. The Intermountain Fire-danger Rating System, *Proceedings, Society of American Foresters' Meeting,* San Francisco, Calif., 1959, pp. 133–137, Washington, D. C.
16  Hepting, George H., and James W. Kimmey. 1949. Heart Rot, *Trees,* The Yearbook of Agriculture, pp. 462–465, U.S. Department of Agriculture, Washington.
17  Hepting, George H. 1970. How Forest Disease and Insect Research Is Paying Off. The Case for Forest Pathology, *Journal of Forestry,* **68**:78–81.
18  Hirt, Ray R. 1933. Blister Rust, A Serious Disease of White Pine, New York State College of Forestry, Syracuse, N.Y. 5 pp.
19  Hodgson, Athol. 1968. Control Burning in Eucalypt Forests in Victoria, Australia, *Journal of Forestry,* **66**:601–605.
20  Hursh, C. R. 1932. The Forest Legion Carries On, *American Forests,* **38**:16–19.
21  Hutchins, L. M. 1949. Diseases and the Forest, *Trees,* The Yearbook of Agriculture, pp. 443–445, U.S. Department of Agriculture, Washington.
22  Irving, George W. 1970. Agricultural Pest Control and the Environment, *Science,* **168**:1419–1424.
23  Keen, F. P. 1949. Pine Bark Beetles, *Trees,* The Yearbook of Agriculture, pp. 427–432, U.S. Department of Agriculture, Washington.
24  Ketcham, David E., Charles A. Wellner, and Samuel S. Evans, Jr. 1968. Western White Pine Management Programs Realigned on Northern Rocky Mountain National Forests, *Journal of Forestry,* **66**:329–332.
25  Martin, J. F., and Perley Spaulding. 1949. Blister Rust on White Pine, *Trees,* The Yearbook of Agriculture, pp. 453–458, U.S. Department of Agriculture, Washington.
26  Phillips, Clinton B. 1960. Fighting Forest Fires from the Air, *Proceedings, Society of American Foresters' Meeting,* San Francisco, Calif., 1959, pp. 137–140, Washington, D. C.
27  Rohwer, S. A. 1949. The Key to Protection, *Trees,* The Yearbook of Agriculture, pp. 413–417, U.S. Department of Agriculture, Washington.
28  Shigo, A. L., and E. v. H. Larson. 1969. A Photo Guide to the Patterns of Discoloration and Decay in Living Northern Hardwoods. U.S. Northeastern Forest Exp. Sta., Upper Darby, Pa. Research paper NE-127.
29  Swingle, R. U., R. R. Whitten, and E. G. Brewer. 1949. Dutch Elm Disease,

*Trees,* The Yearbook of Agriculture, pp. 451–452, U.S. Department of Agriculture, Washington.
30 Tierson, William C., Earl F. Patric, and Donald F. Behrend. 1966. Influence of White-Tailed Deer on the Logged Northern Hardwood Forest, *Journal of Forestry,* **64**:801–805.
31 Tribus, Myron. 1970. Physical View of Cloud Seeding, *Science,* **168**:201–211.
32 U.S. Department of Agriculture, Forest Service. 1958. Timber Resources for America's Future, Forest Resource Report 14, Washington. 713 pp.
33 Waters, William E. 1970. How Forest Disease and Insect Research Is Paying Off. The Case for Forest Entomology, *Journal of Forestry,* **68**:72–77.
34 Wilkins, Austin H. 1948. The Story of the Maine Forest Fire Disaster, *Journal of Forestry,* **46**:568–573.
35 Woods, John B., Jr. 1950. The Rehabilitation of Tillamook Burn, *Journal of Forestry,* **48**(5):362–364.

Chapter 8

# Chemical Ecology

Chemistry contributes to our knowledge of forests because so many functions of plants and animals are carried out by means of chemical reactions or are chemically controlled. Examples are food synthesis, digestion, tissue and organ building, growth, metamorphosis, and reproduction. All involve chemical reactions. Chemicals also play a key role in animal communication, behavior, defense, mating, and rearing of young. Odors attract animals to some foods and repel them from others. The latter may serve as a protective device against foraging herbivores or carnivores. We may think of forest chemistry as embracing the chemistry of forest ecosystems and what keeps them in balance. It also includes the chemistry of wood and secondary forest products as these are transformed into products useful to man. Forest chemistry impinges on the chemistry of health and disease, of predator and prey, of growth, maturity, senescense, death, and decay; and of air, water and soil pollution, and repurification.[2]

To make the subject manageable we shall leave the chemistry of life processes in general to the biochemist, and water purification, sewage treatment, and pollution of air, soil, and water in general to the chemists specializing in these fields. The chemistry of wood and secondary forest products will be taken up in later chapters. There remains the fascinating new field of chemical ecology.[11]

Chemical ecology deals with the chemistry of plants and animals in relation to their physical and biologic environment. The subject is so new that it still has no fixed boundaries nor narrow specialties. This newness is fortunate for the subject has such broad implications and is so complex and little explored that no one can predict where it might lead or what specialties might arise within it. Rather than try to define chemical ecology, or to classify it, a more rewarding effort at this stage may be to review how it came into being, sample some of the relationships that have been uncovered, and try to weigh some of the fruits it seems to hold in store for the talented investigator and for mankind.

## BACKGROUND OF CHEMICAL ECOLOGY

Primitive man learned eons ago that certain naturally occurring salts as well as plant and animal products had a profound effect upon himself and other creatures. Some were deadly poisons, such as strychnine and snake venoms. Others carried him into a dream world. Still others soothed his suffering and helped to cure his ills. As he learned to cultivate plants and domesticate animals he also became aware of their pests. The plagues of locusts, gnats, and flies mentioned in Chaps. 8, 9, and 10 of the Book of Exodus are evidence that men in the times of the Egyptian Pharaohs were well acquainted with insect pests. It was little wonder that man looked for means to control them. He found that many natural products contained poisons for his use. Pyrethrum, rotenone, and such potent inorganic chemicals as arsenicals, lead, copper, zinc, and mercury compounds came into use to control insect and fungal pests. But many of these poisons had two elements of risk: their high toxicity to man and his domestic animals, and their tendency to accumulate in the soil, ultimately poisoning it for further use. Something better was needed.

Two new chemical products played a key role in controlling disease and wound infection during World War II and its aftermath of reconstruction. One was penicillin, a product of the common bread mould. This was found to clear up wound infections and to cure a great number of illnesses. The other was a new insecticide, dichlorodiphenyltrichloroethane, that became known as DDT. The potency of this chemical against insects was noted in 1929 when it saved the Swiss potato crop. Its use during World War II prevented typhus epidemics in North

Africa and Italy. It checked malaria outbreaks and rid humans of fleas and body lice. Later it came into widespread use against insect pests of forests, farm crops, animals, and man. One of its advantages was its long-lasting effectiveness. It was small wonder that both penicillin and DDT were hailed as miracle products. Neither appeared to present any significant hazard to man or his domestic animals. The discoverers of both products received the Nobel Prize.

True, penicillin was not effective against all human disease organisms, nor was DDT effective against all insects. Even some of those for which DDT was at first effective, acquired resistance in succeeding generations. The search went on for other antibiotics and insecticides. Hundreds of soil borne and other fungi were examined and a large number were found to contain antibiotics. These were synthesized and extensively tested. A few were added to the pharmacist's lists of therapeutic agents. Several additional chlorinated hydrocarbons were synthesized and tested. Among them, chlordane, dieldrin, aldrin, methoxychlor, and benzene hexachlorine came into use along with the phosphate parathion and the carbamate, sevin. All were effective against a wide number of insects and many of them were persistent in the soil. Insecticide manufacturers felt that they had good cause for congratulations.

But disturbing side effects began to show up. DDT sprayed from airplanes on forests to kill black flies, gypsy moths and spruce budworms got into lakes and streams where it killed lake trout and other fish and prevented their eggs from hatching. On the Michigan State University campus, robins died from eating earthworms that had accumulated DDT in their fatty tissue from sprayed leaves on which they had fed. When DDT appeared in milk the stage was set for widespread public alarm.

In 1962, barely 33 years after DDT was first recognized as a miracle insecticide, Rachel Carson, with her book *Silent Spring,* lit the fuse that exploded the bomb of public alarm against its use. People became convinced that use of DDT and other broad spectrum persistent poisons must be stopped. Laws were passed in several states and abroad to ban or restrict their use. Miss Carson had also convinced the public that biologic means of controlling pests were available or could be devised. She called attention particularly to insect-eating birds and mammals, natural insect predators, parasites and diseases, and various sanitary measures that keep pests within reasonable bounds in natural forests and grasslands.

Entomologists and others had been exploring these fields, and not without some success. Still, these controls were generally ineffective in coping with widespread insect outbreaks, especially in the case of introduced pests. Moreover the risk exists that once such predators, parasites, or diseases have controlled the intended pest, some of them might attack harmless or even beneficial species. Those discouraged by the ineffectiveness of the predator-parasite-disease approach though were unaware of the full potential of other approaches. This is not surprising, for the instrumental techniques for exploring these new avenues were then in the developmental stages.

## PHEROMONES

Pheromones are chemical substances that various animals and insects emit to communicate messages to others of their kind.[16] Some may be danger signals, such as ants use to mobilize the colony to fight off invaders. Some may be the territorial markers of a dominant male. Some may mark trails. Some may serve as name tags for individuals, such as those that enable a fawn to recognize its mother. Some may be feeding attractants that cause bark beetles to concentrate their attack on a single tree. And some are sex attractants that lure a male to a receptive female. The chief means of conveying messages between individuals of most groups of animals is believed to be chemical communication.

Some pheromones offer promise of being used against insect pests. The female gypsy moth, for instance, is incapable of flight. Entomologists reasoned that she must have some means of attracting males to her. By experiment they determined that the attractant was located in the tip of her abdomen. With the aid of chemists, they sought to collect such tips and to determine the chemical nature of the attractant. Progress was slow until after highly sophisticated instruments became available that made chemical determinations possible from only tiny amounts of material. After several trials the pheromone was identified and synthesized. Early tests have shown that it does attract the males. As of early 1971 extensive field trials had not yet been made, but the hope is that this pheromone can augment or even replace the use of the insecticide sevin and the disease-causing organism *Bacillus thuringiensis* that were the chief methods permitted for widespread application in 1970. Control of populations of this introduced defoliator is highly desirable for it strips hardwood forests bare of leaves over wide areas of the Northeast.

The case of the cotton boll weevil is somewhat different. In this species the male weevil emits a substance that attracts both males and females. The males were collected in large numbers, the secretion extracted from them and identified chemically. It proved to be not a single compound but a group of four separate terpenes. The last of the four was identified and synthesized in 1969. During 1969 and 1970 these terpenes were synthesized in quantity at two different laboratories.[1] Preliminary field trials were made in 1970 with mixtures of the four synthetics. When the weevil populations were low, the mixture of the synthetics appeared to attract as many weevils as did the natural pheromone. But when populations were high the natural product emitted by abundant males seemed to be the more potent lure. Further testing and research will be needed before the efficacy of the pheromone for population control can be established.

The female western pine bark beetle, *Dendroctonus brevicomis*, produces an attractant that has been identified chemically and is named brevicomin. Brevicomin by itself is only moderately effective.[10] But in conjunction with a closely related compound produced by the male, and a terpene (myrcene) produced by the host tree, a very effective mixture results that attracts both males and females. In 1970 these compounds were prepared in amounts suitable for field tests. A region was selected for tests in which the western pine bark beetle had been endemic for over a decade, and was causing substantial annual losses of pines. Traps baited with these compounds were distributed over a 4-square-mile area. More than a million beetles were lured to the traps. At the end of the season no active infestation of beetles could be found within the trapped area. This promising result awaits confirmation by further tests in the years to follow. If these tests yield similar results, depredations by this highly destructive pest may be reduced to tolerable limits.

A first step has been taken in an effort to develop a new control for the European elm bark beetle, which is largely responsible for spread of the Dutch elm disease. In 1970 it was demonstrated that the female did emit a substance to attract males. The task ahead is to collect enough of the substance for chemical identification. Attempts will then be made to synthesize the product, and it will be given wide field tests.

In view of the brief time during which chemical ecologists have been working on insect pheromones the above results seem promising indeed. Proving that pheromones exist in a species of insect pest, collecting the material in sufficient quantity for chemical identification, synthesizing it and giving it field tests, is but part of the story. The complex patterns of response of the insects themselves may also need to be known before an effective population control will be possible. Even in the most promising test to date, that on the western pine bark beetle, as of early 1971 it was still uncertain whether in fact the pheromone-baited traps were responsible for the sharp decline in beetles, or whether other factors, such as buildup of predators and parasites of the beetle, were chiefly responsible. Large numbers of predators were attracted, as well as the bark beetles. A long road of painstaking research may still lie ahead. Evidence of this is available in the complexity of pheromones and the actions they induce.

**Complex Behavior**  Rudinsky[9] describes what appears to occur when the Douglas-fir bark beetle invades its host. The beetle is first attracted by the resin and terpenes of the Douglas-fir in the ratio of 2 females to 1 male. After several hours, during which the female has had time to bore into the bark, the ratio changes from the above to 2 males to 1 female, because of a male attractant that females emit with the frass. After a time this attraction ceases abruptly, and shortly afterward mating occurs. The male stridulates above the hole made by the female, which causes her to emit a substance that masks the sex attractant. Males that do not reach the female first are then free to seek elsewhere for one.

Mammals appear to have very complex systems of pheromones. The beaver is said to have some 100 different ones. A beaver colony is a well-organized community requiring the cooperation of its members in dam building and repair, food gathering, food storage, and house construction. Presumably the pheromones are useful in directing these activities. Canines also have a number of pheromones by means of which they communicate. The extent to which man might learn to use pheromones of wild animals to control population density, to avoid overuse of range, and to protect young trees in plantations, is still a question.

Traditionally man has thought of control of pests in terms of their eradication, at least locally. What he really wishes to do is to reduce to tolerable limits the damage they do to him, his domestic plants and animals, and to the ranges and forests he manages for his welfare. If this goal could be achieved eradication might prove both unnecessary and even undesirable, except perhaps in the case of introduced pests. This is particularly likely to be true with mammals whose presence in the ecosystem has an attraction for man. If their behavior could be controlled, there should be no cause to eliminate their populations.

## ALLELOCHEMICS

Allelochemics are substances produced by one species that affect the health, growth, behavior, and population of organisms of another species.[14] They include repellents; escape substances, such as the ink of cuttlefish; suppressants, such as juglone washed from walnut leaves that may impede the growth of other trees; venoms that poison prey or assailants; inductants, such as cause gall formation; counteractants against venoms and antibodies; and lures to attract prey to the predator. All the above afford adaptive advantage to the organism that produces them.

Another group of allelochemics help the receiving organism. These include the attractants to food, the warnings of danger or toxicity, the stimulants that trigger adaptive mechanisms. A third group, such as wastes, may poison the recipient but have no benefit for the producing organism.

The relations are quite complex. Substances that repel some species may serve as food or food attractants to others. Allelochemics are of wide interest because they determine to a large extent which foods are acceptable and which are not. They are a means by which plants and animals restrict the feeding of others upon them. They also include substances by means of which one plant or group of plants suppresses certain others. A few cases will be cited to indicate the extent of allelochemics that are or may become of interest to foresters and other users of natural vegetation.

**Suppressants** Heavy grazing of California grasslands tends to weaken the grass cover, thus permitting the invasion of shrubs. The two chief shrubby invaders are unpalatable, a mint and a sagebrush. These give off terpenes from their leaves that accumulate in the soil and kill the grass. Eventually the shrubs spread so that little palatable forage remains. However, if a fire burns over the area it not only consumes the shrubs with their terpene laden leaves, but heats the soil and so drives off or consumes the terpenes on or near the surface. Grass and herbs then reinvade and hold the area until heavy grazing again sets off a new cycle. With moderate grazing the grass will hold the land for a long period.[13, 14]

Walnut trees produce the allelopathic substance juglone, eucalyptus produces a toxic oil, false flax weed produces phenolic acids. Other trees releasing allelopathics are found in the genera *Myrtus* and *Ailanthus*. Many shrubs of the hard chaparral also release substances that suppress the growth of other plants. How important these cases of allelopathics may be in forestry is still to be explored.[13] Years ago it was found that certain grasses greatly impeded the growth of hybrid poplars. By experiment it was found that this was caused by some substance given off by the roots of the grass. It has been known since 1941 that various lower organisms, especially fungi, produce antibiotics. What role they actually play in the health of forest trees and in ecologic succession is still to be explored.

The role of mycorrhizal fungi is a particularly fascinating one still not fully understood. Mycorrhizal fungi are those that grow in association with roots of higher plants, notably conifers and other trees. Seedlings of such trees do poorly in soils lacking mycorrhizae. This fact has been known for over a century. Yet even though many competent investigators have given intensive study to the question, the actual relationship has remained obscure. It had been assumed that the tree roots provided the fungus with sugars and other food materials and that the fungus in turn supplied water and minerals to the tree roots. Roots bearing mycorrhiza have few root hairs to take up water. Recently it has been found that mycorrhizae produce antibiotics that prevent root infection by pathogenic fungi.[8] Aside from supplying water and mobilizing sparse basic chemicals for tree nutrition, they probably supply also thiamin, vitamin $B_1$. Went[12] has gone so far as to postulate that in the mid-Amazon Basin mycorrhizae may, as saprophytes, decay leaves and transmit not only sparse mineral nutrients to the tree roots for direct recycling but supply sugars as well from the decomposed cellulose. He offers this as one possible explanation of the fact that, once cleared of trees and surface debris, the sandy soils quickly lose their fertility. Leaching of soil minerals also occurs.

**Bacteria Responses** Bonner[3] reports that the behavior of bacteria also appears to be influenced by chemical stimulants. Some species have a very limited range of substances that they can use as food. When this food is abundant the bacteria multiply rapidly. When it is scarce they tend to produce spores or resting stages that remain dormant until food again is available. The spores then germinate and repopulate the surroundings. However, not all spores germinate at once. The germinating spores appear to give off a substance that inhibits the germination of others nearby. This results in saving a substantial number of spores for a later germination period should unfavorable conditions kill off the active members.

**Insect-induced Plant Responses** This is a fascinating field still to be explored in depth by the chemical ecologist. Wasps of the family *Cynipidae,* in the process of laying their eggs in an oak leaf, hackberry leaf, rose, or goldenrod stem, inject a substance that causes the plant to form a fleshy-tissued gall. The gall protects the young larva and furnishes it with a food supply. The nature of the chemical stimulus provided by the gall wasp is still to be determined. The balsam woolly bark aphid injects some substance that causes elaboration of bark tissue on which it feeds and also induces the stem to form compression wood. The two processes are probably interrelated.

**Possible Use in Control** Some allelochemics offer promise of being used to control populations. Nematodes that feed on potatoes, beets, and other plants furnish an example. The eggs of these creatures may remain dormant for as long as 10 years. Many require specific root exudates to induce hatching. If such exudates could be identified and synthesized cheaply they might be spread through the soil at a time when no roots suitable for food were available. The emerging larvae would then starve, ridding the soil of these crop-destroying pests.[4] Research laboratories are seeking to identify the substances that induce the eggs to hatch. As nematodes may also be a serious pest in forest nurseries, means of controlling them is of concern to foresters.

**Selective Feeding of Insects** Insects that feed only on certain plants are the picture on the opposite side of the coin, for in this case plants must emit some attractant that enables an insect to recognize it as suitable food. This subject, being explored by Dethier, has several interesting sidelights.[4] For instance, do these plants produce specific substances needed by the insects that are not available in other plants?

Feeney[6] found that larvae of the winter moth feed mainly on oak leaves in May when the protein content is high and the tannin low. Moth larvae that feed on late leaves are prevented from using the leaf protein because of the high tannin content. Such larvae do not survive to pupate.

Ant acacia provide protein for the ants that feed on them. The ants in turn protect this acacia from feeding by other small herbivores. Other acacias have cyanogens and other toxic substances to provide protection against herbivores. Flea beetles that feed mainly on plants of the family Cruciferae are attracted by mustard oil. The same oil also attracts a parasite of some aphids that feed on crucifers. Some insects develop the capacity to detoxify materials in plants. This probably results from their high enzyme activity. Insects that feed on but a single plant species show enzyme activity of 14.6; those feeding on but a few plant species show 81.5; while general feeders show 269.4.[6]

## DEFENSE MECHANISMS

Plants and animals have a wide variety of defense mechanisms. Poisonous plants, stinging nettles, irritating oils such as those produced by poison ivy, venom of snakes, scorpions, wasps, and spiders are widely known. Odors of the stink bug and skunk repel birds and animals that might wish to use them for food. Much less well known are a host of other insect and plant mechanisms and products that protect the species from enemies. Eisner[5] presents many examples of defensive substances and of the mechanisms for directing these against enemies.

Odors in general serve both as attractants and repellents. They guide animals to food (Fig. 8-1). They attract pollinating insects to flowers. They probably guide the young back to their nest or den. Some workers believe that salmon, once they enter the main river, are guided by odors back to the brooklet in which they began their life as fry. Here they spawn and

**Figure 8-1** Coyotes and other canines have keen sensitivity to odors by which they both track their prey and convey messages between members of their kind. *(U.S. Bureau of Sport Fisheries and Wildlife.)*

start a new generation on its way downstream to river and ocean, and return.[7]

Allelochemics have wide interest to scientists because of the important role they play in the regulation of ecosystems, and of the implications they have for the evolution of organisms. Their discovery and study have thrown new light on the many substances and mechanisms that plants and animals have developed that enable them to survive in a world full of hazards.

## HORMONES

An entirely different approach to insect control is through upsetting their hormones. Insects have elaborate life histories, some going through six or more larval instars, developmental stages, before the final two—pupating and emerging as an adult. It has been discovered that hormonal substances control these metamorphoses—changes from one stage to another. One is known as the juvenile hormone. This substance was first isolated from the *Cecropia* moth larvae. Once its structure was known it was synthesized. This, when injected into a larva or even when brushed on the outside of it, prevented the larva from changing into a pupa. It was tried on other insect species and found to be highly effective with some but only slightly so with others. Certain analogs of the juvenile hormone were also found to be effective. One of these, that could be prepared in a simple one-step process, proved to be highly effective with every kind of insect tested. Williams[15] found that the substance, named juvabione, was more effective in the early stages of metamorphosis than in the later ones. He reasoned that it might prevent an egg from hatching into a larva. This proved to be the case. Sláma then went back another step and treated a female. Her eggs behaved as if they had been treated directly. Now the task of applying juvenile hormones so as to kill one species and only one became how to apply them to the female before egg laying. Sláma and his co-workers in Czechoslovakia reasoned that the most effective way of doing this was to treat the males and permit them to carry the hormone to the females during mating. They tried this and found indeed that males treated with a hormone analog were effective in sterilizing the females. So a new approach to controlling a specific insect is at hand, provided that practical techniques of rearing and treating large numbers of males with appropriate juvenile hormones can be worked out. This is where the subject appeared to rest in 1969. It is an interesting sidelight that Wilson and his co-workers found materials that acted as juvenile hormones in balsam fir, eastern hemlock, tamarack, and Pacific yew. Juvabione was in fact first isolated from balsam fir.[15]

## THE SEARCH GOES ON

It would be inappropriate to credit Rachel Carson with the founding of chemical ecology. But the public concern she aroused convinced budget officers and legislative bodies that research on ways to control pests without the use of broad-spectrum poisons was in the public interest and would command public support. That it was possible to isolate and analyze hormones and pheromones in such a relatively short time was owing to the use of the nuclear magnetic resonance spectrophotometer, the mass spectrograph, and other highly elaborate and costly instruments that have become generally available for biological research only since 1960. The delay would doubtless have been much longer without the pressure of public concern.

The above successes should not make us sanguine that man's war against insects will soon be won. Rather we should reflect on the sobering thought that insects have not held their place as the most abundant, highly organized form of animal life since Paleozoic times without developing great versatility and adaptivity. It no longer appears likely that man will come up with any one chemical or technique that will control all kinds of insect pests. Man's means of adaptation to unfavorable environment is to change the environment. The insect's means is to change his tolerances or life processes. Either reaction may have far reaching effects on the ecological balance.

**LITERATURE CITED**

1  Anonymous. 1970. Teams Ready Weevil Lure for Field Tests, *Chemical and Engineering News,* **48**(4):40,42.
2  Arnold, Dean E. 1971. Lake Erie Alive but Changing, *The Conservationist* (New York) **25**(3):23–30, 36.
3  Bonner, John Tyler. 1970. The Chemical Ecology of Cells in the Soil, pp. 1–19 in Sondheimer, Ernest, and John B. Simeone (eds.), "Chemical Ecology," Academic Press, Inc., New York.
4  Dethier, V. G. 1970. Chemical Interactions between Plants and Insects, pp. 83–102, in Sondheimer, Ernest, and John B. Simeone (eds.), "Chemical Ecology," Academic Press, Inc., New York.
5  Eisner, Thomas. 1970. Chemical Defense against Predation in Arthropods, pp. 157–217, in Sondheimer, Ernest, and John B. Simeone (eds.), "Chemical Ecology," Academic Press, Inc., New York.
6  Feeny, Paul. 1970. Seasonal Changes in Oak Leaf Tannins and Nutrients as a Cause of Spring Feeding by Winter Moth Caterpillars, *Ecology,* **51**:565–581.
7  Hasler, Arthur D. 1970. Chemical Ecology of Fish, pp. 219–234, in Sondheimer, Ernest, and John B. Simeone (eds.), "Chemical Ecology," Academic Press, Inc., New York.
8  Marx, D. H. 1969. The Influence of Ectotrophic Mycorrhizal Fungi on the Resistance of Pine Roots to Pathogenic Infections. I. Antagonism of Mycorrhizal Fungi to Root Parasitic Fungi and Soil Bacteria. II. Production Identification and Biological Activity of Antibiotics Produced by *Leucopaxillus Cerealis Var. piceina, Phytopathology,* **59**:153–163; 411–417; 549–558; 559–565; 614–619.
9  Rudinsky, J. A. 1969. Masking of the Aggregation Pheromone in *Dendroctonus pseudotsugae* Hopk., *Science,* **166**:884–885.
10  Silverstein, Robert M. 1970. Attractant Pheromones of Coleoptera, pp. 21–40, in Beroza, Morton (ed.), "Chemicals Controlling Insect Behavior," Academic Press, New York.
11  Sondheimer, Ernest, and John B. Simeone (eds.). 1970. *Chemical Ecology,* Academic Press, Inc., New York. 336 pp.
12  Went, F. W. 1970. Plants and the Chemical Environment, pp. 71–82, in Sondheimer, Ernest, and John B. Simeone (eds.), "Chemical Ecology," Academic Press, Inc., New York.
13  Whittaker, R. H. 1970. The Biochemical Ecology of Higher Plants, pp. 43–70,

in Sondheimer, Ernest, and John B. Simeone (eds.), "Chemical Ecology," Academic Press, Inc., New York.
14 Whittaker, R. H., and P. P. Feeny. 1971. Allelochemics: Chemical Interactions between Species. Chemical agents are of major significance in the adaptation of species and organization of communities, *Science,* **171**:757–770.
15 Williams, Carroll M. 1970. Hormonal Interactions between Plants and Animals, pp. 103–132 in Sondheimer, Ernest, and John B. Simeone (eds.), "Chemical Ecology" Academic Press, Inc., New York.
16 Wilson, Edward O. 1970. Chemical Communication within the Animal World, pp. 133–155 in Sondheimer, Ernest, and John B. Simeone (eds.), "Chemical Ecology," Academic Press, Inc., New York.

Chapter 9

# Protecting Soil and Water

People in industrial nations use large quantities of water. An adult may drink 3 to 4 quarts daily. To take care of his other domestic needs, in 1970 a rural person used a daily average of 134 gallons, and an urban dweller used 181 gallons. If to this is added the consumption of water for irrigation, steam-power generation, and industrial processes, in 1970 the daily average total per capita use was over 2,000 gallons. Water use has been increasing rapidly since 1900, when total per capita daily use was but 527 gallons. The water use by years, for five use categories, is given in Table 9-1. From 1900 to 1950 irrigation accounted for approximately one-half of total water consumption. Even in 1969 it accounted for about 40 percent of total use. Steam-power development accounted for 32 percent and industrial use for 21 percent. All uses show a marked upward trend. This is especially true for industrial use and power development. To make a ton of steel requires 32,000 gallons of water, to make a ton of paper requires 28,000 gallons, and to refine a barrel of petroleum products requires 480 gallons. Power generation by both fossil- and atomic-fueled

Table 9-1 Water Use in the United States by Years in Billions of Gallons Daily Average

| Year | Total | Irrigation | Public water utilities | Rural domestic | Industrial | Steam electric utilities |
|---|---|---|---|---|---|---|
| 1900 | 40 | 20 | 3 | 2 | 10 | 5 |
| 1920 | 91 | 56 | 6 | 2 | 18 | 9 |
| 1940 | 136 | 71 | 10 | 3 | 29 | 23 |
| 1960 | 323 | 135 | 22 | 6 | 61 | 99 |
| 1969 | 403 | 157 | 26 | 7 | 84 | 130 |
| 1975* | 450 | 170 | 30 | 7 | 98 | 145 |
| 1980* | 494 | 178 | 32 | 7 | 115 | 162 |

*Projections of future use.
Source: Statistical Abstract of the United States, 1970.

plants requires great quantities of water. It has been estimated that by 1980 as much as one-sixth of the total fresh water runoff from the 48 contiguous states may be needed for cooling power plants. As most of the runoff occurs in the spring, for 9 months of the year power plants will pass the equivalent of half the total runoff through their cooling systems.[5]

Every increase in use of paper, steel, processed food, other basic commodities, and power tends to increase the use of water. The United States is fortunate in having generous supplies. If these supplies are properly managed and distributed, it is believed that needs to the year 2000 and beyond can be met. But this will require restraints on both pollution and profligate use to avoid local shortages. It also means that man must plan carefully to provide the water required by new housing and industrial developments.

## SOURCES OF WATER

Fresh water for human use comes from two sources: that which falls on inland freshwater streams and lakes, and that which falls on land. The latter makes up over 98 percent of the catchment area. Man obtains water from these sources by impounding it in reservoirs, by direct tapping of streams, springs, and lakes, and by boring wells that tap underground aquifers. As man's main source of water comes from the land, the use of land affects the quality and quantity of runoff.

Within the 48 contiguous states roughly 28 percent of the land is in agriculture, 29 percent in open rangeland, and 34 percent in forests and woodlands. The water available for irrigation, municipal, and industrial use comes mainly from the 91 percent of the land of these three classes.

That which falls on cities and highways is usually discharged directly into streams.

Of the three classes, rangelands are the poorest suppliers because they receive the lowest rainfall. Water from farmlands often has a high silt content and may be enriched by leaching of animal manure and commercial fertilizers. Such enrichment often results in an algal bloom that imparts an objectionable odor and taste to the water. Water from farmlands may even be contaminated with animal disease organisms to some of which men are susceptible. Such water requires clarifying and chlorination to make it suitable for municipal and industrial uses. Forests are the most generous suppliers of usable water, in part because they occur on lands that have relatively high rainfall. Forest-covered watersheds are preferred because they deliver water of minimal silt content and are generally free from contaminants harmful to man and to the uses he makes of water. Water from all three classes of land may carry objectionable traces of persistent pesticides, herbicides, and animal poisons that are applied to control insects, unwanted plant growth, rodents, and carnivores.

The prime importance of forests as sources of usable water makes it incumbent upon foresters to understand how forests affect infiltration of water into soil, accumulation and melting of snow, flood runoff, silt content of water, yield of usable water, and moisture content of the atmosphere. To clarify understanding, the several ways which forests influence the water that falls on their canopies will be discussed first.

## HOW FORESTS AFFECT SOIL AND WATER

**The Forest as an Umbrella** Anyone who has sought shelter under a tree knows that a leafy canopy intercepts a substantial amount of rainfall. If he remains long and the shower is heavy he also knows that water begins to drip through from the completely wetted leaves. The leaf area of a forest can be as high as 20 times the area of the land on which the forest stands. All these leaves may intercept water. Some of the intercepted water reaches the ground as drip from the leaves and some from flow of the water down the branches and trunk. As much as 0.02 to 0.10 inch of water may be held on the leaves, whence it is absorbed into the leaf tissue or evaporates into the air.[9]

Interception varies with species. It tends to be greater for conifers than hardwoods because the needlelike leaves may each hold a suspended droplet. During a storm in which net interception for an aspen-birch type forest was 10 percent of rainfall, that of northern hardwoods was 15 percent, that of red pine 29 percent, and that of spruce-fir forest 32 percent.[9] Drip from broadleafed trees reaches the ground in large drops. When these drops have fallen 25 or more feet they reach their terminal velocity and strike the forest floor with greater impact than do smaller sized unimpeded raindrops.

Timber cutting has but a temporary effect in reducing interception of rainfall. The residual trees, new seedlings, and undergrowth rapidly respond to form a leafy canopy over the forest floor.

Beneath the crown and undergrowth canopy is a layer of twig and leaf litter on the forest floor that may be from $1/2$ to 3 or more inches thick. This layer also intercepts rainfall, holding water until the layer becomes saturated, then delivering the water to the underlying soil. Litter may hold as much as 0.1 inch of water per inch of litter depth. Over the course of a year the total amount of water intercepted and evaporated by litter may be appreciable. In a region with 49 inches of rainfall the average evaporation from litter 1 inch deep was found to be 3.0 percent of precipitation. With 3.6 inches of litter depth, the interception loss was 5.3 percent of that which fell.[1]

Total annual interception may amount to as much as 25 percent of precipitation, especially where most rain comes in small summer storms followed by clear weather.[9]

The actual amounts of rainfall intercepted by forest canopies and litter cited above are meant to be illustrative only. They cannot be applied to large forest areas without extensive refinement. It is important to recognize that the forest, in performing the function of breaking the force of raindrops and feeding clear water into the soil underlying the litter, does exact a certain toll in interception loss.

Tree leaves, if cooler than moisture-laden air, condense dew and also collect fog droplets. Some of this moisture may find its way directly into leaf tissue. This is the chief way in which air plants, which have no contact with moist soil, receive water between periods of rainfall.

**Snow Accumulation** Forests have a profound influence on the accumulation of snow and on snow melt. Snow, in contrast to rain, does not fall; it drifts down. The average angle of descent is reported to be 4 degrees from the horizontal. Snow therefore has optimum opportunity to become lodged on conifer branches, especially on those that extend horizontally. Temporary accumulation of snow on conifers is at maximum about 0.2 inch of water equivalent. Much of this ultimately reaches the ground. The loss from evaporation of intercepted snow is estimated to be in the range of 6 to 10 percent of precipitation.[1] Deciduous trees, on the other hand, intercept little snow but they may become glazed with ice during freezing rain—often in sufficient amounts to break branches. Snow depths during winter are therefore greater in hardwood forests than in conifer forests. It is for this reason that deer seek conifer forests when the snow becomes so deep as to impede their travel in open country or hardwood forests. Openings in conifer forests have greater snow depths than beneath the trees.

Much of the snow that lodges on conifers is subsequently blown off, whence it eddies into openings, reaches the ground beneath the canopy, or settles beneath deciduous trees, if any are present. Evaporation of newly fallen snow is low because of its high degree of reflection of solar radiation and because of the low temperatures. Once the snow is covered with dirt and debris, absorbed radiant energy will increase evaporation and melting. Snow accumulation is of high importance in determining the amount of water that will become available for irrigation and other uses (Fig. 9-1).

**Infiltration** Once water reaches the soil it runs off, is stored in small depressions, or soaks into the soil by infiltration. Raindrops falling on bare soil splash mud about, quickly puddling the surface. Small depres-

**Figure 9-1** Measuring snow depth and water content, Teton National Forest. *(U.S. Forest Service.)*

sions fill and overflow, then the water runs over the surface carrying some of the soil with it. The muddier it gets the greater its erosive action becomes. The sheet and gully erosion that results wastes water needed to recharge the soil reservoir, muddies streams, and degrades the soil. Water that infiltrates directly into the soil is much the more useful to man and to the entire ecosystem. Forests provide ideal conditions for infiltration. The lower branches, under vegetation, and above all the resilient forest litter, break the force of falling water drops, thereby preventing soil puddling and erosion. In the soil beneath the litter live a host of organisms that burrow through the soil, mixing organic and mineral particles, aerating the soil, promoting crumb structure, and opening channels for water percolation. Decayed roots add to channels for free water movement so that the soil is wetted in depth. A great amount of surface thereby becomes available to receive the downward-moving water. The water that enters the soil in excess of that which is removed by vegetation gradually moves through the soil to be discharged clear and clean into forest springs and streams.

The quick rise in forest streams that may occur following heavy storms is due largely to the water that falls into the stream channel and along its saturated banks. The turbidity that may accompany such a rise is due mainly from channel and bank cutting.

Except where an impervious soil or ledge rock lies near the surface, forest soils rarely show evidence of overland flow unless the soil has become compacted by roads or logging. Removal of the litter by burning or raking decreases the rate of infiltration. Heavy grazing by cattle or horses will compact the soil, thus reducing infiltration, and may give rise to overland flow with resulting erosion.

Infiltration in a dry porous soil proceeds rapidly with the onset of rain. As the soil becomes wet, infiltration slows to a constant rate. This usually occurs after rain has been falling for 30 to 90 minutes. If rain continues long enough and at a high enough rate the entire soil mantle may become saturated. Thereafter what rain falls must run off.

The effectiveness of forest cover in maintaining soil porosity is outstanding. In New Jersey, 150 inches of cannery waste water was sprayed on a forest during a 10-day period without causing surface runoff or noticeable saturation. By contrast, only 2 inches of water applied to a cultivated field nearby induced saturation and surface runoff.[9]

Frost in soil tends to be of two types: concrete, and honeycomb or stalactite. Concrete frost is almost impermeable, whereas honeycomb or stalactite frost has little effect on infiltration. As the latter type is the most common in the forest it furnishes channels for water to enter the soil at essentially the same rate as in unfrozen soil. Forest soils are also slower to freeze in the early winter because of the protection of the crown canopy and leaf litter. If a snow blanket persists during the winter, soil that may have frozen before the snow fell may become warmed from beneath and thaw completely. It then remains permeable to water as long as the snow depth is sufficient to prevent freezing.

**Evapotranspiration** Plants use enormous amounts of water. The more luxuriant the plant cover the more water it will use as long as the supply remains available. It is estimated that forests use 2,000 tons of water to produce 1 ton of wood. Of the 30 inches of water that falls on the average over the 48 contiguous states, $8^1/_2$ inches flows off to the sea with some 0.4 inch lost by evaporation along the way. About 2 inches is withdrawn by man for irrigation, municipal, and industrial use. Some 19.5 inches is returned to the atmosphere through evaporation and transpiration.[12] The net amount that is used annually by forests in photosynthesis is barely 0.05 inch.

The use of energy is as follows. About 30 percent of the incident solar radiation is reflected by the leaves. Of the 70 percent absorbed, 20 percent is given off as heat to the atmosphere, 49 percent is used to transpire water, and barely 1 percent is used in photosynthesis. Land plants and forests in particular use about one-half of the energy they receive and about two-thirds of the water, as they move water through their roots, stems, and leaves to the atmosphere. They perform all this work to convert less than 1 percent of the energy and 0.16 percent of the water into the products of photosynthesis. Yet it is on this 1 percent of the energy and 0.16 percent of the water so converted that terrestrial life is supported.

Why is it necessary for trees to pump so much water from the soil merely to discharge it into the atmosphere by their crowns? One reason is that trees require nitrogen, phosphorus, magnesium, potassium, and other elements for their photosynthetic, growth, and reproductive processes. They must get these elements from the soil through the sap stream where they are usually present in very dilute solution. Another reason is that they must absorb carbon dioxide from the air into the sap of their leaves. The carbon dioxide is present in the air at low concentration, 0.03 percent. The plant can take up this carbon dioxide only through a moist cell membrane. Much air must therefore come into contact with the moist cell membranes to supply the plant with needed basic material for photosynthesis. To keep the membrane moist, water must be supplied to it, for the air is constantly drying it. Actually the leaf is not so prodigal in use of water as it might at first appear, for it closes its leaf pores, the stomata, at night when light is no longer available for photosynthesis, thus shutting off most of the water flow except for that needed to maintain the turgidity of the cells. Thus plants restore at night the water deficit that may have been built up during the day.[7]

Transpiration also cools the leaves that might otherwise become overheated by intense sunlight and the high temperature of surrounding air. Temperatures of rapidly transpiring leaves have been reported to be as much as 27°C below that of the surrounding air.[9]

**Measurement of Evaporation and Transpiration**  Not all water that falls to the ground returns to the air through plants in the transpiration stream. Some evaporates directly from the wet leaves and stems of plants and from the moist leaf litter and soil. Plant scientists have tried to separate these two means of water loss. This can be done by sealing the roots of plants in closed containers with the stem and leaves protruding. The water loss from such a container is due almost entirely to transpiration and can be determined by periodic weighing of the container. Moist soil can also be exposed in a container and the water loss by evaporation determined by weighing. In both cases water loss is rapid with moist soil and decreases rapidly as the soil becomes dry.

To obtain a more realistic measure of the water cycle as it occurs under natural conditions, as opposed to that of plants in small pots, forest hydrologists have made use of lysimeters. These are large tanks sunk in the soil to ground level. In these tanks, several cubic yards of soil are transferred from the forest, grassland, or field, taking care to disturb the natural soil structure as little as possible. The rainfall that occurs on such lysimeters is carefully gauged as it falls and that which runs off the surface and seeps out from the bottom of the tank is also gauged. Various tree species, grasses, or cultivated crops can be grown in the several lysimeters and their effect on infiltration, surface runoff, evaporation, and transpiration, and on yield of underground water, can thus be determined.

Data on water use by plants in pots or lysimeters must be used with care if one is attempting to project the results to a large watershed. For this purpose natural watersheds themselves must be used, with care being taken to monitor soil moisture, precipitation, hourly runoff, and other factors both before and after applying treatments to the vegetation. The sum of evaporation and transpiration on a watershed underlain by impervious bedrock can be arrived at by measuring the difference between the water that falls as precipitation and that which leaves the watershed as runoff (Fig. 9-2).

The term *evapotranspiration* has been coined to designate this combination of the two. It is this combination that constitutes the loss of water that might otherwise be available for beneficial use. Or, looked at from the reverse side, it is this combination that the vegetation has

**Figure 9-2** Stream-gauging station showing V-notch weir, housing of water-stage recorder, stilling basin, and float well, Apache National Forest. *(U.S. Forest Service.)*

available to support the ecosystem, for it is water that does not enter the streams that supports land vegetation.

**Results from Lysimeter and Gauged Watershed Studies** Plants tend to transpire water abundantly and even prodigally as long as they can obtain it in generous amounts. This has an advantage in draining wet soil and thus permitting air to enter the soil to supply oxygen for root activity and for use by decomposing organisms. As the soil dries, water stresses gradually build up between the root and the soil and within the plant between the roots and the leaves. If the stresses are acute, the leaves can no longer retain their turgor. They wilt. The stomata close, reducing water loss and also reducing photosynthesis. If drying continues long enough to desiccate the protoplasm, the leaf dies. Further drying may kill the entire plant.

On a watershed, plants growing along the stream banks with the water table near the surface may be using water in large amounts, while those above the stream may have their growth reduced due to moisture deficit, and those on the ridge tops and ledges may be dying from lack of water. Yield of water during dry periods from such a watershed can often be increased somewhat by cutting trees and other vegetation from the stream banks.

The fact that almost all plants, even those accustomed to living in deserts, use water freely when it is available,[3] accounts in large part for the differences between discharge of water from forested as contrasted with grassland watersheds. Loss of water from soil by evapotranspiration may be essentially the same whether the soil is tree or grass covered, as long as both have an ample supply. But grasses, being more shallow rooted than trees, exhaust their supply first. Trees continue to use water at their normal rate as long as their roots can tap moist layers. Such moisture then no longer becomes available to feed the streams.

For a similar reason summer stream flow from a forested watershed can be temporarily increased more by clear-cutting the forest in strips or patches than by removing the same volume of wood by cutting trees uniformly over the entire area. The clear-cut areas are able to yield a surplus of water whereas the trees remaining on the partially cut areas are free to use the water of their erstwhile neighbors. Within a few years, not more than 5 years in eastern forests, the tree seedlings and sprouts, together with the shrubs and herbaceous vegetation that follows clear-cutting of forests, and the expansion of roots and crowns on partially cut stands, restore the total evapotranspiration to approximately the level that prevailed before cutting. In dry regions, where the response of vegetation to cutting is less pronounced, or where grass takes over the soil, an increase in water yield may continue for a much longer period.

**Logging Damage** Logging is one forest operation that is likely to cause damage to soil and water. Logging damage to soil as well as to residual trees tends to increase with the degree of cutting and the weight of equipment used. A light thinning or partial harvest cut may cause practically no soil damage. Heavy cutting, especially clear-cutting of mature timber followed by fire on steep slopes, can result in serious soil loss and may give rise to floods. Heavy machinery compacts soil, disturbs litter, and can divert or damage the channels of streams. Foresters have demonstrated that moderately steep slopes can be logged with negligible damage to permanent streams if proper precautions are taken in laying out and draining logging roads, skid roads, and log landings. About 90 percent of the runoff and erosion caused by logging is attributable to roads. Cutting trees does not in itself change the rate at which water infiltrates into the soil, provided the forest floor remains essentially undisturbed. This was demonstrated in a classic experiment on the Coweeta Hydrologic Laboratory, North Carolina.[9] All trees were cut on a small watershed, and the boles, branches, and tops were allowed to remain where they fell. Infiltration remained the same as before and no surface runoff occurred.

**Augmenting Water Yields** From the facts outlined above, it is evident that by manipulating the plant cover man can influence at least to some degree the quantity and quality of water that flows from a drainage area. Generally he can do more to influence the contribution of snow to the annual water yield than he can to influence the contribution from rains. Grass- and shrub-covered land accumulates more snow than forest-covered land, but the snow disappears earlier from open land than from forested land due to early melting and rapid evaporation. Timber cutting can have a marked influence on snow accumulation and on the yield of water from the snow pack. This is important in the western mountains and in the northern regions where the major part of the annual runoff comes from melting snow. It is this that recharges the underground and surface water reservoirs.

Anderson lists six ways in which manipulation of the crown cover can affect snow storage and snow melt in western regions:[1]

1  By cutting openings in the forest, snow water storage can be increased as much as 40 percent.
2  If trees can be left standing in forest cutting so as to shade snow from direct sunlight and long-wave radiation, snow melt can be slowed down by 40 to 50 percent.
3  In maritime climates and also in the eastern United States much snow melting occurs from the latent heat of vaporization of water that is given up when it condenses on the snow. By leaving trees to shelter the snow from the wind, less latent heat and ambient heat is transferred to the snow mass. This may reduce the contribution of snow melt to short-period floods by perhaps 50 percent.
4  By removing trees, snow melt can be made more effective in producing water yield by reducing by 8 inches or more the water that otherwise would be required to recharge the soil reservoir that has been depleted by transpiration of trees.
5  By leaving trees to shelter snow on what would otherwise be windswept ridges, evaporation can be reduced, saving 2 to 3 inches of water per year.
6  By cutting forests so that air that is cooled by contact with snow is trapped, snow melt may be delayed.

Such potential benefits from manipulating the forest cover are substantial and may be well worth the effort required to attain them in water-deficit regions where municipalities and irrigation make heavy demands on the water supply. The reader will realize that not all six benefits are cumulative on any one watershed. Much skill and experienced judgment will be required to attain maximum benefits from forest cutting to increase water yields.

Little can be done by manipulating forest cover to increase water yield from summer rainfall throughout most of the United States. The plant cover, whether it be trees, grass, or field crops, tends to draw down the ground water through transpiration and evaporation so that what falls in summer showers scarcely makes up the deficit. If the showers are heavy, as they often are, runoff from cropland and certainly from bare soil will result, but little water will flow from well-protected soil. Heavy summer showers will also yield water from shallow soils such as exist in northeastern mountain areas. But elsewhere, unless the soil is stripped bare of vegetation, not much can be done to augment summer flow.

**Cloud Seeding** Many years ago Irving Langmuir began experiments in seeding clouds with silver iodide, and later with dry ice, to induce rainfall. He reported positive results.[8] Commercial rainmakers sought to apply his methods to alleviate drought. Many scientists remained skeptical, as no proof was available that the rain produced might not have fallen in any case, or was not at the expense of what might otherwise have fallen on adjacent areas. Beginning in 1968 a series of radar-monitored, statistically designed tests was carried out in Florida. In 1971 it was reported that the tests were conclusive in demonstrating that single-cloud seeding resulted in increased rainfall as compared with the rainfall from similar unseeded clouds. Further controlled experimentation is planned to determine if cloud "mergers" can be promoted to spread the increase over a wide area.[10] If cloud seeding becomes a practical means of augmenting precipitation over wide areas, it will add a new dimension to the management of forests for watershed protection.

## REDUCING STORM AND OTHER DAMAGE BY FOREST COVER

**Avalanches** In steep mountainous country, avalanches or snowslides are often caused by a heavy fall of new snow on top of an old crust. In areas that are subject to frequent snowslides, forests are prevented from growing. Such areas often occur at the heads of very steep valleys and along small side valleys. Limited protection can be achieved by encouraging tree growth and installing other protective measures on the sidewalls. The valley heads can be held only by engineering works.[6]

**Landslides** Landslides can be even more destructive than snowslides. They occur when a great mass of soil lying on steep slopes becomes saturated and slides or flows over the underlying rock surface. An impermeable soil layer may also serve as a base over which the soil slips. Soil, boulders, trees, and other vegetation rush down the slope at high velocity carrying everything in their path. A single severe storm that occurred along the Lehigh River in 1943 gave rise to over 50 small landslides in a 20-mile reach. These descended onto the railway tracks and, together with bank cutting by the river, caused damage of some $500,000. The valley walls in this case were rocky and covered with tree and shrubby growth, thinned by frequent fires. It was observed that very few of the slides occurred on areas that had a good cover of trees. Slides may occur frequently in high mountains where tree cover is lacking or stunted.[6] Clear-cut logging of steep slopes may increase landslides.

**Gullies** Where heavy surface runoff rushes across unprotected soil that has no permanent stream channels, gullies occur. Those that are actively eroding cannot be stopped by the simple expedient of planting trees. The water that causes the cutting must first be diverted or impeded and the soil stabilized to some extent by other measures. Once this has been done, tree growth is possible and permanent stabilization of the gully may be effected.

**Silting** The slow filling of lakes through storm-carried silt and other debris is a geologic process that man can do little to prevent. He can do much, though, to bring about accelerated erosion that may speed the process severalfold by careless cultivation, overgrazing, and other acts that destroy plant cover and expose the soil. A forest cover will essentially stop surface runoff and halt accelerated erosion. Also, by emptying the soil reservoir through evapotranspiration, the forest tends to reduce peak runoff, thereby limiting the volume of water that would otherwise cut stream banks and move debris into lakes and reservoirs.

**Flood Damage** New Year's Eve in 1933 was a time of destruction, sorrow, and death for people who dwelt in La Canada Valley near Los Angeles. Heavy rains fell for $2^1/_2$ days on the steep San Gabriel Mountains, and a flood rushed out of the mountains, carrying silt, sand, and boulders that weighed as much as 60 tons. The water, with its heavy burden, spread over the valley floor, wrecking homes and gardens and blocking highways. As floods go, it was not large nor was the total damage impressive. What was significant about this flood was a contributing cause of it. A few months earlier a fire had swept over the chaparral-covered mountains, laying bare $7\frac{1}{2}$ square miles. Rain fell equally on both burned and unburned slopes, but studies made after the flood revealed that, acre for acre, burned land contributed 50 times as much water to the flood flow as unburned land. The area damaged by flood, 3 square miles, was almost half that burned by the fire.[4]

An even more widespread cause of floods is overgrazing and cultivation of steep mountain slopes. In the Philippines entire villages have been wiped out by flood waters descending from cultivated mountains. The deep-soiled lands at the base of the Himalayas in Kashmir and India are carved up by mile-wide boulder-strewn stream channels that show but a trickle of water during dry seasons. Floods are particularly damaging to soils in dry regions such as the Mediterranean countries, much of Africa and South America, and our own western states. Here the sparse vegetation provides but scanty protection at best. When lands have been heavily grazed or impaired by cultivation, floods almost inevitably follow each heavy storm.

Floods in the United States have caused substantial property damage and resulted in considerable loss of life (Table 9-2). Over the years the number of lives lost has tended to decrease, while the property damage has tended to increase. This has happened even though expenditures for flood control have also increased. Restoration of forest cover, where feasible, has become a regular feature of flood-control programs.

**Erosion by Wind** Forests protect soils against erosion by wind.[11] Forest planting can be used to stabilize sand dunes along the seacoasts and along the shores of our Great Lakes. Dunes, once started moving, can

Table 9-2 Loss of Lives and Damage Caused by Floods in the United States by 5-Year Periods

| Period | Lives lost | Property damage (in millions of dollars) |
| --- | --- | --- |
| 1931–1935 | 368 | 187 |
| 1936–1940 | 607 | 879 |
| 1941–1945 | 346 | 605 |
| 1946–1950 | 306 | 843 |
| 1951–1955 | 502 | 2,507 |
| 1956–1960 | 228 | 877 |
| 1961–1965 | 329 | 1,844 |
| 1966 | 31 | 117 |
| 1967 | 34 | 375 |
| 1968 | 31 | 339 |

*Source:* Statistical Abstract of the United States, 1970.

destroy large areas of valuable agricultural, recreational, and even residential land unless they are stabilized. A spectacular demonstration of dune control was carried out along the west coast of Taiwan. Here the coastal water is quite shallow. This, together with strong winds, favors dune formation. Foresters, by using lath fencing and dune grass, stabilized the dune surface so that Australian pine *(Casuarina)* could be planted. Within 3 to 4 years this formed a closed canopy affording complete protection. Meanwhile, the dune built up in front of the tree barrier so that a new strip of land could be reclaimed from the sea. In 10 years' time 20,000 acres of new land was thus reclaimed from the sea at modest cost. Machine-gun bunkers established by the Japanese to repell landing parties look out over $^1/_4$ mile of forest plantation instead of the open surf.

**Shelterbelt Planting**  The story is told that when President Franklin Roosevelt was caught in a severe dust storm in Oklahoma, he penciled a note asking the Chief of the Forest Service why he could not plant a mile-wide belt of trees from Canada to the Gulf to stop such storms. The Chief was at a loss to answer until one of his advisers pointed out that while one belt of trees would not stop dust blowing, hundreds of them properly placed would go far toward doing so. Thus the Prairie Plains Shelterbelt Program came into being.

The soils, natural vegetation, and results of past tree planting were surveyed throughout the region. Hardy trees and shrubs suited to the various conditions to be encountered from north to south were selected and propagated. Planting designs were made, consisting of shrubs, low trees, and high trees. Contracts with farmers to fence, plant, and cultivate the trees were drawn up. Between 1935 and 1949, 26,873 miles of tree belts were planted and additional plantings were made to protect farm homes and gardens.[11] Shelterbelt planting has also been active outside the prairie states. Wisconsin farmers have planted 5,942 miles; California fruit-tree growers, 2,000 miles; and Indiana muckland farmers, 100 miles.

Shelterbelts trap snow, thereby building up soil moisture in a region of deficient precipitation. They prevent soil from blowing on newly seeded fields. They protect crops from hot, dry winds of summer, and they shelter barns, livestock, and the farmers' homes against driving winter storms. Farmers of Nebraska and South Dakota estimate the cash value of shelterbelts in crop protection to be $43 to $60 a year per farm. Cotton yields in Oklahoma and Texas are reported to have increased from 8 to 17 percent, depending on nearness to the protecting belt. The total protected area extends a distance to leeward of 20 times the height of the trees.

Extensive shelterbelt planting has been carried out in France, the U.S.S.R. and other regions.

## POLLUTION ABATEMENT

Pollution of streams with industrial and municipal waste has rendered many natural supplies unfit for use or useful only after costly treatment. Pollution arises from local sources; hence, its correction has been considered a local and state responsibility. Foresters are concerned with the subject because of the possible role that forests might play in the removal of organic wastes and eutophying chemicals from sewage treatment effluents and overloaded streams.

The capacity of forests to receive and transmit large amounts of cannery waste water to underground reservoirs was mentioned in the section on infiltration. In the process the organic burden carried by the waste water was decomposed. Otherwise it would soon have clogged the soil channels.

At State College, Pennsylvania, another important experiment was carried out. Following secondary treatment, effluent from the municipal sewage plant was sprayed on the forest at the rate of 2 inches per week during an 8-year period. At the time of writing the test is still in progress. The forest reduced the detergent and plant nutrient content of the effluent to approximately the level that prevailed in the natural ground water. Some 80 percent of the forest-filtered water entered the soil reservoir in condition suitable for use in the municipal water system. The other 20 percent was lost by evapotranspiration. Tree growth was markedly stimulated by the water and plant nutrients provided by the effluent. Similar applications were made to land bearing agricultural crops but these proved much less effective than forests during winter. Sludge from the filtering beds of the treatment plant also had a fertilizing effect when applied to the forest. In 1971, large-scale applications of the test were underway by four additional municipalities. The possibility of using forests as a final cleansing agent for waste waters has vital significance for future recycling of water by municipalities and industries.

## RIVER-BASIN PLANNING

Through acts of 1950 and 1965, the federal government offers aid to states for comprehensive planning for river-basin development. Such planning encompasses control of floods, water for municipal use, irrigation, power developments, erosion control, pollution abatement, improvement of streams for fishing and recreation, and general development of forest and other renewable resources. All federal agencies active in a river-basin area are being drawn into the planning groups, and representatives of the several state governments participate.

Foresters become involved in river-basin planning because of the role of forests in soil stabilization, flood abatement, protection of river banks, general environmental enhancement, and recreational use.

Many states are taking steps to improve their water resources. Interstate compacts have been drawn up to develop the resources of river basins. The Delaware River is a case in point. New York, New Jersey, Delaware, and Pennsylvania have, together with the federal government, planned for the future use and development of this basin's water resources. Perhaps of more significance in the long run are the plans being made by individual communities and groups of communities in cooperation with state and federal authorities. Their plans to develop their own water supplies and resources are intimately associated with expanded urban, suburban, industrial, and agricultural development.

One of the first steps in river-basin development is to clear up pollution so that the water is usable for industrial and municipal supply, for recreation, and for navigation. As mentioned above, forests can make a substantial contribution to removing objectionable salts that cause a heavy growth of aquatic vegetation. Interstate compacts are required for major rivers. That the case need not be hopeless in even densely populated regions is borne out by the success attained by the Ohio River Valley Water Sanitation Commission, abbreviated ORSANCO. The achievement of this agency won the American Society of Civil Engineers award for outstanding accomplishment. The story is told in vigorous language by the Commission's executive director Edward J. Cleary in "The ORSANCO Story," published by the Commission in 1967.[2] The achievements on the Delaware and Ohio Rivers, working with state compacts aided by the federal government, and the comprehensive river-basin planning in the Tennessee Valley are ample demonstrations of what can be overcome by governments once they are committed to the achievement of a task.

It needs to be kept in mind that water problems grow in acuity from year to year as population increases, affluence spreads, and industries consume more and more power and raw materials. The momentum of the United States and of the western world is still strongly oriented toward increased consumption of power and material goods. The accomplishments mentioned above should therefore be viewed as only a beginning in the huge task ahead of cleaning up the environment and orienting consumption to products and activities that cause a minimum of deterioration. The success we have today in dealing with such tasks will determine the type of life future Americans can enjoy.

## LITERATURE CITED

1. Anderson, H. W. 1970. Storage and Delivery of Rainfall and Snowmelt Water as Related to Forest Environments, J. M. Powell and C. F. Nolasco (eds.), *Proceedings of the Third Forest Microclimate Symposium, Seebe, Alberta, 1969,* Canadian Forest Service, Calgary, pp. 51–67.
2. Cleary, Edward J. 1967. "The ORSANCO Story: Water Quality Management in the Ohio Valley under an Interstate Compact," The Johns Hopkins Press, Baltimore. 335 pp.
3. Crafts, A. S. 1968. Water Structure and Water in the Plant Body, in vol. 1, p. 30, of Kozlowski, T.T. (ed.), "Water Deficits and Plant Growth," Academic Press, Inc., New York. 390 pp.
4. Frank, Bernard, and Anthony Netboy. 1950. "Water, Land and People," Alfred A. Knopf, Inc., New York. 331 pp.
5. Holcomb, Robert W. 1970. Power Generation: The Next 30 Years, *Science,* **167**:159–160.
6. Kittredge, Joseph. 1948. "Forest Influences," McGraw-Hill Book Company, New York. 394 pp.
7. Kozlowski, T. T. (ed.), 1968. "Water Deficits and Plant Growth;" vol. 1, Development, Control and Measurement, 390 pp.; vol. 2, Plant Water Consumption and Response, 333 pp. Academic Press, Inc., New York.
8. Langmuir, Irving. 1950. Control of Precipitation from Cumulus Clouds by Various Seeding Techniques, *Science,* **112**:35–41.
9. Lull, Howard W. 1964. Ecological and Silvicultural Aspects, section 6, pp. 1–30, in Chow, Ven Te (ed.), "Handbook of Applied Hydrology" McGraw-Hill Book Company, New York.
10. Simpson, Joanne, and William L. Woodley. 1971. Seeding Cumulus in Florida: New 1970 Results, *Science,* **172**:117–126.
11. Stoeckeler, Joseph H., and Ross A. Williams. 1949. Windbreaks and Shelterbelts, *Trees,* The Yearbook of Agriculture, pp. 191–199, U.S. Department of Agriculture, Washington.
12. Water and America's Future, *Proc. of the Soil Conservation Society of America,* Akeny, 1967. 239 pp.

Chapter 10

# Wildlife and Range Management

A visit to the African game parks is the supreme wildlife experience available in the twentieth-century world. To be within 50 feet of free-ranging wild gazelles, impalas, eland, and hartebeest; to cruise up the Victoria Nile among hundreds of crocodiles and hippopotamus; to watch a mother cheetah training her cubs to hunt; to be amused by the kittenlike behavior of a pride of well-fed lions lolling in the shade of a thorn tree; to snap picture after picture of a herd of 50 elephants as the cows and their calves file by within easy camera range; and to see herds of buffalo, wildebeest, and zebra numbering in the hundreds of thousands is to be amazed at the fruitfulness of this relatively dry land.[2]

Must such richness of life be sacrificed to cattle, sheep, and goats as is rapidly taking place? Not, informed scientists believe, if edible meat supply is man's objective. Over much of the range, natural herds can outproduce domestic animals for several reasons.

Domestic animals are selective in their feeding, shunning unpalatable plants. Each kind of wildlife may also be selective, but of different species of plants so that the various kinds together feed from all forage available. Thus domestic animals, if overcrowded, tend to deplete the range and convert grassland to shrubs. The native wildlife, while harvesting more of the total forage than cattle, still tend to maintain the quality of the range. Elephants, for example, are able to break down or tip over acacia trees to feed on the foliage, fruit, and twigs, thereby converting low growing woodlands to grass.

Domestic animals require a readily available water supply. Many wild herbivores do also, but some can go for long periods with limited water and a few, such as the gerenuk, can live on land without free water, supplying its needs from the forage it consumes.

Domestic animals are prone to debilitating endemic diseases to which native herbivores have high resistance. Domestic animals also require protection by man. Wild herbivores are much more able to elude predators and to fulfill all their needs from the range than is domestic stock. Man need only harvest them at a rate that does not deplete the breeding stock, just as he would do with his domestic stock.

Actual studies and game ranching in Africa support the merits of managing the wild herds for meat supply. One large ranch employed professional hunters to remove the wildebeest and zebra from the land. After some months the owner learned to his dismay that the hunters were selling the carcasses for £12,000 per year. This was more than the rancher could hope to net from cattle from the same land. Game farming of Saiga antelope in the U.S.S.R., a species threatened with extinction at the beginning of the twentieth century, has yielded a harvest of some 150,000 animals per year since 1957. The meat and hides are of high quality.[9]

Yet even in the face of such evidence as the above, the range for wild herbivores in Africa is rapidly shrinking because of the pressures of the people on the land for settlement, cultivation, and domestic stock. Moreover the local populations are said to prefer beef to game and so kill few wild animals for food.

When the white man first came to America, it also supported fabulous quantities of wild game. Buffalo, deer, pronghorn antelope, and elk were widely distributed over the continent. Among game birds, ducks, geese, and the passenger pigeon were taken by carloads by market hunters and sportsmen. But as these species became scarce or were eliminated throughout much of their original range and the passenger pigeon became extinct the sportsmen's wrath was vented on the market hunters. Laws were passed to prohibit the sale of game. Later various birds and game animals were given statewide protection. Hunting seasons were established, daily and yearly bag limits imposed, artificial restocking was begun, exotic species of game were imported from abroad for release, and winter feeding was encouraged. Game refuges were also established. Still the game dwindled and hunter frustration from bagless days afield grew. The measures advanced by biologists, sportsmen, and legislators failed to stop the decline in game. Increased restrictive measures seemed to have little effect. A more comprehensive approach was needed.

## WILDLIFE MANAGEMENT

**The Game Management Concept**  Theodore Roosevelt and Gifford Pinchot had applied the concept of perpetuation through wise use to the nation's forests. Could the same concept be applied to game? One man was convinced that it could be and staked his future professional career on this conviction. His name was Aldo Leopold. It was no accident that this man was a forester in midterm of a highly successful career as a U.S. Forest Service administrative officer (Fig. 10-1). Leopold outlined three basic premises for application to game:

1   All outdoor resources should be viewed as one integral whole.
2   Perpetuation through wise use is a public responsibility, and their ownership is a public trust.

**Figure 10-1**  Aldo Leopold, founder of modern wildlife management. *(University of Wisconsin.)*

**3** Science is the tool by which this responsibility can be discharged.

Leopold himself was an ardent hunter and fisherman who sought to preserve these sports. But even more, he was a scholar and philosopher. He systematically kept records of his own days afield and of what he brought home in bag and creel as it varied from place to place, trip to trip, and year to year. He also read avidly and purposefully the writings of others and sought to fit all he learned into a broad pattern of managing land for harvest of wildlife as well as timber on a sustained-yield basis.

Whereas biologists and sportsmen tended to think in terms of deer, elk, and other game species, Leopold focused his attention on their habitat and its capacity to supply the food, coverts, water, and other special needs of wildlife. He gleaned from experience and from records of others that game populations tend to attain a certain density on a given range. The number of breeding adults to survive the winter fluctuated around a mean from spring to spring irrespective of the size of the brood or herd in autumn. This led him to the concept of a harvestable surplus that could be removed from the land each year without affecting the long-term population of the species.

Leopold set about to define species requirements and behavior: nesting areas per breeding pair, tolerances of other species, daily and yearly travel radii, breeding habits, and minimal populations essential for species survival. All influenced or controlled population density. He cogently argued that such measures as closed seasons, daily and yearly bag limits, complete protection of rare species, killing of predators, large refuges, rearing and release of game, importation of exotics, and food distribution in winter were no substitute for improving the quality of the range. He convinced biologists that manipulation of the plant cover and other features of the environment such as watering places, dusting areas, nesting places, coverts for young, and especially appropriate wintering grounds, were far the more rewarding measures to take. He emphasized that these must be well dispersed so that all essentials would be available within the normal cruising range of the species at their time of greatest need.

Leopold demonstrated that the breeding potential of most wild creatures is more than adequate to stock a range within a few years provided the factor or factors that limit population increase are ameliorated, a task often possible at modest cost.

Thus, with the publication in 1933 of Leopold's classic, *Game Management,* biologists, sportsmen, legislators, and public officials were introduced to the modern approach to management of wildlife resources.[7] The University of Wisconsin responded by making Leopold the first professor of game management. Through later writings, notably his charming *Sand County Almanac,* Leopold convinced the general reader of the need for a new ethic toward the life-giving properties of land, and of the satisfactions such an ethic can bring to the individual. Wildlife management as a science has progressed rapidly since Leopold outlined the tasks ahead in 1933. This progress was due largely to the many young people, including Leopold's own sons and daughter, who were inspired by him to follow careers in scientific resource use.

**Characteristics of Wildlife** Wildlife includes all free-ranging vertebrates, fishes, reptiles, amphibians, birds, and mammals. Wildlife must be capable of living and perpetuating its kind in natural environments without the aid of man—either intentional or inadvertent. The Norway rat that depends on man's garbage dumps and farmers' granaries for a livelihood is not classed as wildlife.

Wild creatures differ from domestic in a variety of traits. The wild turkey, for example, differs from the domesticated in greater wariness, a hiding instead of scattering reaction when disturbed, later nesting habits, and a larger brain and thyroid, as well as larger adrenal glands.[5] In wildlife many senses are much more highly developed than in man. The sense of smell is especially noteworthy. The olfactory powers of many carnivores are well known. They are used extensively in stalking prey and in receiving messages conveyed within the species by means of scent pheromones, dealt with in Chap. 8. At night, fawns attract their mothers by pheromones. Pheromones are also used to repel other deer. Beaver, that need to attract others of the colony to repair a break in their dams, emit many different scents, apparently to convey messages.*

The eyesight of hawks, owls, and predatory mammals is especially keen. The hearing of many species is vastly more acute and covers a wider range of pitch than that of humans. Bats and certain aquatic animals have a natural sonar system for avoiding objects and for locating prey. The ability of birds to migrate huge distances and return to their home nest is phenomenal. The capacity of salmon and other anadromous fish to return from the ocean to spawn in the streamlet in which they were hatched as fry seems even more remarkable.

Still, with all their superbly developed senses and alertness, wildlife for the most part have but limited capacity to modify the environment for their comfort and use beyond digging burrows, building nests from readily available material, and making seasonal migrations to take advantage of favorable food supplies and climates. The outstanding exception is the beaver. The beaver fells trees and constructs dams that flood up to several acres of land, killing trees and creating a pond or swamp where none

---

*Author's notes from lecture by Muller-Schwartz, 1970.

existed before. Here the colony builds its house and stores its winter food supply. Even so, in view of the size of the animal, it seems but a modest achievement compared, for example, with the fungus culture of leaf-cutting ants, the use by other ants of aphids for their honeydew, and the house-building proclivities of tropical termites.

**Behavior of Species and Populations** When a species moves into a new territory that is favorable for its development the population grows slowly at first until a good breeding stock is built up. Thereafter the population grows very rapidly until it matches the carrying capacity of the new range. After that the population tends to fluctuate around this maximum. If the range is overused, however, the population will fall drastically until the range recovers, at which time a new cycle of increase occurs. Some species avoid overcrowding of their range by intolerance toward others of their kind. Density leads to fighting, competition for food and coverts, other forms of strife, and lowered reproduction.[13] Some students of animal behavior hold that such strife is taught the young by adults, with this belief based on the observation that young reared in the absence of adults appear to tolerate crowding without aggressive acts. Some species have built-in endocrine mechanisms that lower the reproductive activity when the animals are crowded, thereby keeping the population within bounds.

Species of animals differ widely in their breeding habits, feeding habits, cruising radii, and gregariousness. Differences in these characteristics cause the variation found in range and distribution of species, densities tolerated, hunting pressures withstood, and the influence animals have individually and collectively on their environment. Some animals bear only one or two young annually; others may bear up to ten or more. The ruffed grouse commonly lays 8 to 14 eggs in a nest; the loon only 2. The white-footed mouse may bear four to eight litters of six to eight mice each during a year. The cottontail rabbit may produce three to four litters of four to five young each in a year. The bear produces one or two young every other year.

Species that have large annual clutches or litters of young tend to have high mortality of young. The percentage of young that survive to breeding age is given by Sheldon[13] as follows:

| Species | Percentage surviving to breeding age |
| --- | --- |
| Cottontail rabbit | 8 |
| Bobwhite quail | 13 |
| Pheasant | 15 |
| Ruffed grouse | 13–24 |
| White-tailed deer | 70 |

The capacity of a species to multiply until it occupies its range completely is its biologic potential. This is remarkably high for some game species and accounts for their resistance to hunting pressure and the eruptions in population that sometimes occur. Deer on understocked range may show an annual increase of 40 to 50 percent; elk, 21 percent; raccoon, 50 percent; and quail, as high as 460 percent. For wild ducks, 70 percent of those that reach flying stage have a life span of less than one year. About 60 percent of the ruffed grouse kill is of birds less than a year old. Several other species seem to maintain their populations with about a two-thirds turnover each year.[1, 15]

The distribution of animals in a given area naturally will depend upon how selective they are in their feeding. Animals that eat a wide variety of foods may find abundant forage within a restricted area. Selective feeders, such as a pair of ruffed grouse, must range over some 5 acres to obtain a satisfactory food supply. Other birds, such as wild ducks, geese, blackbirds, and starlings, can be seen feeding in flocks.

Cruising radius is the distance an animal or bird will travel to meet its daily life requirements. All factors essential for its welfare must be found within its daily cruising radius, else the animal will leave the area. It is for this reason that large areas devoted to a single crop—whether this be wheat, potatoes, or spruce—tend to produce a small amount of wildlife, whereas those that are highly diversified, with woodland interspersed with pasture, clean-tilled crops, and waste areas, tend to support dense wildlife populations.

Animal cruising radii for a few species have been determined as follows:

| Species | Radius, miles |
|---|---|
| Bobwhite quail | $1/4$ |
| Cottontail rabbit | $1/4$ |
| Ring-necked pheasant | $2 1/2$ |
| Ruffed grouse | $5/8$ |
| White-tailed deer | 7 |

Some animals tend to exhibit group behavior that is not readily discernible in the individual animals. For instance, there must be limits of density of population in both directions. The animals must be sufficiently numerous to find mates but not so dense as to overstock their range. A population must be of a certain minimum size and occupy certain minimum range if it is to survive. This is particularly true of such animals as the Rocky Mountain sheep that feed in bands and require fairly large areas to maintain themselves. Some animals have special tolerances for others of different species and for various factors of the environment. Others are limited in their tolerance and will drive out populations of other animals or leave themselves. The population should have a proper sex and age-class distribution if it is to remain vigorous and healthy. Some populations can readily be transplanted to a new location, whereas others can establish themselves in a new environment only with difficulty. All these considerations influence what the game manager can do successfully to promote the welfare of wildlife.

**Range Essentials** A wildlife range must provide all the essentials for the life processes of wildlife. It must contain food, coverts, and water sufficient for all species concerned, for both sexes, for all age classes of the various species, and for all activities and life functions of these species. For example, it is not sufficient simply to have plenty of young sprouts as food for deer. There must also be well-distributed water, particularly during the period when fawns are born so that the does need not travel far from the newborn fawns to find food and water.

All essentials of the wildlife range must likewise be well distributed over the range and must be sufficiently close to one another so that an animal can meet its life requirement within its cruising radius. This naturally varies between adults and young and between nursing females and mature males.

Factors that tend to restrict wildlife population are rigorous winters; unseasonable cold or rainy weather, especially during the period when young are born; prolonged drought; failures of nuts, fruits, and forage; accidents; disease outbreaks; predators; parasites; pressure of domestic animals; land clearing for cultivation; swamp drainage; forest fires; and hunting. Man has probably done more to restrict wildlife through his appropriation of wildlife range for his crops, domestic animals, roads, dwellings, and industries, and by starting forest fires and other acts of carelessness, than by all his hunting and trapping (Fig. 10-2).[16]

The interaction of the two factors of biologic potential of species, on the one hand, and range essentials, on the other, determines the game population and the numbers that may be harvested without depleting the breeding stock. On a nonhunted tract of good grouse range, it was found that though the fall population varied from 29 to 55 per 100 acres over a three-year period, the spring population remained constant at 25.[16] In this case the birds are socially intolerant of heavier populations.

Many species, particularly the white-tailed deer, tend to overstock their range, especially where predators are kept in check.[15] This leads to stunted individuals, low reproduction rates, and heavy losses by starvation in winter. It also leads to serious depletion of the food and cover plants that support the herd. Natural restoration of depleted game range is a slow process.

**Developing a Wildlife Management Program** Management of an area for wildlife, like management for timber, requires an inventory of resources. This means determining the types and populations of wildlife

**Figure 10-2** A female blue-winged teal with wing tags and radio transmitter to study nesting and brood movements, Northern Prairie Center. *(U.S. Bureau of Sport Fisheries and Wildlife.)*

present and the condition of the animals. It is equally important to survey the food and cover available to support them.

Biologists have used a variety of approaches to census wildlife. Song birds can be estimated by travelling along census lines, stopping at fixed intervals to record the numbers and species of birds whose songs can be heard. The distance to each songster is estimated and recorded, and from this densities can be calculated. Woodcock can be censused by following along a road or other appropriate line, during the spring at dusk or early dawn, again stopping at regular intervals to record the number of males "peening" and engaging in nuptial flights. Wild geese can be herded and counted during the moulting period when they are unable to fly. Deer can be driven across a road or other open area along which concealed observers can count them. Tracks may be counted after a fresh snowfall. Deer densities may also be estimated from pellet counts.

Game on an open range can be counted from the ground or from aircraft. Live trapping, marking, release, and retrapping may also be used to estimate population densities. Various other methods have been devised. Compared with timber estimating, all these methods are imprecise. Counting flightless geese and large animals from air photos of herds are among the more precise methods of censusing wildlife.

In examining a wildlife range the manager will seek to discover the factor or factors that are limiting the population for each important species. These might be lack of winter coverts, winter food, den sites, places for rearing young, or a population of competing species. Ascertaining and ameliorating these limiting factors will result in the highest population increase for desired species over any comparable expenditure of effort. Ameliorating limiting factors might be done by logging operations, if the timber is mature; by thinning if it is so dense as to exclude low cover for food; by planting if there is an excess of open land; by creating coverts through slash piles or encouragement of shrubs; or by developing a water supply.

In most cases the management of forests for wildlife must be integrated into the management of the land for multiple-use purposes.[12]

**Public Policies and Programs** Migratory birds, including waterfowl, are under federal jurisdiction. All other wildlife is considered the property of the state in which it is located, because wildlife can freely move from one man's land to another's. The owner of the land does not have the right to take wildlife as he sees fit on his own property; he must conform to the game laws established in his state. The only exception is that Indians may take fish and game on reservations at any time. Wildlife on national forests is also considered the property of the state in which the forest is located.

The federal government has jurisdiction over the wildlife on national parks and national monuments. Here the government attempts to maintain an abundant and vigorous population of wild animals.

**Fish and Wildlife Service** The Fish and Wildlife Service of the Department of the Interior is the government's expert agency for management of fish and wildlife resources. This Service is responsible for managing some 317 National Wildlife Refuges totaling 28.5 million acres. Of these, 250 are primarily for waterfowl. Refuges in Alaska are the largest and total some 18.7 million acres. Most of those in other states are less than 10,000 acres in size. The objective of migratory waterfowl refuges is to provide resting and feeding areas along the four main flyways used by these birds during their spring and fall migrations. A refuge need not be large to attract thousands of geese, ducks, and other waterfowl during the migration periods. Other National Wildlife Refuges have been established for rare or threatened species. Big game refuges have been established for grizzly, Alaska brown, black, and polar bear, Dall sheep, bighorn sheep, moose, musk-oxen, elk, pronghorn antelope, and buffalo.

Each refuge is established to protect some primary species or group of species but other resources are also considered. For example, on the

33,000-acre Piedmont National Wildlife Refuge in Georgia, waterfowl management is of primary importance. But ancillary thereto some 380 deer and 7 million board feet of timber are harvested annually. Wild turkey also is hunted on a limited scale.[20]

Some 20 million people visit the National Wildlife Refuges each year. Camping and fishing is allowed on many and controlled hunting on a few.[8]

Other major responsibilities of the Fish and Wildlife Service include research in wildlife management, administration of federal aid to states authorized by the Pittman-Robertson and other acts for the protection and management of wildlife, maintenance of inland fisheries, conservation of salmon, expert service on river-basin development plans, control of predators and rodents, administration of federal statutes for the protection of wild fowl, and cooperation with the American republics in the enforcement of the migratory-bird treaties (Fig. 10-3).[3]

As owner and custodian of roughly one-third of the land area of the United States the federal government has responsibilities for the wildlife that lives on this land. In 1970 the federal government, subject to state and native claims, had jurisdiction over 95 percent of the area of Alaska. Virtually all is wildlife range. This range has been the least modified by human action and the herds of big game the least subject to human control of any in the nation. Among the 48 contiguous states, the 20 western states that have the largest areas of public lands find that from 40 to 48 percent of their big game lives in the national forests and on the lands managed by the Bureau of Land Management. The Public Land Law Review Commission recommended that legislation be drafted to provide for increased cooperation between the federal and state governments in protecting and managing the wildlife resources of the public lands. It recommended that wildlife management be recognized as the dominant use for those federal lands that are critical habitats for dwindling and threatened species of wildlife, whether they be game or nongame species. The Commission also stated that legislation was needed to set forth clearly that ultimate responsibility for determining the wildlife populations the range can support without deterioration rests with the federal government.[10]

**State Conservation Departments** The major activities in game management are centered in the state conservation departments. These departments issue licenses, protect game against illegal hunting and fishing, operate hatcheries, advise legislatures on seasons and bag limits, manage state refuges and public fishing streams and shooting grounds, and conduct research in game production. The last-named activity is often shared with or delegated entirely to the state agricultural colleges. By far the largest number of wildlife specialists are employed by the states.

# WILDLIFE AND RANGE MANAGEMENT

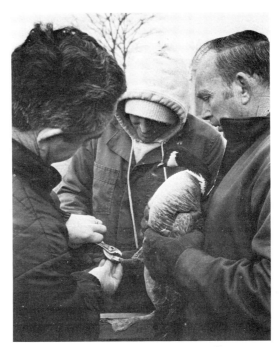

**Figure 10-3** Tagging a rare Aleutian goose to be used for restocking on Amchitka Island, Alaska. *(U.S. Bureau of Sport Fisheries and Wildlife.)*

The states collect the fees for hunting and fishing licenses. It is from such funds, together with the federal supplements mentioned above, that states finance their activities directed toward improving hunting and fishing within their borders. The sportsmen therefore feel that they should have a large voice in determining how fish and game expenditures are to be made. A strong tie tends to develop between the sportsmen and their respective state fish and wildlife administrators.

Only since World War II has the science of wildlife management advanced through scientific research enough to provide a firm basis for administrative decision and legislative action. If this science is to be of avail for wildlife improvement a constant and vigorous flow of information to the sportsmen and public becomes a vital activity of every progressive state fish and game agency. Few state departments operate under more interested and searching public scrutiny.

**Other Agencies** Owners of forests and other wild land have implied responsibility for wildlife. Care in forest management, distributing timber operations over a wide area and over a number of years and avoiding needless destruction of wildlife dens, coverts, and breeding areas are helpful.

The farmer has responsibility because he owns a substantial area of land that produces wildlife. More than other landowners, he may find that wildlife interferes with his activities. Game requires food, and the farmer feels that he has certain rights to the game he has fed. He has the right to control hunting on his land. Farmers have planted waste areas to game food and have taken other measures to promote the welfare of game.

Sportsmen's clubs, more than any other group, influence public policy and contribute funds for wildlife research and management. Through sportsmen's magazines an excellent opportunity is available to interest a wide group of individuals in scientific game management.

The general public also has concern, because wildlife has values that go beyond those realized by the landowner and the sportsman. It is to the American public as a whole that we must look for appropriate legislation to prevent indiscriminate destruction of valuable wild creatures, and for review of proposals made by landowners and sportsmen. The wildlife manager, in turn, has a responsibility to the public to provide accurate, unbiased information on wildlife questions and, particularly, to help the public to appreciate the values inherent in the wildlife resource.

**Values of Wildlife** The justification for spending large sums of money to support government and private activities for wildlife is the value of the wildlife to man. Professor R. T. King, formerly at State University College of Forestry at Syracuse University, has listed wildlife values under seven headings:

1  *Commercial values* include the income derived from commercial fishing, from controlled use of wild animals and their progeny, from sale of furs, pelts, and game as food, and related exploitation of wildlife. To realize such values wild creatures must either be destroyed or confined. This need not necessarily result in reduction of the wild breeding stock.

2  *Recreational values* may be indicated by the expenditures of sportsmen and others in the pursuit of wildlife and in watching it in its natural habitats. This includes expenditures for equipment, supplies, clothing, license fees, transportation, food, lodging, and so on. Some 37 million hunting and fishing licenses were purchased in 1967.

3  *Biological values* are the services that wildlife performs for man in controlling insects, rodents, and other pests; their sanitizing activities, their collection and storage of products man could not as cheaply collect himself, their contributions to soil building and water conservation, and their service in maintaining a balanced population in ecosystems.

4  *Social values* are those that accrue to communities in making the environment more interesting, healthful, and attractive to people through the presence of wildlife. Real estate values may be enhanced by the presence of a wildlife refuge near developed land.

5  *Esthetic values* of natural environments are enhanced by the presence of wild creatures. The study of wild birds and animals has inspired artists, poets, musicians, and photographers to embellish their artistic creations by appropriate use of wildlife subjects. Wildlife interest often leads to a broader concern embracing all renewable resources including environmental quality.

6  *Scientific values* are realized from the study of animal behavior, population phenomena, disease spread in wildlife, and other activities and vicissitudes of wild creatures that help man to understand and appreciate better the world's ecosystems.

7  *Negative values* result when wildlife destroys fences and structures, feeds heavily on crops and fruit, consumes stored grain and hay, damages forest plantations and tree reproduction, preys on domestic stock and poultry, weakens earthen dikes, levees, and dams, or plugs drainage channels, and like actions that result in economic loss. This also includes the costs that men incur to protect crops, forests, and domestic animals from wildlife damage. Wildlife may also become infected with and spread diseases that affect humans and domestic animals, such as rabies and sleeping sickness.

It is the positive sum of all these values that justifies man's efforts to protect and promote the welfare of wildlife. Many of these, though real, cannot be assigned definite dollar values.

**The Task of the Wildlife Manager**  The typical wildlife manager is a state employee with responsibility for the protection and well-being of wildlife over an area of some 30,000 acres or more. The imaginative protector is one who knows his landowners and encourages them to plant game food and cover plants, who addresses sportsmen's clubs on wildlife requirements, who enlists the interests of school children in studying wild creatures and improving nesting areas, and who promotes good sportsmanship among the public. He is, in fact, more a teacher than a warden (Fig. 10-4).

The wildlife manager must be alert to prevent trespass, especially during hunting seasons and during those periods that are critical for wildlife. This means that he must often be in the field during the stormy weather of winter and during the period when young are born in the spring and summer.

A wildlife manager should be a good plant and animal ecologist. He

**Figure 10-4** The African College of Wildlife Management educates wildlife technicians. The college is located at Mweka, Tanzania, on the southern shoulder of Mt. Kilimanjaro.

should understand the management of the resources that the land produces aside from wildlife. The most efficient habitat management often can be effected through a well-coordinated land-use program.

He must be, above all, a man of great patience and of keen insight and understanding.

### RANGE MANAGEMENT

*Wildlife* management is the conscious application of ecological principles to increase the long-range value of wildlife to man. *Range* management is the conscious application of ecological principles to increase the long-range returns to man from livestock, water, and other range resources. Both trace their origin to prehistoric times. Nomadic herdsmen are still to be found in Africa and the Middle East. Here they watch over their flocks as Luke reports the shepherds of Judea were doing when the Angel of the Lord appeared before them to announce the birth of Christ.

Commercial grazing exists in the present day on tropical savannas of Africa, southern Brazil, and northern Australia, and on temperate zone grasslands. Temperate grasslands that are grazed commercially include those of the Andes and southern South America, of Australia and New Zealand, of southern Africa, and of the southeastern and western United States and adjacent Canada and Mexico. Of the four major regions, North

American rangeland is the only one that has nearby prosperous industrial communities that provide a ready market for all the products the range can produce.[4] Though climatically far different from central and eastern Africa, the rangeland of the United States shared the same need as that of Africa to displace a huge population of native ungulates and the predators that fed upon them in order to provide range for domestic livestock. In at least parts of Africa this displacement, as indicated above, has not always led to an increase in the yield of harvestable animal products from the land. Might the same be said of rangelands of the western United States? There is some evidence that this might have been the case. Ernest Thompson Seton, in his *Lives of Game Animals,* gives estimates of the numbers of ungulates that populated the ranges before the extensive slaughter by the white man.[11] His estimate lists 50 million bison, 40 million white-tailed deer, 40 million pronghorn antelope, 10 million each of elk and mule deer, and lesser numbers of black-tailed deer, bighorn sheep, and mountain goats.[11] Stoddard and Smith[14] converted Seton's estimates into animal units equivalent to 1,000 pounds of live weight, roughly that of a cow with one calf. The original game range supported on the basis of these estimates some 67.1 million animal units. In 1959, the comparable figure for cattle, horses, and sheep on the 17 western states exclusive of Alaska and Hawaii, an area roughly comparable with that used by Seton in his estimates, was 50.2 million.

**Development of Range Use in the United States** Horses, sheep, and cattle were brought to the Americas within the first three decades of the discovery of the western hemisphere. In fact Columbus, himself, is reported to have landed horses and sheep in 1493. Many subsequent introductions followed. By 1830 the commercial range-livestock industry began in the western states. Range-livestock number expanded rapidly. The numbers of animals grazing on the 11 western range states by selected years were as follows:[18]

| Year | Cattle, in millions of head | Sheep, in millions of head |
|---|---|---|
| 1867 | 2 | 5 |
| 1885 | 7 | 19 |
| 1905 | 9 | 24 |
| 1925 | 11 | 21 |
| 1954 | 12 | 12 |
| 1959 | 11 | 11 |

Between 1940 and 1969, numbers of sheep on the western range declined while cattle increased. The same trend occurred nationally, sheep dropping from 46 to 18 million head, while cattle increased from 68 to 110 million head.[19]

For a long time there was ample range for all users as cattle and sheep spread over the land vacated by the fast disappearing bison and antelope. But while livestock was increasing, free rangeland was diminishing due to occupation by settlers and grants to states, railway corporations, and individuals. The federal government as landowner exercised no control, so conflict over use became inevitable. So did range deterioration through overuse.

**Grazing on the National Forests** Such was the situation when, in 1905, the Forest Service in the U.S. Department of Agriculture was granted administrative responsibility over the National Forests. Dedicated to wise use of resources for the long-term good of all the people, the Forest Service faced its first major test in regulating grazing on the National Forests. An annual permit system was established with grazing allotments made to users at fixed fees per head of livestock. The stockmen did not graciously surrender prerogatives they had exercised for decades to officers of the Forest Service. The latter often found themselves in the middle between contending users. Halting range deterioration could not be immediately effected. Tlhe momentum of decline was too great. But restrictions on expansion of herds on the national forest were imposed and various practices aimed at range improvement were introduced on an experimental basis.

The stockmen were realistic enough to recognize that in the long run they could not expect to gain their ends by intimidation of homesteaders or Forest Service officers. Moreover, perceptive stockmen realized that regulation of use of the range was necessary to halt deterioration and to allocate use among graziers.

A prolonged series of drought years began in the 1930s. This dealt both stockmen and dry land settlers a doubly cruel blow, for it coincided with the great economic depression with its drastic decline in markets and prices of land products. Even the very soil seemed to rise with the wind in protest, forming huge dust clouds that drifted to the eastern seaboard. Foreclosed farms and ranches were sold for sacrifice prices at sheriff sales. Financial ruin overtook many who had formerly been in comfortable circumstances. Past abuse of western lands had invoked a nationwide penalty that the public could no longer ignore.

The critical deterioration on the open ranges was documented by the Forest Service study *The Western Range*, published in 1936.[17] Irrespective of ownership, topography, or climate, the Service estimated the range to be 52 percent depleted from its virgin condition. The ranchers had little choice but to turn to the federal government for help.

**The Grazing Service** The Taylor Grazing Act of 1934 brought the use of the remaining 200 million acres of unreserved public land under federal control.

The new Grazing Service was established in the Department of Interior. It went through a period of testing with the stockmen such as the Forest Service had experienced. But the machinery was set up for control and improvement. Policies were established only after public hearings. Advisory groups were established so that stockmen had a voice in deciding on both the objectives to be pursued and the methods to be followed to attain them. In fact, their success with the Grazing Service led the stockmen to press new demands on the Forest Service. As a result, the Forest Service discontinued the practice of cutting down on allotments of established ranchers to distribute grazing permits to aspirant newcomers in the business. The Service also allowed the purchaser of an established ranch first chance to take over the grazing privileges of the former owner. Both these practices did much to give the range users a sense of stability that they had not enjoyed before. Other measures followed that led to range improvement and to increasing rather than curtailing the grazing allotments.

During the drought and depression years of the 1930s the range-livestock industry had reached its nadir. The depression ebbed as the decade drew to a close. The long drought also broke. War in Europe brought an increase in food exports. When the United States became involved it became necessary to ration beef and certain other foods. At the urging of the government and with the help of increased prices the stockmen expanded their herds to use the improved forage that resulted from increased rainfall. Prosperity returned to the industry, enabling many operators to accumulate savings that could be invested in ranch improvements. Range management entered upon a new era (Fig. 10-5).

Research by the Forest Service and by the western agricultural colleges had developed management methods that brought about range restoration and improved carrying capacity. The demand for range experts by the Forest Service and Grazing Service encouraged the western universities to offer degree programs in range management. The new techniques were rapidly put into practice on the range.

**The National Grasslands** During the drought and depression years of the 1930s the federal government under the Bankhead-Jones Farm Tenant Act of 1937 bought up submarginal farmlands to retire them from cultivation. Large areas were acquired in the Great Plains area. Most of this was converted to grazing land. In 1950 the Forest Service was

**Figure 10-5** Counting range cattle in Montana. *(U.S. Bureau of Land Management.)*

assigned responsibility for administering some 3,822,000 acres of such lands. These were designated National Grasslands. Regulations and management objectives similar to those on National Forests were followed. In 1968 the National Forests and National Grasslands furnished forage for 1.5 million head of cattle and horses and for 1.9 million head of sheep and goats. The cattle grazed an average of 4.5 months and the sheep 2.5 months on Forest Service–administered lands.

**Importance of Range Management** Every American boy learns about the role that the cowboys and ranchers played in the development of the western states. The livestock industry was and remains today one of the major mainstays of the economy of the region. Grazing occurs on some 1,112 million acres of land in the United States. This includes farm pastures, croplands grazed after harvesting, forested and wooded range as well as native grassland, and semiarid lands. The rangelands of the West and Southeast cover some 915 million acres, 40 percent of the nation's area. In 1959, 87 percent of the sheep and 45 percent of the cattle in the United States were in the 17 western states. Beef cattle from the western range are commonly marketed in the fall of the year. Some go directly to slaughter, but many more are sent to feeding lots in Iowa,

Illinois, and Kansas, where they are fattened for market. The use of the western range, therefore, influences agriculture in the West and in the Middle West and affects food prices throughout the country.[17]

### Grazing Regions of the United States

*The Tall-grass Region* The tall-grass region, or prairie, lies immediately west of the eastern forests and originally extended from Canada to Mexico in a belt 150 to 500 miles wide. This region of the big and little bluestem grasses furnished the main food supply for the great herds of bison. Most of this land is now in field crops. It was the most productive of all range regions (Table 10-1).

*The Short-grass Region* This region lies between the foothills of the Rocky Mountains extending eastward to about the 100th meridian. It reaches far into Canada and southward to the desert regions of Texas, New Mexico, and Arizona. It covers the largest area of all regions. Blue grama and buffalo grasses are characteristic species. This region produces about one-third of the range beef of the nation.

*Pacific Bunch-grass Region* Lying between the Cascades and the Rockies is the fertile Pacific bunch-grass region. Much of the land has been taken over for cropland. The growing period for the grasses is limited to a brief spring period when temperature and soil moisture are favorable. Many nutritious grass species occur in the region.

*The Semidesert-grass Region* Located in the dry Southwest, this region of some 90 million acres is productive because of the highly nutritious and palatable grasses that occur there. Year around grazing is

Table 10-1   Types of Rangelands

| Range type | Area, in millions of acres | Average rainfall, in inches | Area required per animal-unit of grazing capacity per month, in acres |
|---|---|---|---|
| Tall grass | 19 | 20–40 | $3/4$–$1 1/2$ |
| Short grass | 198 | 13 | $2 1/2$–10 |
| Pacific bunch grass | 43 | 10–20 | 4–6 |
| Semidesert grass | 89 | 15–18 | 6–7 |
| Sagebrush grass | 97 | 10–20 | 9–15 |
| Southern-desert shrub | 27 | 5–10 | |
| Salt-desert shrub | 41 | 5–10 | 18–20 |
| Piñon-juniper | 76 | 6–15 | 8–10 |
| Woodland-chaparral | 13 | 10–20 | 2–3 |
| Open forest | 126 | 15–30+ | 6–10 |
| Southeast | 200 | 50+ | 5–30 |

possible where water is available. If overused, the range is taken over by desert shrubs worthless to range livestock.

**Sagebrush-grass Region** Occupying some 96 million acres in the intermountain region between the Sierras and the Rockies, the sagebrush-grass region is one of relatively low productivity. It does provide spring and fall range, which is scarce compared with winter and summer range areas. For this reason overuse was prevalent in the past. Sagebrush is one of the most important forage species of the region.

**The Southern-desert Shrub Region and Salt-desert Shrub Region**
These regions, with 3 to 15 inches of precipitation, extend across southern California, New Mexico, Nevada, and Arizona. Prolonged droughts and high temperature are frequent. The dominant shrub is creosote bush that is almost worthless for grazing. Good forage is available only during brief periods following heavy rains that furnish moisture for the short-lived annual plants. These regions are lowest in forage production per acre.

**Piñon-juniper Region** Ascending in elevation above the desert-shrub region, one passes into the piñon-juniper region. Both piñon pine and juniper form quite open stands. Between the trees are found mountain mahogany, sagebrush, and several kinds of grasses. The region furnishes spring range for animals that winter in the desert-shrub and summer in the coniferous forest. It is intermediate in productivity among the western regions.

**Woodland-chaparral Region** Chaparral is evergreen scrub oak and other evergreen shrubs of which mountain mahogany, snowberry, and manzanita are common species. The type is found in the southern Rockies, the intermountain plateau, and California. Fires have been common in the type and probably account for it not being replaced by forest or woodland. Grasses grow between the shrubs where they are not overdense. The grass, together with the nutritious and palatable shrubs, make it valuable for grazing even though it covers but a relatively small area.

**Open Coniferous Forest Region** The open forest range occurs at elevations above that of the piñon-juniper zone in the Southwest and in the Rocky, Wasatch, and Sierra mountains at elevations above 6,000 to 7,000 feet. Ponderosa pine is the dominant tree as one ascends from the foothills into the coniferous forest. The forest tends to be open grown at the lower limits of its range, permitting grasses to grow interspersed among the trees. At elevations of 8,000 to 10,000 feet Douglas-fir is the more common species, which also tends to form open stands where moisture supply is low. A special feature of the mountain coniferous forests is mountain meadows. These are often found in wet areas that are fed by springs. They support a lush growth of forage highly palatable to livestock and game. Hence they are often overgrazed, especially by elk.

The aspen–Douglas-fir and the spruce-fir zones both have stands of variable density with openings supporting forage plants. Above the spruce-fir zone at 11,000 to 14,000 feet occurs the alpine-tundra zone that provides good summer range for sheep (Fig. 10-6). The coniferous forest region is highly valuable for summer range where forage is lush and green long after it has dried up at the lower elevations. It is here that the grazier encounters the recreationist in the largest numbers and where a growing conflict in use between the two seems certain to bring about increased pressure on the stockmen.

**The Southeast Region** The moderately open longleaf and slash pine and the bottom-land hardwood forests of the Southeast have long been used for grazing cattle and hogs. The latter feed on the pine seeds and seedlings while these are in the grass stage. For about three months in the spring and a shorter period in the fall the grasses are green, lush, and nutritious. At other times supplemental feeding is necessary for good animal production. Rainfall of over 50 inches annually has caused heavy leaching of the soil, leaving it poor in nitrates and phosphorus. The use of

Figure 10-6 Alpine sheep range in Colorado. *(U.S. Bureau of Land Management.)*

protein-rich food supplements has become a general practice. Introduction of Brahman strains into the cattle has produced herds with increased resistance to heat and to tick fever. Because of a year-around growing season and the introduction of many improvements in the care of livestock and pasture, the Southeast has rapidly become a major beef producing region.[14]

**Modern Range Management Techniques** The western cattle ranch has become far more than a rancher's house with attached quarters for cowhands, sheds for saddles, branding irons, and lariats, and a corral. It may now include cultivated and irrigated hayfields, special breeding and calving pastures, and, if sheep are involved, lambing grounds and shearing sheds. Equipment includes heavy disk plows, powerful tractors, and essential seeding and haying machines. Pickup trucks are used for servicing pumping equipment at watering tanks, for salt distribution, and for hundreds of other light duty tasks. Large trucks are used for transport of feed and of animals to distant grazing lands or to rail sidings for marketing.

*Herd Management* Improved breeds of sheep and cattle respond best to careful herd management. This requires constant care in culling slow breeders, low weight gainers, and animals incapable of making good use of the range. Breeding seasons are restricted so that calves and lambs are born at the best time of year for survival and rapid early growth. Those stunted at an early age remain so for life. Good pasture and ample water as well as careful attention during calving and lambing are essential. Calves must be separated from their dams at weaning time. Supplemental feeding on an emergency basis must often be used to tide animals over a prolonged drought or an unseasonably heavy snowfall. Attention naturally needs to be given to salting, providing extra phosphates and protein if needed, as well as providing the usual care given herds managed on farm pastures.

*Forage Improvement* Range-research men have discovered that native grasses and nutritious herbs and shrubs will usually respond well to appropriate treatment. This means, as a basic need, affording the plants an opportunity to accumulate enough reserve food to ripen seed and to tide them over the winter or over a dry period when they can manufacture little carbohydrate. Perennial plants must have enough food reserve in the fall to last through the winter and to support a flush of growth in the spring to provide photosynthesizing leaves. Spring grazing, before the new leaves have restored the plant's reserve food consumed in their growth, weakens the plant throughout the season. Deferred grazing, and rest from grazing during a part of the season, may often be all that is required to start a depleted range on the mend.

Dispersal of animals over the range is essential for full use of forage. This may be done by herding sheep. Fencing, location of watering places and of salt, can induce cattle to make better use of the total range with minimum pressure on the land around the water sources.

Badly depleted ranges may need complete protection from grazing for one or more years. Overgrazing weakens the preferred and generally most nutritious and palatable species. It favors those that are least used by grazing animals, including poisonous species. The recovery capacity of range grasses is such that many can restock a depleted range if it is protected from heavy use and temporarily relieved of the competition of weeds and shrubs by use of herbicides, fire, or other measures. Fire may be needed also to remove toxic terpenes of soft chaparral plants. A striking case is reported of a gaged watershed in Arizona that yielded no runoff from its chaparral-covered slopes. Following a wildfire that killed the chaparral, the slopes were soon covered with grass and the stream began to flow. Thus a range that had yielded very little became productive of both forage and water, in this case without the intervention of man. It is believed that wildfire played a significant role in maintaining grass on the range even in pre-Columbian times.

Reseeding some northern ranges to crested wheatgrass gave some outstanding responses in early trials. It was especially successful on abandoned croplands. Cultivation to destroy unwanted plants followed by seeding and protection for up to 3 years enabled the plants to become established. Such reseeded ranges were able to support as much as three times their former stocking. In the central and southern regions crested wheatgrass did poorly, but seeding with native grasses proved successful. It was found, however, that ranges that had not been used for field crops would often respond almost equally well if given 3 years' protection from grazing. The practice today is to rely more on deferred grazing, rotation grazing, and rest of the range at strategic times to maintain health of the grasses rather than artificial reseeding. Use of herbicides or fire to eliminate competition of shrubs also may be all that is required to restore vigor to the grasses.

Poisonous plants have caused heavy losses to livestock in the past. Stoddard and Smith[14] list 38 species as especially dangerous among 400 that are known to be poisonous. Many of these, if consumed in small amounts, do not cause serious illness. Stockmen have learned how to manage range and livestock so that losses due to poisonous plants are kept within acceptable limits.

**Competition of Wildlife and Insects** Game animals, if overabundant, can degrade a range just as seriously as livestock. Elk, deer, pronghorn antelope, and even mountain sheep are known to deplete their range where hunting pressure and natural population checks are inoperative.

Elk, as cattle, feed on grass when it is available. One cow is estimated to eat as much as 1.9 elk. The elk probably makes use of the most diversified foods of any of the native ungulates. Thus, they not only consume grass but weeds and shrubs that are preferred by sheep. Elk therefore are able to remain in good flesh on all types of forage found from the rancher's hayfields to the Arctic tundra of the mountain tops.

White-tailed, black-tailed, and mule deer feed chiefly on browse but will consume grass and herbs when browse is unavailable. They, as elk, use a wide variety of plants for food, including foliage of conifers, especially that of white cedar and piñon pine. Deer compete with cattle where browse makes up the chief food supply. They also compete with sheep for browse and herbs. One cow consumes as much forage as 7.7 white-tailed deer or 5.8 mule deer. Deer seldom feed in the open areas far from tree or high shrub cover. Where they are protected from normal predation, deer tend to deplete their range.[6]

Pronghorn antelope, in contrast to deer, frequent open range. They feed mainly on herbs and browse rather than on grass. One cow will consume as much as 9.5 antelope. Antelope can do moderately well on range overgrazed by cattle but not on that overgrazed by sheep, as they prefer the same types of food as sheep. Bighorn sheep and mountain goats have been so limited in numbers that they cannot compete with domestic stock (Fig. 10-6).

Probably much more serious than big game animals has been the forage consumed by jack rabbits, prairie dogs, ground squirrels, kangaroo rats, and other rodents. In the Southwest, prairie dogs and kangaroo rats have been estimated to consume 29 percent of all forage and 39 percent of the most valuable range grasses.[14] Columbian ground squirrels may consume daily forage equivalent to 17 percent of their body weight. Rabbits are abundant throughout the west. At peaks in population cycles they make heavy demands upon forage. All these and other forage-eating rodents are controlled on National Forests and other public rangelands by the federal Fish and Wildlife Service working in close cooperation with the state game authorities. Such efforts have almost eliminated prairie dogs from much of the range. The campaign against prairie dogs has also brought the black-footed ferret to the verge of extinction, as it feeds principally on prairie dogs that it catches in their burrows.

Outbreaks of herbivorous insects may strip a range clean of green vegetation. Grasshoppers and Mormon crickets have been the most damaging but others may cause serious forage losses, at least locally. Grasshopper defoliation and twig debarking of sagebrush have been known to kill out this shrub over the area affected. These insects are controlled by distributing poisoned bait. One insect has proven a boon to the stockmen. It is a small beetle, *Chrysolina gamellata,* that feeds on the exotic poisonous weed St. Johnswort. First used in Australia to control this noxious weed, it has proven also effective on the western range.

**Predators** Wolves, coyotes, mountain lions, bobcats, and bears all fed on wild range dwelling animals before the introduction of cattle, sheep, and goats. Some of these predators found the more readily caught domestic animals easy prey. So stockmen and government began warfare on these predators. Skilled hunters, trappers, and bait distributors were employed. The campaign proved so successful that the wolf and the grizzly bear have been almost completely eliminated from the range. The bobcat, never a serious menace, has been greatly reduced in numbers. Mountain lions have become only occasional marauders. The coyote remains the only serious predator, occasionally taking sheep and young calves. The coyote's normal diet is native rodents. These, with carrion, make up over 60 percent of the daily food. Deer form 14 percent of the diet and domestic livestock but 7 percent.[14] The coyote has been extensively hunted, trapped, and poisoned to protect domestic livestock. This has also reduced its take of rabbits and other forage-consuming rodents and of game animals. Whether the net effect of this war on the coyote has been positive or negative considering its cost and the cost of control measures for rodents is still a matter for conjecture. It is worthy of note that the Public Land Law Review Commission in its report of 1970 expressed its conviction that predator control programs should be eliminated or reduced on federal public lands in furtherance of wildlife management objectives.[10] This conviction is shared by many people. Distribution of poisoned bait was discontinued by executive order in 1972.

The western rangeland as a whole is reported to have improved substantially since 1935. This improvement has probably been more marked on the National Forests than on the lands administered by the Bureau of Land Management—mainly, perhaps, because of the heavier precipitation in the mountains where the forests are located.

With improvement of the range and with the decline in sheep grazing has come a corresponding increase in wild ungulates, especially of deer, elk, and pronghorn antelope. Some biologists and range managers are of the opinion that cattle fit into the range ecosystem occupying the position vacated by the bison. The occupancy of the range with intermingled herds of domestic animals with game species may lead to a healthier range for both, provided the numbers of both are properly controlled, disease is held in check, and the animals are well distributed. The availability of adequate winter range appears to be the limiting factor in game population.

**Range Management Economics** To the middle-western farmer the cowboy riding range to tend herds of 50 or more cattle and the herder with his band of 400 to 1,000 sheep appeared economical in use of labor. And so it was in the past. But rising labor costs with the general increase in affluence have outstripped the rise in price of beef, lambs, and wool. This has forced the stockmen to decrease their use of labor and increase the use of machinery. This in turn has meant an increase in size of an economical operation. In 1930 the average cattle ranch in the 17 western states had 185 head, by 1950 this ranch had 200 head, and by 1970, 350 to 450 head. The corresponding values of the ranch were $35,000, $85,000, and $220,000. One man could still handle the stock except for help needed during peak work periods. During the same 40-year period the increase in size of sheep ranches rose from 850 head in 1930 to some 3,000 head in 1970. Ranch values increased somewhat more rapidly than herd numbers.

In the more popular areas of the western states the price of land has been inflated by extension of metropolitan areas, by increase in numbers of summer homes, and by the urge of many people of means to have a ranch of their own. As a consequence ranches changed hands in 1970 for values as high as $1,000 per cow unit or even higher. This was deemed to be twice normal values. In the opinion of experienced stockmen, ranchers obliged to borrow money to set up in business at such prices would face a doubtful financial future.

Sheep raising may require even larger investments for an economic sized unit than one for cattle. The large area of land required, the relatively low prices for lambs and wool, but most of all the unavailability of experienced herders and the considerable labor required for herd management and shearing, have led since 1930 to the pronounced decline in sheep numbers on both the open range and farms.

**Range Use in Other Lands** The English and Scottish moorlands afford medium-quality open range for sheep, cattle, and horses. In the main, the number of animals is adjusted to the productivity of the range. The damp, cool weather, as well as the carefully bred stock, is conducive to splendid wool production. Practices vary widely with local laws and customs. In some places the moor is broken up by heavy plows, cultivated, fertilized, and sown to good pasture grasses, thereby greatly increasing the carrying capacity.

In the Mediterranean region, a relatively dry climate and limited natural resources have led to heavy overuse of the grazing lands. Over much of the Mediterranean region, grazing by sheep and goats has broken down land terraces built by ancient people and has led to serious erosion. Strenuous efforts are currently being made to rehabilitate the entire region. Spain has taken the lead in the West, and Israel in the East. The programs include torrent control, irrigation, restoration of rangeland, forest planting, and improved agricultural practice. Astounding transformations are being effected where concentrated effort has been expended.

The Valley of Kashmir in India is one of the loveliest regions of the world. Walled in by the Himalayas, this 5,000-foot-high valley has a delightful summer climate and a deep loess soil unmatched in fertility. Its fame has been known for ages. Dotted with lakes and well supplied with streams of clear mountain water, it should be a paradise for mankind. But it no longer is. The nearby hills are grazed barren by sheep and goats. The mountain lands are overgrazed by cattle. The unprotected soil washes into the stream channels that cut deep gullies in the rich loess soils. Erosion has proceeded so far that many channels have been clogged with boulders. Gradually such streams build up boulder plains a mile or more wide. Unleashed natural forces tear away the rich soil. Meanwhile, hungry mouths multiply as modern science checks early mortality. Elsewhere in India the same problem of overgrazing and soil erosion persists, in part because little effort is made to control cattle population.

Seeking to improve the management of the world's grazing land is one of mankind's most urgent tasks. It will not easily be achieved because man must be supported while the improvement is taking place.

In some developing lands economic factors appear to be working toward lessened pressure on rangeland. Turkey, for example, has universal military service for youth. Sons of shepherds are taken from their fathers' flocks to cantonments where they are taught to read and write, and learn to operate and repair motor trucks and other mechanized equipment. Use of such equipment is growing rapidly in Turkey. It would seem unlikely that many such youths, after leaving military service, will set aside what they learned and the well-paid 8-hour day it offers to chase sheep and goats over maqui-covered mountains on a 24-hour day responsibility.

**OUTLOOK FOR WILDLIFE AND RANGE MANAGEMENT**

Two forces are becoming increasingly potent in influencing the future use of wild land. One is the rapidly growing urge for an unpolluted environment. To achieve this man is turning extensively to use of natural biological and ecological processes to maintain a healthy balance. The highly vocal exponents of environmental health are aided in pressing their demands for public action by the millions of concerned citizens and the recreationists that throng public lands.

The other force is economic. *Intensive* management of lands generally pays off better than *extensive* management. This fact has caused eastern farmers to abandon millions of acres of hard-won hill fields and pastures, most of which return to forest. Similar economic pressure is inducing foresters to practice intensive silviculture, and livestock men to improve herds, manage them intensively, and make heavy investments for range improvement. It has also led to the growth of feed lots in the West where stock are fed on hay and feed concentrates.

It seems inevitable that, just as much low-producing land has been abandoned for farming, so much low-value rangeland will prove no longer to be profitable for stockmen. The good ranges will be used intensively, supplemented by fenced pastures, hay, and food concentrates. Lands low in productivity will be left largely for use by wildlife and recreationists. Some may prove to be much more productive in such use than for the grazing of domestic livestock. The largely privately owned southeastern range is being rapidly expanded and can be counted upon to support increasing numbers of livestock. Management of herds is intensive, and the loss, if any, of animal production in the West will be offset.

As of 1970 the outlook for the western range was far brighter than it had been since the 1930s. One well-informed stockman has stated, "We seem to be well on our way toward making the western range more productive than it has ever been before."

**LITERATURE CITED**

1 Allen, Durward L. 1954. "Our Wildlife Legacy," Funk & Wagnalls Company, New York. 422 pp.
2 Brown, Leslie. 1965. "Africa: A Natural History," Random House, Inc., New York. 300 pp.
3 Chapman, Oscar L. 1950. "Annual Report of the Secretary of the Interior (fiscal year ended June 30, 1950)," U.S. Government Printing Office, Washington. 411 pp.
4 Clawson, Marion. 1950. "The Western Range Livestock Industry," McGraw-Hill Book Company, New York. 401 pp.
5 Hewitt, Oliver H. 1965–1966. Characteristics of Wildlife, *The Conservationist,* 20(3):3–5, 35.
6 Humphrey, Nyles. 1966. A District Manager's Dilemma—Wildlife Habitat and Multiple Use Management, *Proc. 46th Ann. Conf. W. Assoc. State Game and Fish Comms.,* pp. 212–217, WR 131:11 [U.S. Bureau of Land Management, Malta, Mont.]
7 Leopold, Aldo. 1933. "Game Management," Charles Scribner's Sons, New York. 481 pp.
8 Murphy, Robert. 1968. "Wild Sanctuaries: Our National Wildlife Refuges—A Heritage Restored," E. P. Dutton & Company, Inc., New York.

9. Ovington, J. D. (ed.). 1963. "The Better Use of the World's Fauna for Food," Hafner Publishing Company, Inc., New York.
10. Public Land Law Review Commission. 1970. One Third of the Nation's Land, Government Printing Office, Washington. 342 pp.
11. Seton, Ernest Thompson. 1909. "Lives of Game Animals," vol. 3, "Hoofed Animals," Literary Guild of America, Inc., New York. 780 pp.
12. Shafer, Elwood L., Jr. 1964. Deer Browsing of Hardwoods in the Northeast: A Review and Analysis of the Situation and the Research Needed, Northeastern Forest Experiment Station, Upper Darby, Pa. (U.S. Forest Service Resource Paper NE-33.) 37 pp.
13. Sheldon, William G. 1969. Wildlife Dynamics in New York's Fish and Wildlife Resource Biology, Management and Use, State of New York, Department of Conservation. P. 17.
14. Stoddart, Laurence A., and Arthur D. Smith. 1943. "Range Management," McGraw-Hill Book Company, New York. 547 pp.
15. Taylor, Walter P. 1954. "The Deer of North America," The Stackpole Company, Harrisburg, Pa. 668 pp.
16. Trippensee, Reuben Edwin. 1948. "Wildlife Management," McGraw-Hill Book Company, New York. 479 pp.
17. U.S. Department of Agriculture, Forest Service. 1936. "The Western Range," 74th Cong., 2d Sess., S. Doc. 199. 620 pp.
18. U.S. Department of Commerce, Bureau of the Census. 1959. "U.S. Census of Agriculture 1959," Washington.
19. U.S. Department of Commerce, Bureau of the Census. 1970. "Statistical Abstract of the United States," Washington.
20. U.S. Department of Interior, Fish and Wildlife Service, Bureau of Sport Fisheries and Wildlife. 1970. Piedmont National Wildlife Refuge. Even-Aged Forestry Management, Washington. 12 pp.

# Part Four
# Forest Management

Chapter 11

# Appraising Forest Resources

Purchasers and sellers of forest land, tax assessors, estate appraisers, timber buyers and sellers, lending agencies, loggers, federal and state officials engaged in managing forest lands, and public and corporation budget officers are all concerned with how to determine the value of forest resources. Determining timber volumes and growth rate are dealt with in books on forest measurement.[1,3] Forest-appraisal methods are dealt with in books on forest management and valuation.[2,4]

One of the first steps in an evaluation process is to determine the units of measurement to use to which a price per unit may be applied. These have been established in precise terms for land area, timber volume, and annual growth increment. They have been established for forest soil and site productivity, site index, in terms that admit of ready field application to acceptable degrees of accuracy. Precision diminishes progressively as attempts are made to assign units of measure and dollar values to water yields, forage yields, wildlife, soil protection, recreation, contributions of forests to environmental quality, and esthetics. This does not imply that such values are unreal nor that they are ignored by forest owners and purchasers. They may even provide the chief incentive for ownership. But until a means of quantifying them in acceptable units is developed, a different approach than the above must be used to determine value. Such an approach will be introduced later in this chapter.

Units of measure vary widely from region to region. Two major systems are in use: The English and the metric. The English system is used in the United Kingdom, the British Commonwealth nations, and the United States. The metric system is used throughout Latin America and Europe, and by the United Nations Food and Agriculture Organization. The confusion growing out of attempting to convert from one system to another could be eliminated if all nations could agree to adopt a common system. The metric clearly has advantages over all others. In 1970 the United Kingdom was in the process of converting to the metric system. The U.S. Department of Commerce has asked the professions and major industries of the United States to study the task of such conversion here. The forestry profession has complied by appointing a committee to make such a study. A preliminary report was made in 1970. In the meantime, foresters who work abroad in nations using the metric system will be obliged to think in terms of both the metric and the English systems. If conversion is carried out in the United States, all its foresters must become conversant with both systems for many years to come.

**ENGLISH AND METRIC UNITS OF MEASUREMENT COMMONLY USED IN FORESTRY**

It is assumed that the reader has general familiarity with the basic units used for measurement of length, area, volume, temperature, and weight in both the metric and English systems.

Additional units of measure are defined in the respective chapters. Some of the units that foresters commonly use in the United States, the United Kingdom, and the British Commonwealth nations are given with their metric equivalents in Table 11-1. Foresters obliged to make rough conversions between English and metric units in the field keep in mind certain approximate equivalents such as the following:

1 chain and 20 m
1 acre and 0.4 ha
1 sq ft per acre and 0.23 $m^2$ per ha
14 cu ft per acre and 1 $m^3$ per ha

**Measurement of Land** A timber inventory involves measurement of land, measurement of trees, and conversion of tree measurements into units in which wood products are bought and sold.

## Table 11-1 Commonly Used Units of Measurement Applied in Forestry with English and Metric Equivalents

| English | | Metric |
|---|---|---|
| *Units of length and distance* | | |
| 1 yard (yd) | = 3 feet (ft) | = 0.9144 meters (m) |
| 1 chain (ch) | = 66 ft | = 20.116 m |
| 1 mile (mi) | = 80 chains | = 1.6093 kilometers (km) |
| 49.712 ch | = 0.6214 mi | = 1.0 km |
| *Units of area* | | |
| 1 square foot (sq ft) | | = 0.0929 square meters ($m^2$) |
| 10.7639 sq ft | = 1.1960 sq yd | = 1.0 $m^2$ |
| 1 acre (a) | = 10 sq ch | = 0.4047 hectares (ha) |
| 2.471 a | | = 1 ha = 10000 $m^2$ |
| 1 section | = 640 a | = 259 ha |
| 1 township | = 36 sections | = 103.24 square kilometers |
| 4.36 sq ft per acre | | = 1 $m^2$ per ha |
| 1 sq ft per acre | | = 0.23 $m^2$ per ha |
| *Units of volume* | | |
| 1 cubic foot (cu ft) | | = 0.0283 cubic meters ($m^3$) |
| 35.3145 cu ft | = 1.3079 cu yd | = 1.0 $m^3$ |
| 1 board foot | | |
| (bf) | = 12 cu inches | = 196.5 cubic centimeters (cc) |
| 424 bf sawn lumber | | = 1 $m^3$ |
| 1000 bf sawn lumber | | = 2.36 $m^3$ |
| 14.29 cu ft per acre | | = 1 $m^3$ per hectare |
| 1000 bf per acre sawlogs | | = 11.2 $m^3$ per hectare |
| 1 acre foot | = 43,560 cu ft | = 1233 $m^3$ |
| 1 cu ft per second (cfs) | | = .0283 $m^3$ per second |
| *Approximate equivalents* | | |
| 1 cord (cd) | | |
| round wood | | |
| (4 × 4 × 8 ft) | ≅ 90 cu ft | = 2.55 $m^3$ |
| 0.39 cd solid | | |
| round wood | | ≅ 1.0 $m^3$ |
| 221 bf | | ≅ 1.0 $m^3$ round wood |
| 6.25 bf lumber | ≅ 1 cu ft sawlogs | |
| 1000 bf per acre of | | |
| sawlogs | | ≅ 11.2 $m^3$ per ha |
| 1.0 cord (solid) round | | |
| wood per acre | | ≅ 6.2 $m^3$ per ha |
| *Units of Weight* | | |
| 1 ton (t) | = 2000 pounds (lb) | = .9072 metric tons (mt) |
| 1 ton per acre | | = 2.24 mt per ha |

The instrument used in the United States public land surveys for measurement of distance was Gunter's chain of 66 feet or 4 rods. This is the measure implied in land deeds when the term chain is used.

Land in the eastern United States was divided up by a number of different survey systems. Large units were referred to as towns, townships, plantations, patents, purchases, tracts, and various other terms. These in turn were divided into lots. Where two surveys failed to join, a gore occurred. None of these units had a fixed size, nor did the lot lines of different surveys all have the same directions. Properties are described in deeds and other documents by reference to tracts and lots, by abutting land ownerships, or by a metes and bounds description referenced to a known point. An example of a metes and bounds description might read as follows:

> All of that property located in the Town of Mooers, County of Clinton, New York, described as follows: beginning at the Southeast corner of lot 180 Nova Scotia Refugee Tract, thence N8°E 20ch to a corner marked by a stake and stones, thence N82°W 20ch to a blazed maple tree, thence S8°W 20ch to a pipe set in a stone wall, thence S82°E 20ch to the point of beginning, containing in all 40 acres be the same more or less.

The original land survey always governs, regardless of how accurate it may have been, unless subsequent adjoining owners agree to a new survey.

To simplify division of lands in the public domain, the U.S. Public Land Survey System was inaugurated. Meridian lines were run north and south and parallels east and west. From these the land was divided into townships 6 miles square containing 36 sections of 640 acres each. Sections could be further subdivided by lines parallel to boundaries. Provision was made for correction lines to account for the earth's curvature. This system prevails generally on all lands north of the Ohio River and those west of the Mississippi River. The land laws specify how the land shall be surveyed, how the survey lines are to be marked by blazed trees, and how section and quarter-section corners are to be monumented and witnessed.

Land is measured by horizontal distances. In hilly or mountainous land it is often more convenient to measure the distance along the ground, then measure the angle of slope, and convert the slope distance to horizontal distance. The angle of slope may be measured with the vertical arc of a transit or by a clinometer that reads the slope in percent. A simple trigonometric calculation gives the horizontal distance. The Gunter chain may be purchased with a topographic trailer that is graduated for slope corrections.

For rough mapping and estimating timber on tracts that have well-marked section lines or boundaries, foresters often use a hand compass and approximate the distance by pacing, checking their position every mile or so. For precise work the modern transit and other precision instruments are used and an experienced land surveyor is employed.

Land area may be computed from a metes and bounds survey. Areas of properties can be readily determined if they are made up of sections or lots of known area.

**Measurement of Trees** Tree diameters are measured at $4^1/_2$ feet from the ground using tree calipers, a tape calibrated to read diameters directly, or special instruments designed for the purpose (Fig. 11-1). Tree heights are determined by standing at a measured distance from the tree and determining the angle or slope—in percent—of the lines of sight to the top and base of the tree. If the horizontal distance is 100 feet and the slopes to the top and base are respectively $+.60$ and $-.10$, the height of the tree is $100 \ (.60 + .10) = 70$ feet. Often a forester measures the merchantable height, which, in the case of pulpwood, might be to a top diameter of 4 inches. Various instruments, known as hypsometers, have been constructed for measuring tree heights. As an index of stand density, the forester has found useful the cross-sectional area of all tree stems

**Figure 11-1** Using a diameter tape to measure a large loblolly pine, Ocala National Forest, Florida. *(U.S. Forest Service.)*

standing on an acre. This is computed from the diameter, measured at 4.5 feet from the ground. The expressions basal area per acre in square feet or per hectare in square meters are in common use.

**Conversion of Tree Measurements to Volumes** The board foot is a board 12 inches square and one inch thick or the equivalent thereof. It is the unit by which lumber is sold on the United States and several other markets. The unit is imprecise as used because it is applied both to rough lumber and to the same lumber after dressing, which reduces its width and thickness substantially. A board $2 \times 6$ inches by 12 feet contains $2 \times \frac{1}{2} \times 12$ or 12 board feet. After seasoning and planing it will be about $1^{5}/_{8}$ by $5^{1}/_{2}$ inches by 12 feet, but it is still called a $2 \times 6$—12 and sold as 12 board feet.

Logs and lumber are generally bought by the thousand board feet. Sawlogs and veneer logs are usually measured for diameter at the small end or at the center and then the total length is determined. These measurements are converted to board feet by use of a log rule. A log rule is a table giving the board foot contents of logs by diameters and lengths. Several are in use, the most precise being the International Rule. Unless otherwise stated, all references to board feet of logs or standing timber in this book imply the International Log Rule $^{1}/_{4}$-inch sawkerf.

Fuel wood and pulpwood are measured by the standard cord made up of a stack $4 \times 4 \times 8$ feet. The solid cubic foot contents of a cord of pulpwood will vary with the size and straightness of the component pieces. As used by the United States Forest Service, one cord is the equivalent of 90 cubic feet of solid wood. Variations from the standard length of pieces of 4 feet are common. Fuelwood is sold by the 16- and 24-inch cord, pulpwood by the 50-, 52-, and 100-inch cord. It is also sold by the cunit, which is 100 cubic feet of solid wood. Much modern purchasing of pulpwood is by weight, which avoids the necessity of trimming to standard lengths, is more quickly determined at the mill yard, and serves as a better index of the value of the wood to the mill. Crossties, posts, poles, shoe-last blanks, baseball bat blanks, and a host of similar products are sold by the *piece*. Piling and long poles may be sold by the linear foot. Mine timber may be sold either by the piece or by weight. Short-length veneer bolts may be sold by the cord. Prices tend to be higher for the larger sizes and longer lengths of products. Practices vary considerably from region to region.

In countries using the metric system, volumes of logs, pulpwood, and other products are expressed in cubic meters of solid wood substance. Obviously they thereby eliminate much of the above complexity.

To simplify conversion of tree measurements of diameter and height into units by which timber is bought and sold, foresters have constructed volume tables. By diameters and heights, these show the tree content in board feet, cubic feet, cords, cubic meters, or special units or products. It is then a simple task to compute volumes from numbers of trees in each height and diameter class. Separate volume tables have been constructed for different species and for different degrees of taper in the tree stem, as well as for different log rules.

**Determining Annual Timber Growth** Timber yield or increment involves volume, area, and time. It is expressed as cubic feet per acre per year (ft$^3$/acre/yr), board feet per acre per year (bd ft/acre/yr) or cubic meters per hectare per year (m$^3$/ha/yr). The yield of a specific forest tract would be given in thousand board feet per year ($10^3 \times$ bd ft/yr) or thousand cubic meters per year ($10^3 \times$ m$^3$/yr).

The ideal way to project timber increments is from records of past harvests over a rotation from the same forest or from one closely similar to it. Such information is available for many European forests. It is also becoming available from the harvests of rapidly growing genetic stock on man-made forests. It will be a long while before such information is accumulated for the slow-growing timber types of North America. This is so because on only a few properties are systematic efforts made to maintain the necessary records.

The use of modern electronic data processing has made feasible a direct mathematical approach to timber growth projection. The three major factors determining growth for any species are stand density, which can be expressed as basal area per acre; site index, expressed as the average height of dominant trees at 50 years of age; and stand age. A mathematical model is developed to account for the influence of these three factors on growth and mortality. Such a model for natural stands of loblolly pine had the form:

$$\text{Net growth} = a + b\,(BA) + c\,(BA)^2 + d\,\frac{1{,}000}{AS} + e\,\frac{BA}{S}$$

where $BA$ = basal area
$A$ = stand age
$S$ = average site index at age 50
and $a, b, c, d,$ and $e$ are constants

The constants $a, b, c, d,$ and $e$ were determined by multiple regresion equations in which actual field data from each of 202 sample plots were substituted in the above equation for basal area, age, site index, and net growth. In this case, data were collected from 1,552 acres on which successive 100 percent inventories had been made at 5- to 12-year intervals. A computer was used to work out the values of the five constants. Three of them, *a, c,* and *e,* bore negative signs. The final equation accounted for 68 percent of the variation in stand growth.[1] The method is designed to cope with stands of variable density, which is the type most commonly encountered in natural forests.

Considerable mathematical sophistication is required to develop and test appropriate regression equations by means of a computer and to determine the value of constants for different species. But once this has been done a computer can quickly work out the individual values for an increment table. Such tables are usually the work of a forest biometrician.

A third approach to increment determination is that of stand projection. This involves reconstructing the stand as it existed 10 years earlier, using permanent sample plot measurements or increment-core measurements with measurements of the trees that died during the decade. (An increment core is a radial cylinder of wood cut from the stem by a special borer.) It is assumed that the stand of today that is similar to the reconstructed stand will grow during the next decade at the same rate as the sampled stand grew. Forest increment tables based on this method are available for many areas in the United States.[5] The method is weak in that growth cannot be projected with confidence beyond an initial 10-year period without collecting new field data. Otherwise the errors increase in a geometric ratio.

The continuous forest inventory method is used in nationwide forest surveys and on large industrial holdings. It requires use of permanent inventory plots of fixed area. On these, each tree is recorded on data processing cards by distance and azimuth from the plot center, together with its species, diameter to 0.1 inch, merchantable height, and whatever data on vigor, crown class, and bole quality may be desired. Plots are remeasured at 3- to 10-year intervals and the changes due to mortality, growth, and timber cutting are determined. From such information the gross and net increment can be determined for the intervening period. A minimum of 100 plots should be established to get reliable growth projections. Where timber type and forest site qualities vary considerably, many more plots will be needed. The establishment and maintenance of continuous inventory plots is time-consuming and expensive. It is, however, one of the more reliable approaches to growth determination.

The most reliable method of determining growth is to make periodic inventories of the forest every 10 to 20 years. Net growth is then equal to current inventory, plus the volume removed by cutting, less the volume standing at the beginning of the period. This method also is expensive as will be apparent from the following section.

## MAKING TIMBER INVENTORIES

Forest inventories may be made for properties of any size—from a small farm woodlot to a large industrial forest, or even for a state, nation, or the world. The methods used and the object of the inventory will vary with the size of the property, the accuracy needed, and the amount of money available for performing the task. If the objective is to assess the value of a property, then the inventory must determine the value of the land apart from the value of the timber. If the property is primarily valuable for the growing of timber, the value of the land depends upon its productivity in timber growth. The value of the merchantable timber depends upon the species, size, and quality of the timber, and the ease with which it can be logged. It follows that the first step in making a forest inventory is to decide on the objectives, then to plan an inventory that will achieve these objectives.

**Planning Forest Inventories**  Making a forest inventory involves a considerable investment as it requires systematic data collection over the entire property. It should be carefully planned to assure completeness, accuracy, and minimum cost. The first decision concerns the precise purpose or purposes the inventory is to serve. If it is to form the basis for a comprehensive forest management plan, the following information may be needed:

1 Timber volumes by species, size-classes, and timber quality.
2 Areas of major timber types, age-classes, and site indices.
3 Annual timber increment through growth.
4 Probable costs of logging operations.
5 Existing and proposed road system.
6 Proposed forest subdivisions.
7 Features of special monetary or scenic value, such as gravel deposits, ponds, streams, and mountain peaks.

The first step, then, in planning an inventory is to list the various items that might be covered and to eliminate those that are unnecessary or too costly.

The second step is to decide on the accuracy needed, as this will influence the intensity and cost of the inventory. The intensity of inventory needed to attain a desired degree of accuracy increases with the variability of the forest stands. Variability can be determined by collecting data from representative plots and determining their standard deviation from the mean. From this a simple statistical computation will indicate the intensity of sampling needed to attain the desired degree of accuracy. The intensity needed might vary from 100 percent for a 20-acre woodlot to 0.5 percent or less for a property of 50,000 acres. For forests of a high degree of complexity, such as those in the tropics, sampling intensity may be adjusted to that necessary to achieve the desired accuracy for volumes of major species.

Decision must also be made on the units of measurement to be used, the tables needed, and the office computations necessary to convert field measurements to usable summaries of information. It is desirable for office computations to be carried out as field work progresses so that changes in intensity or nature of field work can be made promptly if needed.

Aerial photos of the forest to be inventoried may be used to prepare base maps and to delineate road systems, forest types, and age-classes. If areas of forest types and age-classes can be accurately delineated on aerial photos, sampling intensity can be reduced. Only after the above decisions have been made is it possible to plan the field work so as to achieve maximum efficiency. This requires familiarity with statistical procedures and with the costs of the various operations, as well as the mathematical skill to achieve the desired information at minimum cost. The detailed outline of field work is then prepared.

**Collecting Field Data** An initial field task is to determine and mark the boundaries of the property and of any major subdivisions for which separate information is desired. Field crews then collect the information called for, recording it on data cards or sheets provided by the survey planner. This usually involves running compass lines across the property, stopping at regular intervals to record tree diameters and heights by species on plots, commonly circular, of $1/5$ acre or more. Whatever additional information the inventory supervisor specifies is also obtained, such as stands in need of early treatment, open areas to plant, road layout, logical compartment boundaries, protective measures needed, and other essentials for a comprehensive management plan. The inventory plan may require that a certain percentage of the field plots be established for continuous forest inventory.

It is important that the work performed on each plot be carefully done, for errors introduced by careless measurement of trees and plot areas will be magnified manyfold in the total inventory. Unless measurements of plot radii and trees are checked frequently, errors tend to become cumulative rather than offsetting.

A number of short cuts are used by foresters working with large properties. Instead of measuring all trees on a plot, the foresters may estimate diameters by 2-inch diameter classes. Heights need be measured for only a few representative trees (Fig. 11-2). With relatively uniform timber, it is even possible to make acceptable inventories directly from aerial photos using tree heights or crown diameters and crown closure as indices of stem volumes.

**Data Processing and Preparing Reports**  Field data are transferred to maps and to data processing cards in the office unless such cards have been used to record information in the field. Computer programs are prepared to summarize timber volumes by species, size-classes, timber quality, and by major forest subdivisions, site qualities, timber types, and age-classes, or whichever of these as are needed. The computer then computes the desired information summaries. The summaries are studied and an interpretive report is prepared—often with maps, charts, and other visual aids to help the reader to get a good understanding of the property inventoried. Dollar values may be applied for bookkeeping use. Informa-

**Figure 11-2**  Using a Relaskob to measure tree heights and basal areas in a red pine stand, Chippewa National Forest, Minnesota. *(U.S. Forest Service.)*

tion on annual increment will certainly be included. Such data are essential for the preparation of a timber management plan for the property.

### TIMBER INVENTORIES FOR COUNTIES, STATES, AND NATIONS

To obtain the timber volumes and growth increments for very large areas, techniques such as those outlined above can be used. But first the timber survey team will get all the information it can from state forests, national forests, industrial forests, and other large holdings on which surveys may have been made. These are then updated. Next, a survey to sample the remaining area is designed, bearing in mind the accuracy desired for the total. Aerial photographs are extensively used for determining forest areas and timber classes. Sample plots are established on the ground, some of which will be recorded in detail for subsequent remeasurement. Additional permanent plots may be established for subsequent surveys. Accuracies attained on a nationwide basis are believed to be within 2 to 3 percent of the combined timber volumes.

### FOREST INVENTORIES FOR CONTINENTS AND FOR THE WORLD

Periodically the United Nations Food and Agriculture Organization (UN-FAO) makes forest inventories for the continents and for the world. It first obtains all possible information from the individual nations. Many nations, such as Finland, Germany, the United States, and Canada have reliable data. These are supplemented with whatever information the UN-FAO may have collected for individual countries. These data are then adjusted and compiled, taking into account all sources of information available. From this the totals for the continents and the world are projected. As but 55 percent of the world's forests had been covered by timber inventories by 1971, worldwide summaries are imprecise—especially for remote regions such as the Amazon and Orinoco basins of South America. However, statistics for the accessible forests from which most of our timber must come—up to the year 2000—are highly reliable and form a sound base for timber resource planning for at least two or more decades ahead. By that time substantial additional areas will probably have been covered.

### THE STATE OF THE ART

Before leaving our present subject, we should consider the current state of the knowledge and practice of forest measurements. This is one of the first phases of forestry to be accurately quantified. Mathematics and mathematical statistics of considerable sophistication are now used in designing large-scale forest inventories. Growth parameters are being recognized and used in mathematical models to simulate forest growth. The growth responses to be expected from various forest treatments can then be determined by programming a computer to test such models with data from forest stands. A whole family of new information can thus be placed at the disposal of the decision maker to guide him in his forest operations planning.

While the scope of precise raw data on the reaction of forests to treatment remains limited, our capacity to use what we have with versatility and confidence has been increased enormously since 1950, when foresters began to use computers.

## THE APPRAISAL PROCESS

Forest properties may be appraised on three bases—their liquidation value, their long-term investment value, and their current market value.

The *liquidation value* is assumed to be the current value of the land without trees, plus the value of the standing trees of commercial sizes and quality. Trees below merchantable size are assumed to have no value. This method assumes that the presence of small trees or stumps on the land is a matter of indifference to the prospective purchasers.

The *long-term investment value* of a forest property is based on what its average net monetary yield may be and what it is projected to be in the future. This requires taking into account the annual growth increment of the forest, and the sale value of products and services less all expenses for taxes, protection, management, and administration. This net is then capitalized at the going rate of interest plus an allowance for risk. If the general economy is undergoing inflation that would result in higher prices for stumpage in the future than at present, this may be taken into account.

Under this method a forest yielding a net revenue of $5 per acre per year with a 10-percent rate for interest and risk would be worth $50 per acre.

Both these approaches are theoretical. The *immediate sales value* is what an average interested buyer would be willing to pay, and a willing seller to accept. This might be only half the theoretical liquidating value because of the costs involved in transferring property and selling the timber. It might also be much more than the liquidation or investment value if the buyer were interested in purchasing the land for recreational use or for a summer estate. Land appraisers base their judgments on what timber land is currently selling for, rather than on what it might yield the logger or the long-term investor. Also, forest properties as such are not traded as industrial stocks on a large exchange. Hence estate appraisers may set the appraisal at what could be expected on a quick sale.

For timber growing, a property of 20,000 acres is worth more per acre to the purchaser than are 100 properties of 200 acres each. This is so because the costs of making purchases vary directly with the number of transactions. It costs little more to get an abstract of title for a 20,000-acre property than for a 100-acre property. It may indeed cost less, for the larger property is likely to have changed hands fewer times. Negotiations with the seller may take no more time. Legal fees may be no greater, assuming the title is clear. Examining the property on the ground, surveying, and estimating the timber will be much less per acre for the large than for the small property. Potential buyers of large properties generally recognize all these advantages.

As will be brought out later, to assure profitable independent operation a forest property should be in the neighborhood of 20,000 acres. Only with such a property does the owner have sufficient timber yield to support a sawmill or other significant-sized processing plant. Very few properties of this size or larger can be found in the United States, except for those already owned by government, or large forest industries. Consequently, changes in ownership are likely to occur as mergers involving stock transfers rather than as outright purchases. This complicates the task of arriving at a fair market value for such properties.

**MEASUREMENT OF WATER**

Water that falls in the forest is measured in inches or millimeters. That which accumulates in reservoirs or snow packs may be measured in acre-feet. Each acre-foot equals 43,560 cubic feet. The flow of water in a stream is determined by diverting the flow through a calibrated weir. It is usually expressed in cubic feet per second (cfs) or cubic meters per second ($m^3/s$). Water yields may be expressed in terms of cubic feet per second per square mile of catchment area (cfs/sq mi). The analogous expression in metric terms would be cubic meters per second per square kilometer ($m^3/s/km^2$).

Stream flow tends to increase sharply during or immediately after a storm, then gradually decreases to normal flow. It is therefore common practice to monitor stream flow by means of a water stage recorder with a float in a stilling well that stands at the level of water flowing through the weir. The area under the stage recording tracing can then be integrated to determine the total flow per day, storm period, or month.

Various devices and methods are used to determine rain and snow interception by forest canopies. Standard rain and snow gauges may be operated with one set in the open and another set under a forest canopy. The difference in average catch between the two is an indication of the amount intercepted. For precise determinations for research purposes a number of refinements are introduced, including use of a battery of gauges, and collars around tree stems to intercept flow down the stem to the ground.

## MEASUREMENT OF FORAGE

The farmer measures his hay crop in tons and his yield in tons per acre. Range forage and browse may be determined by clipping and weighing the palatable portions of grasses and other herbs and the succulent portions of woody vegetation. This is usually done only for research purposes. The stockmen and range managers use the term animal-unit-month by which they mean the amount of forage that is required to support a cow with calf in good flesh for one month. The yield of a range may be expressed as animal-unit-months per square mile. It is also sometimes expressed in terms of the number of acres required to support one animal-unit for a year (Table 10-1). Forage value is recognized by the fees private range owners charge per grazing animal by month or year.

Value of range livestock is determined from their value at the feed lot or stockyards.

## WILDLIFE

Wildlife populations are determined by a variety of means, as mentioned in Chap. 10. The standard ways are to count the animals at water holes, on the open range, or when they are driven across a road or open area. The population is expressed in terms of the range they use, and its density is given by numbers per acre or square mile. American foresters and game managers usually have a far from precise census of the game by species on their forests. Most European foresters, on the other hand, know their wildlife population fairly accurately, and even know many of the trophy animals as individuals. The European forester can always answer the question of how many game animals he has per 100 hectares. Good reasons exist for European foresters being well informed on game population. Game there is the property of the forest owner and its taking and sale brings in considerable revenue. The forester supervises hunting, determines the annual harvest of game, and in some cases designates the specific animal a hunter may shoot. The European forester has responsibility for a much smaller area than his American counterpart, and

European forests generally have limited areas of thicket where animals may hide. Winter feeding is often practiced.

In Europe monetary values can be ascribed to forest game because of the substantial fee forest owners receive annually from hunters for the privilege of hunting, and the price they obtain from sale of carcasses the game hunters have killed. Forest owners in the United States may charge for the privilege of hunting on their land but such fees are generally modest and few landowners impose them.

Attempts to assign monetary values to the recreational, biologic, social, scientific, and esthetic values of wildlife are largely speculative. They are inseparable from these values for the forest as a whole.

**RECREATION AND ESTHETICS**

Recreational use of forests and parks is expressed in visitor days. Day visitors are often listed separately from those who remain longer, at a campground or lodge. Various efforts have been made to get expressions of visitor satisfaction. So far this determination is still largely in the research stage, although length of stay is generally correlated with satisfaction, as is amount of money spent at a particular resort or camping area. With the spread of privately operated fee campgrounds it may be possible to get better quantitative information on visitor satisfaction.

The value of esthetics can be estimated in part by the entrance fees collected for access to scenic areas. An approach still in the beginning stages is the scenic easement. The easement fee is paid by government to landowners for keeping their land open and pleasing in appearance. It is being explored with special interest in many places where emphasis is placed on maintaining a quality environment in the neighborhood of thickly settled sections.

Another index of esthetic and amenity values is the cost the public is willing to pay or to impose on governments and industries to clean up pollution, littering, and other types of environmental degradation. Such costs are substantial and are being increasingly incurred and imposed. The expenses of establishing and maintaining parks, historic sites, natural areas, and wilderness are all incurred mainly for educational, recreational, scientific, and esthetic purposes. The public appreciation of such amenity values seems likely to increase rapidly during the years ahead.

**VALUE TO THE OWNER**

The timber value of a forest is emphasized above, for it is this value that the owner can expect to realize upon sale. The actual value to the owner is the satisfaction he realizes from the property. This may result from a stream or pond, a lofty lookout point, trees of unusual size or character, wildlife, the quietude and seclusion, or response of the property to his care. These are largely personal values, yet they are certainly just as real to the owner as the pleasure of owning a pedigreed horse, a fine motor car, or a fast boat.

Approximately 4 million people own the 300 million acres of forest land held in small properties. Many of these owners appreciate amenity values and it is these values that determine largely how the owners will react to proposals to cut their timber for commercial use.

Owner values are clearly recognized by the land appraiser because they soon become reflected to some degree in the prices people pay for forest land, or demand for it before they will sell it. Such prices, just as those of real estate in general, are much higher for properties that have features of outstanding esthetic quality and that adjoin neighboring properties of like quality than they are for properties that have been abused. Until more precise appraisal techniques are worked out for recreational and amenity values of forests, the going price of properties serves as the most reliable measure of their market value.

Man cannot place meaningful prices on all forest values any more than he can on other highly cherished life values. Yet such values are among those that add most to the joys of living and that are defended with the most zeal and determination.

**LITERATURE CITED**

1. Avery, T. Eugene. 1967. "Forest Measurements," McGraw-Hill Book Company, New York. 290 pp.
2. Chapman, Herman H., and Walter H. Meyer. 1947. "Forest Valuation," McGraw-Hill Book Company, New York. 490 pp.
3. Chapman, Herman H., and Walter H. Meyer. 1949. "Forest Mensuration," McGraw-Hill Book Company, New York. 522 pp.
4. Davis, Kenneth P. 1966. "Forest Management," 2d ed., McGraw-Hill Book Company, New York. 519 pp.
5. Feree, Miles J., and Robert K. Hagar. [undated] Timber Growth Rates for Natural Forest Stands in New York State, State University of New York College of Forestry in Syracuse, Technical Bulletin Publication 78. 56 pp.

Chapter 12

# Harvesting Timber Crops

Timber harvesting is generally referred to as logging or lumbering. It includes those activities involved in felling, delimbing, and cutting trees into manageable lengths and transporting the usable parts of the main stem and branches to a timber processing plant.

**LOGGING AND FOREST RENEWAL**

When white men and women first came to America their initial task was to erect shelters and clear land for crops and villages. Lumber manufacture was started soon thereafter. Logging then became a step in the land development process, and lumber export a means of obtaining needed supplies and equipment from abroad. Standing timber was superabundant and occupied land needed for crops and pasture. Little thought was given to future timber supplies. As a result logging in the United States traditionally aimed at but one objective—getting out the timber at the lowest cost possible. Independent loggers and their equipment manufacturers had to 1972 taken few significant steps to outgrow this tradition. In fact almost every major development leading to lower logging costs has increased logging damage. Introduction of animal skidding, tractor logging, cable logging, tree length skidding, and, most recently, machine harvesters seems to have progressively increased the damage to residual trees that might have formed the basis for a second cut. Logging has at times created high fire hazards, prepared the way for outbreaks of insects and diseases, obstructed natural drainages, compacted the soil, induced erosion, and left the forest increasingly unsightly. Compared with what well-managed forests produce, the yield of most cutover lands has been low—and drastically so where hot fires raged through the slashings, destroying forest humus with its supporting organisms. Is it any wonder that logging came to be regarded by conservationists and other concerned people as essentially a destructive process?

Timber harvesting has, therefore, an additional task to perform. This task is forest renewal. For the future of the forest and of all people and life that depend upon it, this additional purpose is much more important than getting logs to the market cheaply. Until loggers and their equipment developers recognize this and act accordingly, they seem destined to operate under the menacing cloud of public disapproval and distrust. It is in their interest as well as that of foresters, landowners, forest workers, and timber processors that this cloud be lifted. Feasible systems for combined timber harvest and forest renewal have long been used in Europe and others are being applied by some forestry industry companies in the South and the West. There is hope that analogous systems can be devised for nationwide and worldwide use.

## DEVELOPMENT OF TIMBER HARVESTING AND TRANSPORT

Timber harvesting, because of the weight and bulk of logs handled, has been recognized as hard, dangerous work. The Greeks referred to hewers of wood and bearers of water as menial workers or slaves. Traditionally logging involves the following operations: felling, delimbing, bucking, skidding, loading, and transport, together with such supporting activities as planning, road building and maintenance, equipment maintenance, and, if required, logging camp construction and operation.

**Felling**  Early man used stone axes to fell trees and to fashion the trunks into useful articles and structures. With the coming of the Bronze and Iron Ages, metal axes replaced those made of stone. The metal saw came into use in Egypt and the Fertile Crescent between the years 3000 and 2475 B.C.[4] The ax and hand pulled saw were the tools used to fell timber from that time until the close of World War II. They are still in wide use in many places. Considerable judgment and skill are required to fell a large tree in the direction desired in order to facilitate log removal and to avoid the log's lodging against other trees or damaging those to be reserved for a later cutting. A skillful timber faller can drop a tree almost anywhere he wishes unless the tree is decayed, leans badly, or has a one-sided crown. Even then considerable latitude exists as to where the tree can be dropped. For example, wedges may be driven in the saw cut to start the tree tipping in the desired direction.

*Delimbing* Limbs must be cut flush with the stem of the merchantable portion. Delimbing is a one-man task, formerly done with an ax.

*Bucking* Bucking is the term used for cutting the tree stem into logs or pulpwood bolts of desired lengths. It was generally performed with a one-man or two-man crosscut saw, though axes have been used. In the Amazon Basin in 1967, the writer saw logs as large as 6 feet in diameter that had been severed with an ax.

Since 1945 the power chain saw has been rapidly displacing the ax and the hand-pulled saw for felling, delimbing, and bucking trees.

*Skidding* Moving logs from the stump to the landing where they are loaded on cars or trucks is termed skidding. Many methods have been used, as are implied when referring to hand, animal, cable, or tractor logging.

**Hand Logging** Much logging in prehistoric and early historic times was probably hand logging carried out by men alone, unaided by draft animals or machines. We see this illustrated in early carvings on tombs and buildings. The trees were felled and cut into logs with axes or saws and the logs were then dragged or rolled by hand to water or to the place of processing and use. Some products, dugout canoes for example, were carved from the log where the tree fell and were then dragged to water. Hand logging was the first method used on the Pacific Coast of the United States. The huge trees were cut into logs of a size that could be rolled down a slope to water by use of prypoles, jacks, and ropes attached to horse-powered windlasses. From there the logs could be floated to the sawmill.[6]

Primitive as hand logging may seem to most of us today, it was still in use in remote regions as late as 1970. The author saw it used in Kashmir in 1954, and in Mexico and the Upper Amazon in 1967.

**Animal Logging** Oxen, horses, mules, camels, and elephants have all been used directly or indirectly for logging. Oxen were used to log timber in the South, New England, and other parts of the United States as late as the 1930s, mostly on small farm operations. The author saw them in use in the mountain forests of Italy and Turkey in the 1960s. Slow moving and patient, and using the simplest form of harness, oxen have their advantages for skidding logs. They probably cause the least damage to the forest of any present-day logging method.

Horses and mules were extensively used in the past and are still used in various places. A well-trained skidding horse can skid a log over considerable distance without the aid of a driver. For heavy logs a team of horses may be used, but a team is less maneuverable than a single horse or a yoke of oxen.

Both camels and burros may be used to transport logging supplies and gear, and to carry out small logs, posts, or planks whipsawed at the stump.

Elephant logging is the most impressive of animal logging. An elephant can skid heavy logs along the ground to a landing. Here, directed by his mahout, and using his tusks and his trunk, he can roll the logs together and pile them. He can also lift them on a wagon or truck or, if they are too heavy, he can load them by cross-hauling as a team of horses would do. The well-trained elephant is man's supreme beast of burden for timber harvesting. These animals are used in India, Bangladesh, Burma, Thailand, and other Asiatic countries.

No matter how powerful or well-trained animals may be, they are no match for power skidders and motor trucks in terms of heft of logs they can handle and length of working day they can be used. Moreover, animals require usually more man-hours of care per animal-hour of productive work than a machine requires of maintenance per hour of productive work. As machine output of useful work per hour greatly exceeds that of an animal, the latter have largely been retired from logging in high-wage industrial countries.

**Cable Logging** Cable logging has long been in use in Europe for logging steep terrain. In its simplest form it may consist of a wire or a cable line fixed at the top and the bottom of a promontory or ridge. Small logs or pulpwood are suspended in a sling hooked over such a cable and allowed to slide to the bottom. Efficiency is improved by using a carriage that rides on the fixed cable. The carriage has a working cable to which logs are attached, winched up to the carriage, and then hauled to the top or allowed to descend to the bottom. This is the principle of the Wyssen system as developed in Switzerland. Much more elaborate cable systems were developed in the United States and Canada where timber is large and wages are high. Two of these systems are described below.

The *skyline system* has a cable suspended from two tall spar trees on which a carriage rides with backhaul and working cable for attachment of logs. This system has been used to remove logs from steep ravines and also to "swing" logs from one landing to another in West Coast logging.

*High-lead logging* is the most commonly used cable system in the western United States. In this system the main haul cable feeds from the logging winch drum through a main block fastened at the top of a tall spar tree or portable spar. From there it is led to the waiting logs by the backhaul line. Short cables, called chokers, are looped around logs and hooked to the main haul line. This line is then winched in to the spar tree with the free ends of the logs skidding along the ground. If an obstruction is encountered the cable tightens, lifting the logs over it. At the spar tree the chokers are detached and returned for another turn of logs. The spar tree with the pile of logs constitutes a landing where skidding terminates and logs are machine loaded onto trucks or rail cars for transportation to the mill.

The above paragraphs describe cable logging in its simplest form. Actually the entire cable system, that constitutes a setting, appears to have a maze of cables, blocks, and supporting network to keep the systems taut and in free-working order.

A more complete description of cable logging is given in books on logging.[2, 5, 6] For complete detail the reader is referred to the western Logging Handbook and literature of equipment manufacturers.

**Tractor Logging** The most widely used logging system in the United States in 1971 was tractor and truck logging. This varied from the very simple—using a wheeled farm tractor to skid, and a farm truck to haul pulpwood and small logs—to the more complicated—using heavy rubber-mounted tractors called skidders and heavy logging trucks, such as are used on the West Coast. In between was the modest sized rubber-mounted skidder capable of handling up to seven logs with butt diameters of 20 inches or less. The present-day log skidder is a specially built four-wheel-drive tractor, articulated (i.e., hinged) in the center, with wide wheelbase, short turning radius, and a winch that feeds the logging cable through an elevated fairlead. Logs are attached to the winch cable by choking cables or chains. Instead of the winch and elevated fairlead, skidders may be equipped with a rotating boom hinged in the center and known as a knuckle boom. Pneumatic tongs mounted on the end of the boom are operated from the cab, enabling the driver to pick up and bunch logs for skidding. A great advantage of the rubber-mounted skidder is the speed at which it can move—up to 20 miles per hour along a graded skidway. Special rubber-mounted skidders have been developed for use in swampy ground. These skidders have large, low-pressure tires that provide traction and flotation. The skidder as used in the East usually skids tree-length logs. These are cut into sawlog and pulpwood lengths, at the landing. As logs are increasingly being purchased on the basis of grade, knowledge of these grades and of how to cut the logs from the stem so that maximum price will be realized is essential for profitable logging. Bucking at the landing permits assigning a skilled man to mark the log lengths to be cut from each tree-length stem.

The above discussion gives a brief general outline of skidding systems. The task of the logger is to select the particular system and equipment that best fits into his operation. If clear-cutting is to be done, any one of the several skidding methods may be used. If partial cutting is required, cable logging by the high-lead or ground-skidding systems is unsuitable. The same is true of timber harvesting machines. Animal or tractor logging are both adapted to partial cutting if proper care is taken in laying out skidways and if trees are carefully felled and skidded. Loggers need to keep in mind that felling and skidding are the two operations that damage residual trees and, together with logging roads, cause the most erosion and other damage to the site.

**Timber Harvesting Machines** The latest proven system of logging used in 1971 was that of the timber harvesting machine. Development of these machines was begun in the U.S.S.R., Canada, and the United States soon after the end of World War II. Early models could sever a tree at the stump, load it onto the rear of the harvester, then move to another tree until a full load was aboard the machine. The machine was then driven to the landing where the tree branches were removed and the stem was cut to log lengths with chain saws. By 1971 the following additional machine types and systems had been developed. In one system, one machine performs the functions outlined for the above machine, while at the landing a second machine takes over to delimb and buck the stems to pulpwood lengths. Such a system is used in Canada. The Buschcombine performs all these operations at the stump, carrying the pulpwood bolts with it to successive trees until a full load of bolts is acquired, whence it proceeds to the landing to transfer its load to a logging truck or to loading pallets. Other harvesters perform delimbing, topping, felling, and log bunching. A skidder delivers the bunched logs to a slasher saw for bucking and loading on hauling equipment. Other machines cut the tree into pulpwood lengths at the stump and stack the wood for prehauling to a loading area. It is possible with a proper combination of machines to harvest pulpwood when the ground is covered with snow up to 7 feet deep, as the machine operators do not have to leave their cabs.[2, 3]

Harvesting machines (Fig. 12-1) work well only where the timber is dense and of uniform size and species, the ground conditions are

**Figure 12-1** Mechanized tree harvesting in Maine. Trees are limbed, topped, and felled by a Beloit harvester, skidded to a landing by a rubber-mounted skidder, and hauled in tree lengths to the slasher that cuts them into 4-foot lengths. Scott Paper Co. operation.

suitable—that is, not too sloping, stony, or swampy—and where 24-hour operation is feasible. As of 1971 timber harvesters were not machines to be recommended for developing countries nor for handling large-sized western timber or hardwoods.

Development of timber harvesters and of mechanized tree fellers, delimbers, toppers, slashers, loaders, and various types of skidders was occurring so rapidly in the late 1960s and early 1970s that machines in use in 1972 are likely to be outmoded within a 5-year period. For detailed information on models the reader is referred to current publications of logging associations, equipment manufacturers, and others dealing with logging machines and equipment.

**Methods under Experimental Development** The logger naturally dreams of a flexible system that can move freely above the forest, drop a line to the ground, lift a log or tree bodily above the tree tops, and carry it to a landing. Two possibilities have been tried—helicopter logging and balloon logging. The helicopter is expensive in operation and maintenance for such a mundane task as logging, although it is extensively used in fire-fighting and other forestry tasks. The balloon has many of the drawbacks of the skyline system in that ground anchorage must be used, and any wind results in erratic behavior of the balloon and the lines and logs suspended from it. It has, however, become operational in several locations in the mountainous West, and there are those who have faith that its difficulties can be reduced so that balloon logging can become widespread. It should be the least damaging to the environment of any method of moving logs from the stump to the landing.

**Loading and Hauling**  At the landing, logs were formerly placed on a rollway whence they were rolled by hand or cross-hauled by horses up pole skids onto trucks or freight cars for hauling to the mill. By 1972 mechanical loaders had largely replaced such methods. The smaller and simpler machines included small powered conveyors used for pulpwood, forklifts, and swinging booms for logs. The heel-boom loader was popular for loading heavy logs (Fig. 12-2). A modification is the versatile, one-man-operated knuckle boom equipped with pneumatic-operated tongs to pick up the logs and place them on the truck where desired.

Timber has been transported to sawmills and pulpmills by wagons pulled by animals, by overhead cable ways, by railways, by motor trucks

**Figure 12-2**  Mechanical log loader at landing, Warm Springs Indian Reservation operation. *(U.S. Bureau of Indian Affairs.)*

and tractor-trailers, and by floating in flumes and streams. Portable sawmills have also been built that could be set up at log landings in the forest. Somewhere in the world, each of these methods was still in use in 1972. The modern motor truck has replaced most others in the United States, though timber floating was still in use in parts of Maine in 1972.

Logging trucks are of many sizes and capacities. Those used on the West Coast will haul up to 70,000 pounds, usually in the form of logs 32 feet long. Others will haul up to 80 to 200 tree-length logs of small size. Smaller equipment is used in the South and North but here also tree-length logs are being hauled increasingly. Even the relatively small log trucks used in the North will haul up to 10 tons. Specially designed trucks have a real advantage over conventional trucks that have been converted for log hauling.

Any motor truck requires a good road to move heavy loads rapidly over long distances. Road construction and maintenance therefore becomes a major task of truck-haul logging where a good network of roads does not already exist. The cost of such construction is usually deducted from the price paid for stumpage. If roads are to be constructed well enough to withstand heavy hauling they should be of a permanent type. For example, most silt that gets into forest streams during logging operations comes from logging roads. Therefore, these roads need to be carefully laid out to minimize erosion. Water from ditch lines should be diverted so that it does not discharge directly into streams.

**PLANNING AND LAYOUT**

The present-day logging operator has each of the five major systems outlined above to consider along with almost countless modifications of each. Somewhere in the world he will find use for each of the five, although his actual choice will usually be restricted to one or two of them. Success and profit depend upon wise choice of a system to fit local situations with which the logger must deal.

An independent logger's first task is to find markets for the several types of products that can be obtained from the forest property he is to log. The greater the number of products and species he plans to deliver to mills, the more flexible his operation must be. Planning therefore begins at the delivery points for products and proceeds backward to the forest from which they are to be removed. The distance, roads available and their condition, and products to be delivered will largely determine the loading and transportation equipment needed. They will also influence the size of landing he must have to sort and store the several products while he is accumulating enough for a load. Ordinarily, and especially in the summer, he will want to move his logs as quickly as possible from the landing to minimize log inventory and to avoid log deterioration.

The logger must also plan his operation within the scope of credit resources available to finance it. Generally speaking the more complex the machine he uses, the more necessary it becomes to provide it with ideal operating conditions. He may well find it advisable to forego some labor-saving devices in order to attain greater flexibility of operation and lower capital input.

From the above it is clear that the plan for a logging operation must fit the regional and local conditions for which it is made, and that it also must be kept within the financial resources and skilled manpower available to perform the task. A large western logging operation requires heavy equipment, large investments, and highly skilled manpower. The eastern logger has a much smaller operation with medium-sized equipment and crews. However he is likely to have several types of products to be removed. Even a small operation of only 1 million board feet of logs and 1,000 cords of pulpwood may mean handling 14 species of logs, with 3 grades for some of them, and delivery to 12 separate markets—each having its own specifications. But if the logger is to obtain the best prices for his products it will pay him to seek out these markets and prepare to deliver the material in truckload lots to each of them.

It is no simple task to maintain a steady flow of logs and other products to diversified markets and still meet the limitations imposed by conditions within the forest. Advanced layout of roads and skidways, locations of the major concentrations of special products, timing of specific product harvest, and systematic followup of areas believed to be logged out are necessary to keep the operation moving smoothly and to avoid leaving trees marked for cutting or logs already cut in the woods.

The logging foreman must understand the cost of the several operations. It is his obligation to plan each phase so that the total cost will be minimized. He must lay out the operation so that it moves in an orderly fashion without bottlenecks and without one crew's endangering or being in the way of another. Coordinating all these activities and supervising them is a difficult task calling for special knowledge and administrative skill.

Logging planners must give consideration to other limitations on logging of which the following are examples:[2]

1   Weight limits of bridges and roads.
2   Limitations on timber floating.
3   Safety codes, workmen's compensation laws, accident prevention.

4 Labor recruitment and training.
5 Wages and hours laws.
6 Fire control and slash disposal laws.
7 Logging camps and their operation, with repair parts and shops if required.
8 Labor relations and union contracts.

Some of the most sophisticated planning and programming of operations is to be found in the forestry operations, including logging of the pulpwood forests of the southern United States. Here a single product, pulpwood, harvested on a short rotation, with clear-cutting, site preparation, and planting as the silvicultural system, would seem to simplify planning—and it does. But it also makes possible a multitude of choices. How much land should the company own and operate as opposed to simply purchasing wood from small landowners thereby building up local good will? What types of services and contracts should be used to assure a wood supply from such private lands? How can the best silviculture be introduced and carried out on lands the company looks to for wood supply but does not own? What system or systems of logging and programming will prove best overall for company wood procurement? With these and other limitations on control how can the forester demonstrate to corporate management that his plan will result in minimized costs over the long run? But this is the type of service that is being demanded of foresters today.

The forester who is planning logging operations in a developing nation with tropical hardwood timber has an even more perplexing task. There may be hundreds of species, few of which are known to international trade. There may be no equipment-supply houses with spare parts for his machines, or replacements if a machine becomes disabled. The labor force may have to be trained to operate the machines. Foreign exchange may be difficult to obtain for needed supplies and equipment. Roads, communications, transportation, banking, and other services may be limited. Few experts, if any, can be called in if something goes wrong. A man in such a position must be a long-range planner and in addition an imaginative improvisor to meet the many emergencies that may arise. It is no easy task to introduce a highly sophisticated industrial program in a culture only entering on the machine age. But it is from such countries that much of the quality timber of the future must come, and major United States and European companies are planning accordingly. Developing countries also have opportunities for establishing forest industries of their own, so the task is not one to be ignored.

## WASTE IN HARVESTING TIMBER

Logging has traditionally been a wasteful operation. Where timber is abundant, only choice species and individual trees are harvested. Trees too small to cut are often injured or broken over. Logs themselves may be damaged or broken by careless felling or while being skidded or river driven. Logs may also be overlooked in the forest, damaged by insects or fungus, or may sink during river driving.

Fortunately, in lands that have developed economies it has become good business as well as good forest practice to keep logging waste to a minimum. Because of the ease with which trees can be severed with the power chain saw, it has become profitable to harvest the usable parts of trees considered to have been culls at the time of former harvests. The use of pine and hardwoods for pulp manufacture has made it profitable to harvest crooked, limby, and small trees, as well as the tops of sawlog-sized trees.

Perhaps even more important has been the insistence of foresters on careful woods operations, and the gradual realization of loggers themselves that they can have more stumpage to cut if their work is not wastefully done. The growing use of clear cutting in mature timber is eliminating the damage that might otherwise be caused to residual trees.

All the above considerations have led to progressive reduction in logging waste during the period from 1945 to 1970. The growing concern of the forest industries to provide for their future supplies on lands tributary to their mills—irrespective of ownership—has led them to pursue nonwasteful practices. To this can be added the pressure of public opinion that is warning the logging industry that esthetics and other environmental values are not to be ignored.[1,7]

## INTEGRATION WITH FORESTRY PRACTICE

When logging is finally recognized as but one stage in the total operations involved in managing forest lands, wasteful processes should gradually cease (Fig. 12-3). Emphasis then will be placed on minimizing total costs of all operations over the rotation, whether the benefits from them be currently realized or be in the form of future upgrading of the total environment. Much damage can be corrected most readily if action is taken immediately after the logging operation. This is particularly true of steps needed to assure early establishment of the next timber crop. Road and trail damage, disruption of drainage channels, removal of unsightly debris, clean-up of trees damaged in logging, and active steps to leave the

**Figure 12-3** Selectively cut forest of western conifers, Colville Indian Reservation, Washington. *(U.S. Bureau of Indian Affairs.)*

land in a tidy, manageable condition will minimize the costs of future forestry and assure the public that the land is being given long-range care. Fortunately many large timber companies were following just such practice in 1970. And their foresters were being given a hearing in logging congresses on how to carry out silvicultural measures economically. No one in the business thinks it out of place to see in his logging journal articles bearing such titles as: "Proper Use and Management Will Insure Viable Forests," "Pre-Commercial Thinning in Douglas Fir," or "Timber Production and Aesthetics Are Compatible."

Of great significance, too, is the fact that some timber companies have begun to organize their own subsidiaries for recreational land development. The men who have these responsibilities have access to top management and can put pressure on those responsible for logging to conduct their work so that the environment will remain pleasing after logging is completed.

Logging badly needs to earn a new public image in America and many men in the business now realize this. This new image is needed not only to reduce public criticism of loggers but also to attract enterprising young men into the business. For with all its hazards, hard work, capital requirements, and attention to detail, logging still remains one of the avenues by which a man of ability, enterprise, and desire for independence can build his own business and make a fair livelihood at it.

**LITERATURE CITED**

1   Applied Forest Research Institute. Timber Harvesting Methods and Equipment of Today and Tomorrow—Needs, Goals and Limitations in New York State, AFRI Misc. Report No. 2, March 1969, State University College of Forestry at Syracuse University, Syracuse, N.Y.
2   Bromley, W. S. (ed.). 1969. "Pulpwood Production," 2nd ed., 1969, The Interstate Printers & Publishers, Inc., Danville, Ill. 259 pp.
3   Crebbin, Peter A. 1969. Tree Length Operation in Lodgepole Pine, *Loggers Handbook,* **29**:21–23, 44.
4   Makkonen, Olli. 1969. Ancient Forestry, An Historical Study, Part II. The Procurement and Trade of Forest Products, *Acta Forestalia Fennica,* **95**:1–46.
5   Simmons, Fred C. 1951. Northeastern Loggers Handbook, U.S. Department of Agriculture Handbook No. 6, Washington. 160 pp.
6   Wackerman, A. E., W. D. Hagenstein, and A. S. Michell. 1966. "Harvesting Timber Crops," 2nd ed., McGraw-Hill Book Company, New York. 540 pp.
7   Westveld, R. H. 1968. A Challenge for the Wood Products Industries, *Journal of Forestry,* **66**:471–474.

Chapter 13

# The Business of Growing Timber

A forest manager is employed to operate a property in accordance with the wishes of the owner. If the owner is the public, these wishes may encompass many things: watershed protection, public hunting grounds, grazing, recreation, environment protection, and timber production. If the owner operates a wood-using plant, he will need to meet all or a part of his timber requirements from his forest. Return on investment may be a minor concern to owners who are primarily interested in recreation or hunting, or to wood-using industries whose main source of income is from processing rather than growing timber. An individual private owner may have mixed motives for owning forests. If income production ranks high among his objectives he will probably be obliged to look to timber as his main source. Leasing camp sites, hunting rights, recreational use, and selling water have all been used as a means of obtaining income from forest properties. The total income from these sources to forest owners who have not been engaged in the resort business as such has been negligible compared with what has been realized from selling timber. In this chapter, timber growing as a business will be explored.

The manager of a retail store can obtain from departments of commerce and other sources information on capital, floor space, inventory, and turnover for a given business volume. How the budget should be distributed among such costs as rents, advertising, labor, store supplies, and miscellaneous expenses may also be learned. The Agricultural Extension Services can advise a prospective dairy farmer on size of herd, land area and quality, and the crops he should grow. He is given feeding schedules for dry stock and milkers. Equipment needs and overall capital expenditures can be estimated. Suggested work schedules and appropriate accounting records are also offered. These serve to minimize business risks.

Such comprehensive business guides are seldom available to the man wishing to make a business of growing timber. On very few forest properties in America have the necessary records been kept for preparing such guides. Careful studies have been made of the economic returns to be expected from specific operations such as planting, thinning, pruning, and other practices, but it is quite another matter to synthesize this information into meaningful guides that can be applied to operating a forest as a whole. Local differences in timber quality, growth, value, and production costs are so great that detailed guides to forest business are unlikely to appear soon. In recent years forest economists have given increasing attention to developing methods of analysis to apply to specific forest operations. These have been tested against such field data as have been available. Such approaches have yielded some rather startling information. For example, private individuals have probably invested more money in tree planting than in any other specific forestry activity. They have been encouraged to do so by foresters and by public subsidies. In some areas owners have been well rewarded for tree planting investments. But over much of the northern United States tree planting seems to have been far less remunerative than other forestry options the owner might have pursued. Webster[14, 15] found that in Pennsylvania, thinning young hardwood pole stands yielded highest returns in terms of investment costs. To achieve the same returns from planting scrub oak lands would require an expenditure 75 times as large. This very fact, that returns in relation to cost vary so greatly and that few reliable analyses of such variability in response have been made, points up the high risk involved in forestry investment. It also suggests that potentially high returns may be realized by those who can judge astutely the basic business factors involved and are willing to venture into a new field of investment.

**Basic Factors in Timber Growing**  A producing timber property has certain basic features analogous to those to be found in a manufacturing enterprise. Climate, topography, and soil are the timber grower's basic physical factors governing the supply and availability of his raw materials. As such, they are analogous to the location and building that houses an industrial plant. The soil supplies two raw materials: water and mineral elements essential for tree growth. These may be available in ample amounts or in limited amounts that restrict rapidity of growth. The third raw material, carbon dioxide of the air, is so widely distributed that it never need give concern to a forest manager. Sunlight supplies the energy for manufacturing woody substance from minerals, water, and carbon dioxide. Temperature and the efficiency of chloroplasts determine how fast these raw materials can be converted into wood products by the trees. The forest manager, once his location has been set, has no feasible way to modify climate, topography, or the physical character of the soil. All he can do is adjust his operations so as to take the best possible advantage of these basic factors.

**The Producing Machinery** The producing machinery of a forest is the growing stock, that is, the trees of all sizes and ages that occupy the soil. If this growing stock is well distributed over the land, if it is likewise well distributed as to age, and if it is made up of valuable species and well-formed units, it will produce regular crops of high-value products. If the growing stock is sparse or scattered, if it is composed of a high percentage of low-value species or poor-quality units, it will produce a low volume and low-value products. Such a forest is analogous to a factory using improperly arranged, worn out, and obsolete machinery. A fire may just as effectively destroy the producing machinery of the forest as might a fire in an industrial plant.

One striking difference exists between the forest and the industrial plant. The trees that grow the wood are both the producing machines and the end product. This fact has confused many people unfamiliar with forest management. An industrial owner would scarcely expect his factory to be productive if he sold off his machinery, allowed fire to gut his factory, and invested no new capital to replace it. Yet some people seem to expect forests to do so.

**Forest Valuation** It was stated in Chap. 11 that a forest property might be evaluated on three bases: its liquidating value, its value as a going business, and its market value. The market value, that at which comparable forest properties are currently bought and sold, may bear little relation to either the liquidating value or the value of the forest as a timber-growing investment.

If a man pays too much for his forest property he may run into financial difficulties because of the interest he must pay or forego. Also, a high purchase price may invite an increase in the assessed valuation and the taxes the owner must pay.

Assessments are based by law on the highest use of the land. Land that has attractive features, such as lakes and streams, may be assessed much higher than its value as a timber-growing investment.

An investment in a forest property may be considered as a hedge against long-term inflation. In general, landowners have benefited from inflation, particularly those whose property lay in areas of increasing population and economic activity. Many forests lie elsewhere. The general exodus of people from rural to urban areas has led to abandonment of farmland and an increased tax burden to remaining landowners. In such cases the costs of ownership may increase as rapidly as long-term inflationary trends. Unless the owner can harvest enough timber regularly from his land to cover current costs by a substantial margin, he may find that his forest offers only an illusory protection against inflation.

Historically lumber prices have advanced far more rapidly than the prices of all commodities, and the prices of standing timber have behaved similarly. But wide fluctuations occur. Local price fluctuations are even more erratic. In fact during periods of depressed prices small sawmills may be forced out of business or may cease to operate for extended periods. The paper industry has generally been more stable than the lumber business, but it has not always provided a dependable market for wood. New technologies have changed the demand markedly. For example in the 1940s and 1950s aspen pulpwood stumpage brought $3 to $4 per cord in northern New York, whereas other hardwoods were not accepted at all. In 1970 hardwood stumpage brought $2 to $3 per cord and aspen was rejected entirely, though it could be marketed in Canada and elsewhere. The development of the neutral sulphite pulping method for hardwoods accounted for the shift. Aspen was suitable for such a pulping process but needed a milder cook and gave a lower yield of pulp per cord than did the denser woods such as beech and maple. Hence aspen fell into disfavor with local pulp manufacturers.

A market for diversified products tends to minimize such risks. Still, the fact that many timber-processing businesses are small adds to the instability of any one of them. All such factors influence the chances for success in the timber-growing business and hence affect the price that should be paid for a forest property.

Bare forest land that must be planted and tended for 40 years or more before merchantable products can be harvested has very low value for timber investment. In fact, it may have only a negative value for the private investor because the costs and risks involved may exceed the prospective return. Such land, if extensive, can be restored to forest only by the slow process of nature or at public expense. In the latter case, subsidiary values inherent in productive forest land would be expected to offset the short-time deficit in return.

**Selecting Land for a Timber Investment** The individual or corporation seeking to obtain a profit from managing forest land for timber must take great care in selecting an area of land. Basic factors to consider are available markets now and in the future; climate; topography; soil productivity; present stocking of timber trees, and the species, size, quality of stems, and distribution of age-classes; local taxes; accessibility to public roads and the ease of road construction; prevalence of destructive insects and diseases and the measures available for their control; prospects of supplemental income from leasing hunting rights, from mineral or gravel deposits, or from leases for recreational purposes; attitude of local people toward forestry; and willingness of timber operators to use cutting practices that will lead to building up the value of the growing stock rather than depleting it. As timber growing of necessity is a long-term investment from which, in view of the risks involved, only modest returns are to be anticipated, the owner should make sure that his title is clear and that the property has been free from trespass and adverse possession. A lawyer can examine a title search and a forester or surveyor can establish the boundaries and note any cases of serious trespass or adverse possession. With favorable reports on these, the owner can feel assured that the property is really his and that no claims will be made upon it from creditors or former owners or users.

**Tending the Forest-producing Machine** Operating the forest factory so as to produce efficiently high volume and quality of timber is applied silviculture. Silviculture means making sure that the land owned is well stocked with growing trees. It involves eliminating the ineffective timber producers—by thinning, improvement cuttings, release cuttings, and other operations—so as to concentrate the growth on those trees that will develop the highest quality of timber.

**Organization of the Timber Factory** The forest manager's job is to organize the forest for efficient use of soil, growing stock, labor, and equipment. He decides what roads, trails, skidways, and other improvements are needed. He locates his protection equipment so that the producing plant can be readily protected against fires and trespass. He decides on an orderly arrangement of forest compartments so that the road and skidway systems will be used to best advantage, thereby minimizing his costs of operation (Fig. 13-1). He attempts to adjust his

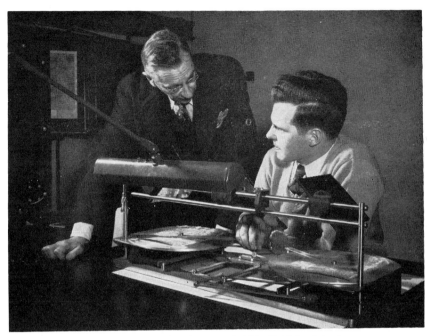

**Figure 13-1** Foresters reduce the time required in field work by preparing preliminary maps of forest types and condition classes from aerial photos. A stereoplotter is being used in the photo. *(C. Wesley Brewster.)*

growing stock so as to provide a regular yield of products, year by year.[4] He times his timber-cutting operations to take advantage of good markets and curtails selling when markets are low. He knows that he can store his inventory on the stump and have it increase progressively in value through growth, and he also knows that there is a limit on how long he can allow his inventory to accumulate without having it congest his factory so that total production slows down. Generally, he seeks to manage his land intensively for high-value products, thereby reducing the relative importance of fixed costs, such as roads, boundary maintenance, prevention of trespass, and land taxes, that increase directly with size of ownership. He invests first in those forestry measures that bring the highest return on costs. He is almost certain to find that some of his lands—steep, rocky slopes and thin or swampy soils—cannot return a yield on a forestry investment. He will either eliminate such lands from his holdings or recognize that they must be carried but will spend as little as possible on them.

The forest manager also decides what equipment he requires to manage his forest most effectively. This includes personal transportation, protection equipment, logging equipment, road-building and maintenance equipment, and record-keeping equipment. He should give special consideration to what equipment he should own and what he should lease or contract for. As he will want his woods work done as efficiently as possible, he will be alert to new equipment developments and may well find it expedient to pioneer in new equipment use.

**Forest Labor** The forest manager employs forest labor to protect his forest and to keep it in good growing condition. He also requires labor to prepare his products for shipment; that is, to fell, limb, buck, skid, and haul his logs and other forest products. He must decide how many workers he can employ permanently, and the extent to which he can depend on temporary labor or contractors to speed up his operations during times of good markets.

**Markets** One of the major tasks of the forest manager is to find good markets for all the different types of products his factory produces. The forest factory produces a large amount of medium- and low-grade products and a relatively small amount of high-grade products. Some forests, in fact, produce almost no high-grade timber, and even the best of forests, having all age-classes represented, will produce a large volume of small-sized timber that is difficult to market. This fact immediately confronts the forest manager with the necessity of deciding how far he should enter into the timber-processing field in order to provide a market for his products. He may decide to operate a small portable sawmill to convert to rough lumber the low-quality logs that will not pay transportation costs to a permanent mill.[3]

It is in the marketing field that forest managers encounter their most serious difficulties. Nationally, the markets for forest products tend to get progressively better decade by decade, but this may not be true locally. Local markets depend on the presence and economic prosperity of local wood-using industries. Timber is so costly to transport that a good market for fireplace wood in New York City means nothing to the Adirondack timberland owner.

The forest manager must always balance carefully the cost of buying equipment against renting it. If the forest manager has full-time use for such equipment, he may well afford to own it.

Decisions on direct operation as opposed to contract operations will be determined in large degree by the size of operation and the wisdom of incurring costs for machinery, a permanent labor force, and the clerical and supervisory aid needed to manage efficiently a dispersed operation. Generally, the further the landowner is able to go in timber processing, the better the market he is able to provide for his products. But his marketing task becomes increasingly complex and the capital tied up in inventory burgeons as the variety of products to be sold multiplies.

**Owner Logging** The best way for a forest owner to assure that logging operations on his property are carried out at the most propitious time and in a manner that minimizes damage to residual trees and site is to do the job himself or have it done under his own supervision. European foresters have traditionally been responsible for logging operations and look askance at the American practice of having such work done by outside contractors.

The capital required to enter the logging business in the East or the South in 1972 was considerable, some $25,000 to $35,000. The western logger would have an investment several times this amount. Capital requirements could probably be reduced by one-third by purchasing the equipment secondhand or on a time-payment basis. Credit with a local bank will probably be needed to maintain solvency during work stoppages caused by bad weather, equipment breakdowns, slack markets, or other causes.

Obviously the owner must be willing to employ a foreman with the necessary skills for successful logging. He must assume the risks involved if he is to perform his own logging operations. It is though the best way to assure integration of his timber-growing business and best overall use of land and capital investment.

**Record Keeping** The forest manager obviously cannot know whether his operations are financially successful unless he keeps accurate records, including an inventory of his land, growing stock, improvements, and equipment. He should keep separately his costs for taxes, protection, general management, logging, sale of products, and timber processing. Costs of unit operations reveal which of his various operations are making money and which are losing money, and form the basis for intelligent action to produce profits.

## CREDIT FOR FORESTRY BUSINESS

Because income from sale of timber may fluctuate widely year by year, credit may become a pressing need. Credit can be safely extended only with assurance that the operation to be undertaken will yield ample returns to repay the loan. Once a business is well established, credit may be used extensively, provided this leads to a marginal interest return above whatever rate the owner must pay.

The long-term growth rates of forest stands are not high—from 3 to 6 percent for northern hardwoods and up to 10 percent for southern pines. At certain stages in the life of a tree or a stand, however, the value increase is very rapid. For example, a 60-year-old hard maple, growing in diameter at a constant rate, might be 12 inches in diameter. It would contain about 50 board feet and would be worth but 50 cents if, indeed, it were merchantable at all. At 90 years, it would grow to 18 inches in diameter, contain 200 board feet, and be worth about $6. This is a 12-fold increase in value in 30 years. Likewise, a 60-year-old northern hardwood stand might bring but $10 per acre if it was sold for pulpwood, but if it were held 30 years more, it could have 5,500 board feet worth $110. It is from such knowledge that the forest investor will seek to benefit and that will justify his use of credit.

Forest financing also involves forest insurance and insurance required of those who employ labor. Adequate fire insurance is difficult to obtain at reasonable rates. Large forest property owners require no insurance against small fires but do require insurance against large fires that might destroy the major portion of their forest.

Insurance against severe damage by windstorm, ice, snow, insects, tree diseases, and animals is even less obtainable. All these are potential sources of serious loss, especially to young growth. The owner's best insurance is intensive management that will provide for prompt salvage of whatever losses occur. A hurricane, such as swept over Georgia, Alabama, and Mississippi in 1969, or a huge fire, such as burned over 220,000 acres in Maine in 1947, can destroy a forest business for the lifetime of any individual owner.

The timberland owner, particularly if he engages in logging or any other timber operations, must carry Workmen's Compensation insurance and Social Security for his workmen. Basic knowledge of business law and contracts is essential.

## BUSINESS ACUMEN

The overall success of the forest manager will depend, in the last analysis, upon his business acumen. This means the skill with which he is able to manipulate his forest, his growing stock, his forest labor, his marketing and sales, and his use of credit and insurance so as to maximize financial returns from operating the property. The forest manager must keep an adequate reserve of working capital so that his operations can be flexible. Forced liquidation of valuable growing stock when markets are badly depressed can be disastrous.

The greatest financial hazards to the business of timber growing are market fluctuations caused by business and building cycles. Timber markets may virtually vanish for 2 or more years at a time. Building activities expand very rapidly during the ascending phase of the business cycle. Building cycles show higher peaks and lower troughs than industrial production. The index of building activity in 1918 at the close of the war was approximately 25; by 1925 it had risen more than 10-fold, to 260. During the same period business activity, as measured by industrial production, increased only from 60 to 90. By 1932 building had dropped from index 260 back to 20, and industrial production, which reached its peak in 1929, dropped from 115 to about 55.[17] Lumber production appears to fluctuate less violently than the building cycle but more violently than the business cycle.[6]

**Long-term Trends in Prices** Long-term trends in American business enterprise have been marked by an increase in quantity and quality of output and a decrease in labor requirement. This has frequently meant a decline in price or at least a decline in man-hours of labor required to purchase the products. This, in turn, has been possible by increasing greatly the productivity of labor through modern production techniques, standardized design, power machinery, and applied research. Living standards have risen thereby. This trend has characterized the manufacture of such products as chemicals, steel, automobiles, household appliances, and power equipment.

This has been somewhat less than true of the lumber industry. Long-term prices for lumber and standing timber have been strongly upward. During the 100-year period from 1870 to 1969, lumber prices increased seven times as much as the average prices for all commodities. Lumber prices doubled or more than doubled during the Civil War and during each World War. After the close of each war, prices receded but tended to hold some of the gains. A 50 percent increase in softwood lumber prices occurred between January 1968 and March 1969. Again, with the passing of a tight situation, the price receded to but 13 percent above the January 1968 level.[6, 8]

Such sharp fluctuations as mentioned above do help to raise the price of standing timber. Yet even in 1970 sawlog stumpage prices bore no direct relation to the cost of growing the timber. They resulted from the competition of buyers and were not imposed by timber growers as such. Sharp price increases can have an adverse effect, for they tend to discourage the use of lumber. In fact lumber has long been at a disadvantage, compared to other construction materials, because of the above-mentioned upward price trend. However, this price relation seemed to have changed[9] in the 1960s (see Chap. 18).

One of the major causes of increase in lumber prices has been the increase in the cost of labor in logging and sawmill operations without a corresponding increase in output per workman. Where trees are scattered and small, it has been impossible to maintain quality of product at the same price as in former years.

The long-term outlook for the business of forestry has other favorable features. The U.S. Forest Service projects that by the year 2000 the total commercial wood demand may be as high as 2.4 times that of 1970.[7] The above projections include the following increases for the decade 1970 to 1980: sawlogs from 46 to 64 billion board feet, veneer logs from 9 to 15 billion board feet, and pulpwood from 81 to 121 million cords. Such a growing market for forest products, if realized, would seem to assure a favorable outlook for the business of forestry—provided costs for taxes, labor, and supervision do not outrun the advantage offered by the increased market.

Another factor favorable to forestry is that, in computing taxable income under Internal Revenue laws, timber growth accumulated since purchase of property may be considered a long-term capital gain. Such laws are of course subject to change at the discretion of Congress.

**Case Studies of Forest Investments** Some of the principles outlined above can be illustrated from records kept by men who purchased forest tracts for investment.

*Case 1.* Mr. A in 1956 purchased 311 acres of abandoned farmland of which 244 acres was stocked with trees. His costs for land, title search and legal opinion, deed recording, surveying, and timber inventory totaled $6,485. He was able to rent pastureland for $35 per year. The property had not been logged for 20 years prior to purchase. In 1964 he sold 248,000 bd ft of sawtimber and 330 cords of pulpwood for a total of $4,156. In 1969 he made a reinventory and projected the results to the year 1971 from growth data collected at the time of reinventory. The total sawtimber had increased from 404,000 bd ft in 1956 to 1,336,000 bd ft even though 248,000 bd ft had been cut. The pulpwood, mostly conifer, was unmarketable in 1971 and so was omitted from the value change. The value change in inventory at 1964 prices was $13,191. In 1971, expenses for taxes, sales administration, reinventory, custodial fee, and other fees totaled $7,848. The net return of $10,024 for the 15-year period represented a simple interest rate of 10.3 percent.

*Case 2.* Mr. B purchased a 191-acre property in 1961 for a total cost of $4,200. He had an income from sales of $4,851, an inventory gain of

$2,190, and total costs of $2,805 to the year 1971. His net gain per year was $385—equivalent to a simple interest on investment of 9.2 percent.

## SOURCES OF INFORMATION ON FORESTRY BUSINESS

The best source of information on the business of forestry is an experienced consulting forester who is familiar with the local conditions that affect timber growth rate, markets, stumpage values, land prices, and operating conditions. Additional sources are state and federal foresters, forestry schools, forest experiment stations, and county agricultural extension services. Large corporate owners should be in a position to employ their own foresters. The small forest owner should depend mainly on consulting foresters and on those employed by state and federal agencies to assist woodland owners. A number of good source books are available. Three deal specifically with the management of small properties.[10-12] Other books deal with specific subjects pertaining to the business of forestry: forestry economics,[5,16] forest management,[4] forest measurements,[1] and logging.[2,13]

## LITERATURE CITED

1  Avery, T. Eugene. 1967. "Forest Measurements," McGraw-Hill Book Company, New York. 290 pp.
2  Bromley, Willard S. (ed.). 1969. "Pulpwood Production," 2d ed., The Interstate Printers & Publishers, Danville, Ill. 259 pp.
3  Brown, Nelson C., and James Samuel Bethel. 1958. "Lumber," 2d ed., John Wiley & Sons, Inc., New York. 379 pp.
4  Davis, Kenneth P. 1966. "Forest Management: Regulation and Valuation," 2d ed., McGraw-Hill Book Company, New York. 519 pp.
5  Duerr, William A. 1960. "Fundamentals of Forestry Economics," McGraw-Hill Book Company, New York. 579 pp.
6  Hair, Dwight. 1958. Historical Forestry Statistics of the United States, U.S. Department of Agriculture, Forest Service, Statistical Bulletin 228, Washington. 36 pp.
7  Hair, Dwight, and H. O. Fleischer. 1970. Meeting Growing Demands for Wood Products, U.S. Department of Agriculture, Forest Service, Washington. 22 pp.
8  Hair, Dwight, and Alice H. Ulrich. 1970. The Demand and Price Situation for Forest Products 1969-70, U.S. Department of Agriculture, Forest Service, Miscellaneous Publication No. 1165, Washington. 79 pp.
9  Josephson, H. R. 1969. Substitution—A Problem for Wood?, U.S. Department of Agriculture, Forest Service, Washington. 11 pp.
10  Shirley, Hardy L., and Paul F. Graves. 1967. "Forest Ownership for Pleasure and Profit," Syracuse University Press, Syracuse, N.Y. 213 pp.

11  Stoddard, Charles H. 1958. "The Small Private Forest in the United States," Resources for the Future, Washington, D.C. 171 pp.
12  Vardaman, James M. 1965. "Tree Farm Business Management," The Ronald Press Company, New York. 209 pp.
13  Wackerman, A. E., A. S. Mitchell, and W. D. Hagenstein. 1966. "Harvesting Timber Crops," 2d ed., McGraw-Hill Book Company, New York. 540 pp.
14  Webster, Henry. 1960.Timber Management Opportunities in Pennsylvania, U.S. Department of Agriculture, Forest Service, Northeastern Forest Experiment Station, Station Paper No. 137, Upper Darby, Pa. 37 pp.
15  Webster, Henry H. 1965. Profit Criteria and Timber Management, *Journal of Forestry,* **63**:260–266.
16  Worrell, Albert C. 1959. "Economics of American Forestry," John Wiley & Sons, Inc., New York. 441 pp.
17  Zivnuska, John A. 1949. Commercial Forestry in an Unstable Economy, *Journal of Forestry,* **47**:4–13.

Chapter 14

# Forest Administration

The function of forest administration is to use the knowledge and talents of forest scientists and other specialists to plan and put into practice in the forest the management program that best meets the wishes of the owner and the needs of society. The administrator must perform this function with the financial and technical resources that the owner makes available to him.

## SETTING GOALS

An administrator's first task is to establish management goals in conformity with the owner's wishes. Whether the owner be an individual, corporation, state, or the federal government, the objectives of ownership may be ill defined or difficult to achieve on the forest property available. Using all his expert knowledge, the forest administrator should therefore set forth what he believes to be attainable objectives, thereby helping the owner to clarify his goals and to decide how to allocate the financial resources necessary to achieve them.

To gain an understanding of the task of a forest administrator, let us consider a hypothetical example. Let us assume that a company purchased a 100,000-acre tract of southern pine land to supply wood for a

400-ton-per-day integrated pulp and paper mill, using sulfate pulp. Mill construction was begun while the company's chief forester was still engaged in a comprehensive timber inventory. The branch manager realized that he would have to buy wood to supplement what the company lands could produce on a sustained-yield basis. After completion of the timber inventory, he learned that company land could supply annually only some 54,000 cords under an intensive sustained-yield management program. Mill requirements would be 132,000 cords. The chief forester recommended that company forest holdings be increased by some 25,000 acres, which he estimated would yield an additional 13,000 cords. This would meet half the mill needs over the first decade. The remainder would be bought from small landholders and contract loggers. This recommendation was approved and the chief forester was directed to prepare a plan of operation to meet mill requirements, including land purchase, timber management, company logging, and wood purchasing.

### PLANNING AND BUDGETING

To continue with our above example, the chief forester decided to establish four forest districts of approximately 25,000 acres with a qualified professional forester in charge of each. In addition he would need two professionals to buy land and wood from private landowners and contract loggers. With supporting personnel his total organization was as follows:

| Personnel needed | Breakdown | Subtotal |
|---|---|---|
| Professional personnel | | |
| Chief forester | 1 | |
| District foresters | 4 | |
| Wood procurement foresters | 2 | |
| | | 7 |
| Technicians | | |
| Logging foremen | 6 | |
| Nurseryman | 1 | |
| Maintenance supervisor | 1 | |
| | | 8 |
| Labor | | |
| Loggers | 54 | |
| Maintenance men | 2 | |
| Part-time labor for field planting and nursery work, man-year equivalent | 8 | |
| | | 64 |
| Office and clerical personnel | | 5 |
| Total | | 84 |

Having planned his organization, he proceeded to prepare capital and operating budgets for the first year's operations. The capital budget included items for land purchase, nursery site purchase and development, nursery machinery and equipment, maintenance-shop building and equipment, district foresters' headquarters and equipment—including pickup trucks, logging machinery, trucks and equipment, field-planting equipment, and road construction and maintenance machinery. A brief statement was prepared to justify each major item in the budget. Equipment was specified by type, size, manufacturer, and model number.

The operating budget carried, in addition to salaries, a brief job description for each professional and technical employee, setting forth the experience needed and the duties to be performed. Salaries were fixed in accordance with company policy. Wages for loggers and temporary workers were listed at current rates for the region. Separate items were included for fuel, lubricating oil and grease, spare parts, and the various other needed supply items. These were listed separately for office, shop, nursery, and forestry activities. An item was included for official travel. The operating budget also carried items for interest and taxes on land, and depreciation of equipment. The latter would be used to build up a capital reserve for equipment replacement. A minor budget item was for preparing sites for recreational use by the public. The forester justified such expenditure on the basis of the goodwill it would engender. The annual operating budget totaled $2,375,000.

The preparation of such a work plan and budget involved careful selection of each major item of equipment to be sure that it would be fully adequate to meet local operating conditions. The entire plan included sketches of layout for each district and its headquarters and for nursery and shop. Some 6 months' time was spent in preparing it.

After careful review the branch manager endorsed the work plan and the capital and operating budgets and forwarded them to the home office for approval. After review by the vice-president in charge of operations and the president, approval was granted. The branch manager was then authorized to proceed with the program.

## STAFFING AND ORGANIZING A FORESTRY DEPARTMENT

The chief forest officer's first staffing task was to select his professional personnel. To obtain a list of candidates he sent a description of his proposed organization to forestry schools, the Society of American Foresters, and others in a position to recommend candidates. From their recommendations those candidates deemed best qualified were each sent a letter outlining the company's plans. Application forms were enclosed. Applicants responding were screened and recommendations were sought from former employers and others. Promising individuals were invited to the headquarters for interview by the chief forester, local personnel officer, and branch manager. Successful candidates were asked to report for duty at times convenient to both the company and the appointee. A similar procedure was followed to select the nurseryman. Each district forester was asked to recommend logging foremen. These were interviewed by the chief forest officer and personnel manager and appointments were made. It was made clear to each foreman that the district forester was his supervising officer. A list of candidates for shop foreman was provided by the personnel officer and the final selection was made by the chief forest officer.

To clarify organizational lines, duties, and responsibilities, each professional forester and forest technician was furnished a job description. He was also furnished a copy of the company's personnel policies statement, which included information on safety regulations, what to do in case of accident, company insurance and retirement programs, sick and annual leave privileges, and related items of concern. It also contained a statement from the company president, setting forth broad company objectives relating to personnel and relations with the public.

The chief forester personally saw to it that each new appointee met the others in his organization and that the professionals became acquainted with other company officers through whom they must work, including the branch manager. He offered his services and those of others in helping the new appointee to find acceptable housing and become acquainted with those who were to work with him. Such niceties cost little and do much to build esprit de corps and elicit willing cooperation.

## TRAINING, WORK ASSIGNMENT, AND SUPERVISION

The immediate task of a new employee is to learn his work assignment and the conditions under which it is to be carried out (Fig. 14-1). In our example, each job description was written in broad language, terminating with the clause "and perform other duties as assigned." In cases where no job description is supplied, the new employee might himself prepare one and ask his superior if it covers what is expected of him. It is clearly the duty of the superior to explain what duties the new employee is expected to perform, to acquaint him with pertinent organization policy under which he must work, and to furnish him the necessary instructions to discharge his responsibilities. When a change in assignment or an unfamiliar task is to be undertaken, additional instruction or training should be provided. It is especially important that the superior delegate the authority to take necessary action along with the assignment of

**Figure 14-1** Forest Service ranger instructs foreman training class. *(U.S. Forest Service.)*

responsibility. Delegated authority should not be impaired or withdrawn unless the responsibility also is withdrawn. Such might occur in a case of incapacitating illness. A district forest manager cannot administer his district effectively unless he has authority commensurate with his responsibility (Fig. 14-2).[4]

The chief forest officer of a corporation must know what is happening in the several districts. He keeps informed of this by the reports of the district foresters, the mill records of wood delivered, the reports of his own staff officers, and his personal visits to the district for inspection of work performed and underway. This affords an opportunity to commend the man for tasks well done, to question him on ways to facilitate the task or improve performance, and to explain new procedures and policies. It also affords an opportunity to suggest better methods where performance may not be up to company standards. The chief forest officer or his staff representative will be slow to make a point of minor deviations as long as overall performance is fully up to standard. Initiative and innovation are to be encouraged, not suppressed.

A supervisor should seek to know his professional employee well so that he can aid him in appropriate ways to achieve his goal in life. This needs to be done without the employee feeling that his privacy is being invaded nor his life being directed. Such well-considered help can be one way of eliciting high-level employee performance and satisfaction with his position. It is also important to let each employee know how he stands with the company or agency. The feeling of security so imparted enables a

**Figure 14-2** A western National Forest Ranger station. *(U.S. Forest Service.)*

man to perform his tasks with confidence and to look for additional ways to be of service.

The more able the professional employee, the greater becomes the obligation of his superior officer to help him develop his technical knowledge and administrative skills. One approach is to give him challenging new assignments that enlarge his thinking and mature his administrative skills. The late Dean Henry S. Graves of Yale University expressed it in these terms: "The way to develop young men is to give them new responsibilities just as rapidly as they gain the competence to discharge them."

## REPORTING

As the work year draws to a close the chief forest officer prepares his annual report. It will cover income from and costs of timber harvested and delivered to mills, timber purchased and harvested from noncompany lands, wood delivered from company and other lands, cutover and open land planted, land purchased and sold, wildfires that broke out and damage caused, changes in timber inventory resulting from yearly operations, and a projection of timber growth for the future. It will show land management costs for taxes, forest protection, silviculture and reforestation, roads, surveying and boundary marking, timber inventories, and other information that will acquaint the chief executive officer of the company with the contribution the forestry department has made to the company's business.

The chief forest officer will not fail to include a section on public use of company lands for fishing, hunting, picnicking, and camping—including the reaction of local people to the privilege of such use. Other significant reactions of both local and distant public to the company's land and forest policies should also be reported, for this can be of high importance to top-level administrators.

The chief forester will be sure to include in his annual report any details of progress on new activities outlined in the preceding year's plan, and for which new funds were made available. Other material to be included are personnel changes, community services performed by forest officers, innovations introduced, and savings effected.

An important part of the report will be the plans for the ensuing year and the budget required to achieve them. This will include both the costs and expected returns for new activities. It may also include an estimate of income that would be foregone if the expenditure were not made.

## DIRECTING A FORESTRY DEPARTMENT

Setting goals, planning, budgeting, establishing an organization, recruiting and training personnel, making work assignments, supervising and reporting are all means by which the chief forest officer directs the work of his department. So is decision making. But more than these are involved in directing. In fact the worth of an administrator is not determined by the number of decisions he makes personally. It may depend more upon the number he does not make, but that his subordinates make.

Directing includes setting the tone of the organization. Professional men work with eagerness and devotion when they have a significant role in shaping the plans and commitments of their department. They will make decisions with confidence when they know they have the needed facts and that their superior will back them up. In forestry especially it is necessary to delegate much decision making to the district forester on the ground. In controlling a fire it is imperative to do so. Each officer in a forestry department should have ready access to his chief to discuss with him both company business and any troublesome personal problems with which he needs help or at least sympathetic understanding.

Almost every administrative officer's task involves a burdensome amount of routine that must be performed. Journal and report reading to keep informed is indispensable. Men of education and talent are expected to master routine. But their chief responsibility is to give creative thought to broaden activities, develop new processes and products, reduce routine, and increase output of goods and services. Broadly speaking it is their task to conceive of new ways in which the department, company, or agency can serve society. It is the skill with which the administrator mobilizes effectively the creative thinking power of his organization that makes it an inspiring place in which to work.

Two development engineers from a leading chemical company once came to a college to make a pilot plant test of a new process. Much to their dismay, the process proved unworkable. Attempting to console them, a college officer said, "Well, you will have to go back to your department head for a new assignment." "That," replied one engineer, "is what we may not do unless we want to get fired. We must develop a better idea."

Creative thought is catalyzed by presenting challenging new situations to an alert professional team. It is by such means that an administrator builds an organization noted for high-level innovative performance. It requires tolerance and skill to develop innovative talent because innovators propose queer ideas and upset routine. A forest service officer once said of an innovative colleague, "Raphael Zon is a man with lots of ideas. Two out of three are completely impractical but the third is a corker. Anyone who can bat .333 plays in the big leagues." An organization that does not make good use of its men such as Zon had better take a sharp look to its future.

### USING OUTSIDE TALENT

"No man is sufficient unto himself," nor is any corporate forestry department, state forest service, or the national Forest Service. Each must take advantage of the developments of the profession and of the scientific and intellectual community at large. This is done by following scientific, trade, and other literature, by participating in professional and association meetings and in discussions with colleagues in other organizations. Visits to other operations and inviting selected guests to visit one's own activities can be helpful. Employing consultants to cope with tasks beyond the capacity of local talent is another way to keep an organization alert and progressive.

No functioning head of a large corporation or government agency can afford to be unconcerned about the attitude of the public toward the product it merchandises or the services it provides. This is particularly true of the Chief of the U.S. Forest Service, as the public protests and court actions begun in the late 1960s attest. Such interference can sometimes be forestalled if important issues are discussed with an advisory board made up of responsible representatives of the several public groups that may be affected. Reporting to the public on what an agency or corporation is doing and why may dispel many misunderstandings that could impair public relations.

## THE ADMINISTRATIVE STAFF

The chief administrative officer of a large corporation such as the International Paper Company or of a large government agency such as U.S. Forest Service must operate through an administrative staff. The functional roles played by staff members and the manner in which the chief executive organizes and uses his staff may vary considerably from one administrator to another, even within the same corporation or public agency. Some administrators prefer and seem to function best with a second in command. In corporations in which the chairman of the board of directors serves as chief administrative officer, the president of the company fills this function.

Whether or not there is a second officer, there is almost always an administrative and policy council or staff. This would normally include the heads of major organization functions. In a corporation the officer responsible for administering the work of each major division might serve, such as the executive vice-presidents for manufacturing, sales, policy, and administration, together with the chief counsel, chief fiscal officer, and possibly chief of public relations.

The U.S. Forest Service administrative council consists of the Chief, Associate Chief, and the deputy chiefs for Program and Legislation, National Forest systems, State and Private Forestry, Research, and Administration. The chief generally meets weekly with his council to discuss important legislative, policy, personnel, public relations, and other issues. To keep all staff members well informed, and to share routine administrative tasks, the six men take turns in serving weekly or longer terms as acting chief for discharge of routine business requiring the signature of the chief or acting chief. Each deputy chief also has his administrative staff, made up of the major divisions of his branch. One or more of these may be called in when the chief and his deputies need their expert knowledge. Regional foresters and directors of forest experiment stations may be called in when issues with which they are familiar must be decided.

Because of the confidential nature of much that takes place in meetings of the top administrative council, the group is generally kept small and closely knit so that high-level personnel changes and other issues of utmost importance can be debated vigorously without fear of information leaks.

In the function as a member of the chief's administrative council, each man is an arm of the chief and not merely a representative of his branch.

Many decisions made in the chief's council may need to be cleared with key line officers before action is taken. Usually this would be done in advance of final council debate. Directors of divisions, regional foresters, and directors of forest experiment stations may meet once annually with the administrative council to discuss broad policy issues and new programs to be undertaken.

The chief administrative officer along with his council are in effect the top decision-making arm of the organization and in a sense function as the institution they head and direct. From the above it becomes clear that the top executive is not primarily a decision maker himself but is rather the leader of a small executive group whose function it is to consider alternatives, debate issues, and arrive at decisions for the organization as a whole. When urgent issues arise on which a consensus cannot be reached, the chief administrative officer is obliged to make the final decision. Occasionally outside confidential advisors or consultants, or men who have recently retired from high-level administrative posts within the organization, may be consulted. The final decision must in such cases be made by the chief administrative officer.

Whenever an important decision is made in the Forest Service, whether by a field officer, a member of the chief's office, or by the Chief himself, the Chief is ultimately responsible for the decision and must assume the burden of support or of rescinding it if circumstances justify the latter action (Fig. 14-3). The same, of course, applies to the chief administrative officer of any corporation or government agency.

**ADMINISTRATIVE AND MANAGEMENT THEORY**

Recognizing the importance to mankind of the work of administrators and executives, a number of scholars have sought to develop management theory that executives could apply to lighten their burden of decision making and improve its accuracy. The American Society of Management holds annual meetings and publishes periodic reviews on the subject.[1]

Management is defined as the process of getting things done through

**Figure 14-3** Edward P. Cliff, Chief, U.S. Forest Service, 1962–. Actions taken and policies established today have far-reaching consequences to American forests for decades to come. *(U.S. Forest Service.)*

and with people operating in organized groups.[3] One approach to improving the art of management is to study various cases to determine principles and useful practice.

A second school of thought holds that management is leadership and that managers must learn how to motivate people. Its background subject is psychology.

The sociologic approach to management advocates study of the system of cultural interrelations in the organization and how to work with them.

Economic theorists tend to view the executive as primarily a decision maker. These have tried to study how executives make useful decisions.

The mathematical approach goes a step farther and introduces operations research, mathematical simulation, and systems analysis to work out the decision that optimizes returns on investment.[6]

One of the newest approaches is the ecological. This views the enterprise as operating in a more or less complex environment in which outside agencies as well as inside ones impinge on goals set and actions undertaken.

In meeting his responsibilities the modern executive will draw upon his knowledge of administrative cases, his understanding of people in the abstract, and certainly of the key members in his organization—including shop stewards and other group leaders. He must be a decision maker and will certainly seek to use his staff and sometimes expert consultants to collect and digest information, clarify the points at issue, and explain outcomes to be expected from a series of possible actions. To the extent that quantitative data exist or can readily be obtained he will seek to use mathematical simulation and other techniques to aid in decision making, and he will not ignore the environment in which his enterprise operates and the way in which his decisions may affect it favorably or unfavorably.

But for all the aids now available, the fact remains that the top administrator must operate with a considerable element of uncertainty. Decisions amenable to mathematical resolution are made by lower echelons. One of the marks of good administration is the extent to which decisions are made at the place where the greatest degree of competence can be brought to bear on them.[5] Among a top executive's most important decisions is selecting the appropriate man to carry out each major task. Such decisions must be made largely on subjective evidence. Top-level administrators must make use of all the pertinent information they can obtain. Yet with all the theory, principles, and techniques, high-level administration remains an art that each individual must acquire for himself.

## APTITUDES FOR SUCCESS IN ADMINISTRATION

Ordway Tead[8] lists as the most important qualities demanded of the administrator (1) abundant physical and nervous energy, (2) a sense of purpose and direction, (3) enthusiasm, (4) friendliness and affection, (5) integrity, (6) technical mastery, (7) decisiveness, (8) intelligence, (9) teaching skill, and (10) faith. Howard Smith[7] states that the dominant urges of the good executive must be (1) to deal with his fellow men on the basis of concern for their welfare as well as his own, and (2) to climb to leadership on the basis of demonstrated worth and with the approval and respect of his fellows. The necessary administrative qualities are consequently those of character, personality, and ability.

Cleeton and Mason[2] found that successful executives made substantially higher scores on tests of health and drive, judgment of fact, reaction to human qualities, and leadership than did salesmen, technical workers, and college students. Clearly many traits and abilities are important for success in high-level administration.

## FINDING ONE'S PLACE IN ADMINISTRATIVE WORK

Obviously, not all men should aspire to become high-level administrators. Many will find their greatest usefulness and satisfaction in serving in staff positions with responsibility for the development of certain technical phases of forestry.

Even men who aspire to leadership should realize that administrative posts are few and competition keen. Many fortuitous circumstances may stop their career short of their life's ambition. In both public and private careers, men in advanced administrative positions may sometimes be called upon to take a courageous position that antagonizes powerful people who can block their advancement. There are as rich satisfactions to be had from being a member of an effective team as there are from being its leader.

## CAREER DEVELOPMENT

Opportunities to perform high-level administrative work are obviously more abundant in large than in small organizations. Many organizations that follow a practice of promotions from within establish career ladders for aspirants to high-level administration. Certain steps on such a ladder can rarely be omitted. In the U.S. Forest Service, for example, those of district ranger, national forest supervisor, and either regional forester or deputy chief in the central office are considered of key importance. The objective is to give potential administrators broad experience in posts where their mistakes will cause little damage to the entire organization.

Opportunities for advancement in administrative work are greater in organizations that are growing rapidly, in organizations that handle a wide variety of tasks rather than solely routine work, and in those that are concerned primarily with action programs rather than with research.

Young men desiring to advance in administrative work should follow recognized rules of effective conduct. Loyalty to the organization and to superiors and subordinates is indispensable. Friendly consideration for co-workers, and readiness to accept responsibility and to discharge it fully will gain a reputation with superiors. Participation in nonofficial tasks and community service, if important, helps to develop leadership.

A planned course of study can be helpful in acquainting an aspirant administrator with subject areas that may prove valuable to him. Postgraduate study in public or business administration forms a good beginning. Acquaintance with the broad principles of operations research will open the way toward use of this potentially powerful tool in exploring the possible outcomes of different courses of action.

Readings in these fields and in those that broaden one's sensitivity to human qualities, sharpen judgment of fact, and analyze qualities of leadership, will probably prove helpful.

Men who cannot or will not accept heavy responsibilities are unfitted for top-level forest administration. However, for those who are willing to devote themselves to the objectives for which their organization is working and who are temperamentally fitted to lead and to carry the load, it is the field that offers the largest monetary compensation and a rich opportunity for worthwhile service.

**LITERATURE CITED**

1  American Management Association. 1957. Leadership on the Job: Guides to Good Supervision, New York. 303 pp.
2  Cleeton, Glenn W., and Charles W. Mason. 1946. "Executive Ability: Its Discovery and Development," The Antioch Press, Yellow Springs, Ohio. 540 pp.
3  Gulick, Luther. 1937. Notes on the Theory of Organization, Papers on the Science of Administration, Institute of Public Administration, Columbia University, New York. 195 pp.
4  Kaufman, Herbert. 1960. "The Forest Ranger, A Study in Administrative Behavior," The Johns Hopkins Press, Baltimore. 259 pp.
5  Simon, Herbert A. 1957. "Administrative Behavior," The Macmillan Company, New York. 259 pp.
6  Simon, Herbert A. 1960. "The New Science of Management Decision," Harper & Brothers, New York. 50 pp.
7  Smith, Howard. 1946. "Developing Your Executive Ability," McGraw-Hill Book Company, New York. 225 pp.
8  Tead, Ordway. 1935. "The Art of Leadership," McGraw-Hill Book Company, New York. 308 pp.

Part Five

# Forest Products

Chapter 15

# Wood: Its Nature and Uses

Men throughout the ages have expressed their admiration of trees in literature, art, and music. Shade and ornamental trees that embellish the environment of homes, villages, and cities seem to acquire an individuality. When such trees die or have to be removed men accustomed to their presence are saddened. Naturalists and scientists have shown their appreciation by trying to understand the life processes of trees. Understanding the nature of a tree can serve as an introduction to understanding wood, for woody tissue is what gives a tree its dominant place in the plant world.

## THE NATURE OF A TREE

A tree's main stem, branches, and twigs support its food-manufacturing leaves above those of competing vegetation and channel water and soil nutrients to them. The roots serve as anchorage and as absorbers of water and minerals. The leaves use the energy of sunlight to synthesize cellulose and lignin that forms the basic structural material for roots, stems, and branches. Such are the general functions of the primary elements of the tree. But how these functions are performed is a much more intricate story that reveals increasing complexity as our knowledge expands.[2]

Stems and branches need to be flexible enough to sway with the wind without breaking, yet stiff and resilient enough to return to their original position when the wind passes. When lower branches, due to shading, no longer produce enough food to form fresh sheaths of tissue extending back to the main stem, they die, become dry and brittle, and are easily broken off.

The tree stem must be vertical and the crown weight symmetrically distributed lest the tree fall of its own weight. Also, the wood needs to be distributed along the stem in such a way as to provide the greatest resistance to the forces of wind and gravity. The shape of the bole, the flare at root collar, and at major branches, are the result of growth responses to the mechanical stresses that must be withstood. That trees modify their stem structure in response to stress can be experimentally demonstrated. By what mechanism they do so is unproven. It has been suggested that stress causes piezoelectric gradients in the crystalline cellulose molecules of developing cell walls that may cause the flow of the growth-stimulating auxin and photosynthetic products from the crown to the places of greatest stress.[4]

If a tree leans it will tend to right itself by developing short, heavy-walled cells known as compression wood on its lower side and cells with gelatinous walls called tension wood on its upper side. Because wood is stronger in tension than in compression, the leaning tree becomes oval in cross section. Lumber sawed from such trees tends to shrink longitudinally, causing it to bow or twist upon drying.

Efficient use of cellulose and lignin is attained in the woody stem by building hollow, rectangular, elongated cells cemented snugly to adjoining cells by lignin. This imparts a high degree of stiffness.

The woody stem and branches must also serve as pipelines to transport abundant water and minerals to the leaves. In conifers this function is performed by the hollow supporting cells, the tracheids. In broad-leafed trees there is a separate water-conducting system made up of tubular vessels. The vessels are thin-walled and so must be supported by thick-walled woody fibers.[2]

By use of three abundant elements—carbon dioxide, water, and sunlight—the tree develops a marvelously complex structural material—*wood*. If nature did not supply this in great abundance, man himself would probably seek to develop a product of similar properties. A simply written, beautifully illustrated description of wood, its properties, and its uses, by Harlow, appeared in 1970.[2]

## USEFUL PROPERTIES OF WOOD

A modern textbook on wood technology lists nine ways in which the physical properties of wood equal or excel those of other structural materials:

1  Wood can be fashioned into various shapes by simple hand tools or machines.
2  It can be joined with nails, screws, bolts, and connectors, using simple tools. It can be fastened with adhesives to produce a joint as strong as the wood itself.
3  Wood has high-dimensional stability to temperature changes, and in the direction of the grain to moisture changes. As plywood it can be made stable to moisture changes in its two principal directions.
4  Wood is highly resistant to corrosion by sea water and weak solutions of acids or alkalies.
5  Wood does not corrode. If kept dry it endures indefinitely.
6  Although it is combustible, wood in large sizes heats up slowly and loses its strength gradually when exposed to fire. By contrast, a steel beam of comparable load-bearing capacity heats up quickly and uniformly and collapses suddenly at temperatures that generally prevail in burning buildings.
7  Wood has excellent insulating properties against sound and heat. When dry, it also has excellent insulating powers against electricity. Compared with wood the heat conductivity of brick is 6-fold as much; of glass, 8-fold; of concrete, 15-fold; of steel, 390-fold; and of aluminum, 1,700-fold.
8  Wood has excellent rigidity and strength, weight for weight, compared with other construction materials.
9  Wood has intrinsic beauty in grain, color, and texture, each in infinite variety, for no two pieces of wood are exactly alike in appearance.[6]

Aside from these advantages of wood for structural purposes, it has additional useful features and qualities:

1  Wood is well distributed, generally abundant, and relatively cheap throughout the moist temperate and tropical regions of the earth.
2  It has an absorbent surface suitable for holding glue, paint, and stain.
3  Its elasticity is such that it is suitable for long bows and for stringed, percussion, and woodwind musical instruments.
4  Wood has high shock resistance that makes it the preferred material for railway ties, flooring, hammer and other tool handles, golf club drivers, shuttles and picker sticks used in weaving, and other items that require shock resistance.[3]

5   Wood is tough rather than brittle, though this quality varies over a considerable range with growth conditions and species. Under stress wood tends to give warning by sagging, cracking, and splintering before breaking. This quality is highly valued where wood is used as a supporting structure in mines and elsewhere.

6   Wood is readily disposable by burning or decay when it is no longer useful.

### DISADVANTAGES OF WOOD

Among wood's disadvantages are:

1   The presence of knots, spiral grain, tension or compression wood, checks, and other defects that lead to its overspecification in use.

2   Its tendency to shrink and swell across the grain with moisture changes.

3   Its tendency to split along the grain.

4   Its rheology—tendency to flow under prolonged pressure—that causes stressed joints and striking tool handles to loosen.

The properties of wood vary widely from species to species, within species by genetic strain, and by the growth conditions that prevail in the habitat, including the competition of associated trees. Wood properties can therefore be modified substantially by silvicultural treatment. Generally speaking the silvicultural practices that tend to improve growth, such as thinning, pruning, and fertilizing, tend also to improve wood quality for structural use. This is achieved by maintaining relatively uniform growth rates and by favoring high-density wood. The relationship is complex and involved.[1, 4]

### HOW WOOD IS FORMED

Wood is made up of tiny cells that are formed in the two actively growing regions of the tree: the tip regions of root and stem, and the cambial region. Elongation of stems and roots occurs only at the tips. Here, behind a cover of protecting tissue, lies the apical meristem, or region of dividing cells, and here are formed the cells responsible for all increase in length of stem and root. The apical meristem gives rise to all cells that make up the primary stem tissue. Buds that develop into branches, leaves, or flowers are formed at the tip.

The apical meristem also gives rise to the cells that form the cambium a short distance back from the growing tip. The cambium is also a region of rapidly dividing cells. It forms between the primary xylem (wood) and phloem (bark) in the newly formed stem and gives rise to all secondary tissues found in the mature tree stem. The cells cut off on the inside of the cambium become woody, thereby giving strength to the main tree stem. Those cut off on the outside develop into the bark that protects the cambium and main stem and that conducts fluids. Near the outer surface of the young stem is another bark-forming region called the cork cambium, which forms corky tissue. As the stem grows, the cork cambium is pushed outward, ruptures, and dies. New cork cambia are formed inside by young bark cells. This process is repeated continually as the tree grows.[1, 5] The expansion of the main woody cylinder causes rupture of the outer bark, which eventually is sloughed off.

The wood on the tree stem is laid down each year in a sheath that extends from the root tips to the stem tips. This sheath completely surrounds the stem and, once formed, becomes lignified (woody) and develops into permanent tissue. Except during the year of its formation, the sheath does not itself increase in diameter or length. Each year a new sheath is formed on top of the old and thus the tree increases in girth. Each year also the actively growing stem tips form new cells that extend the height of the stem. Each new sheath is longer than its predecessor by the amount of total elongation that occurred during its year of formation.

## GROSS STRUCTURE OF THE TREE STEM

The mature tree stem from the center out contains a pith, a region of soft spongy cells that has little lignification and therefore adds little to the strength of the tree stem. Outside the pith are the various sheaths of wood. When the stem is cut in cross section, we see these sheaths as rings, called growth rings. They extend from the pith to the cambial region that lies between the wood and the bark. Outside lies the bark, which may be made up of alternate layers of corky and fibrous tissue. Bark tissue in itself adds little to the strength of the tree stem.

If a small portion of the tree stem is broken, it tends to splinter, for the main elements of the stem are much longer than they are thick and are arranged parallel to the tree stem. This gives the tree stem its strength to resist bending and breaking as it sways in the breeze, or bears its burden of leaves in summer, or snow and ice in winter.

Between the pith and the bark there is often one zone of darkly colored wood and another of lightly colored wood. The dark wood is spoken of as heartwood and the light-colored wood as sapwood. This difference in color is due in part to the deposition of various chemical substances in the heartwood of the tree. The light-colored sapwood forms the main channel for movement of water and minerals up the tree trunk to the branches and leaves. Being highly permeable to water, sapwood on posts and other wood products exposed to attack by fungi decays readily. The dark-colored heartwood of most trees conducts little water, because the conducting vessels are either plugged by thin membranes, called tyloses, as in white oak, or impregnated with resin, gums, or other substances that impede water movement, as in longleaf pine. The heartwood functions chiefly as a supporting column for the crown. Generally it is considerably more decay-resistant than sapwood.

A cross section of the tree stem may also reveal a series of fine lines extending radially from the bark toward the pith. These are called rays. They are very prominent in oaks and inconspicuous in aspen and, therefore, help one to identify kinds of wood. The ray cells serve as a channel for movement of materials across the growth rings and as storage cells for reserve foods. Their function will be better understood after a description of the cellular structure of the tree stem.

### Microscopic Structure of Wood

*The Cellular Structure of a Coniferous Stem*  When a very thin section of a coniferous woody stem such as white pine is prepared and examined under the microscope, the cellular makeup of the stem becomes evident (Fig. 15-1). With low-power magnification the differences between springwood and summerwood that mark the annual rings of temperate-zone woods become evident. The cells formed in the spring are large and have large central openings; those formed in the late summer are flattened and have narrow openings in the center. A cross-sectional view will reveal here and there large channels lined by thin-walled cells. These are resin ducts that transport the resin up, down, or across the tree stem. The major-supporting elements of the woody stem are tracheids, which are small- to medium-sized, heavy-walled cells. When these cells are viewed in a section cut in a radial direction, it can be seen that they have small circular markings, known as bordered pits.

Running at right angles to the tracheids are the ray cells. These consist of ray tracheids and thin-walled ray cells. Since the thin-walled ray cells are weaker than tracheids, they are responsible for the wood's splitting along the line of a ray. Other types of cells in a pine stem are those that secrete resin. These are found both in the rays and in the vertical stem tissue.

The tracheids are no longer living cells nor do they perform any living function. Many of the thin-walled cells may also be dead, but those that secrete resin are living. Other living cells enter into the storing and transformation of foods manufactured in the tree leaves.

*The Cellular Structure of Hardwoods*  The term "hardwood" is applied loosely to all forest trees that do not belong to the group known as

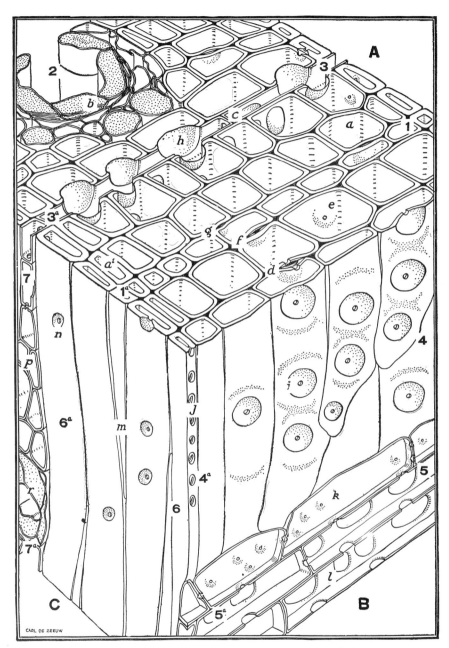

**Figure 15-1** Schematic diagram showing cellular structure of northern white pine wood. Face A, a cross section showing a resin canal 2, tracheids a. Face B, radial section. Face C, tangential section. *(From Brown, Panshin, and Forsaith, "Textbook of Wood Technology," vol. 1.)*

conifers. A more appropriate name is porous woods, because these stems, in contrast to those of the cone-bearing trees, are made up of small wood fibers with very narrow cavities and of large thin-walled connected tubes that serve as the main conductive channels for movement of water and other materials up and down the tree stem. These show up as pores and are very conspicuous on a cross section of the wood. Their presence can easily be demonstrated by taking a short length of red oak or some other stem, placing one end in water, and blowing on the other end. A series of minute bubbles will disclose the passage of air through the pores.

The hardwoods are divided into two groups: the ring-porous group and the diffuse-porous group. In the ring-porous species the springwood pores are very large, whereas the summerwood pores are small. In the diffuse-porous woods the difference in size of pores is far less noticeable. When sections of the hardwood stems are examined under the microscope, it is found that the wood is made up of long, heavy-walled wood fibers that bear pits, vessel segments that serve as the conducting members for transport of water, and other thin-walled cells some of which serve as gum-canal cells and others of which enter into the transformation and storage of food (Fig. 15-2). Rays may be very broad and conspicuous or narrow and inconspicuous. They may be made up entirely of thin-walled cells or of radially oriented thin cells and vertically oriented tracheids. As in the conifers, they serve to transport material across the tree stem. The living cells within the main woody stem transform plant foods, changing sugars into starches and, later on, starches into sugar. In this way food stored during late summer can be utilized during early spring for rapid growth of new plant tissue.

**Wood Identification** The various types of cells, their shapes, and their arrangements serve as features for identifying wood. Most common woods can be readily identified by the gross features that are easily discernible to the naked eye or under a hand lens. A few woods can be identified only by a study of their microscopic structure. Here, the types of cells that are present, their markings and shapes, the number and types of pits, and the presence or absence of gums, resins, and various other features in the wood have definite value in deciding upon the kind of wood in question.

**The Structure of the Woody Cell** The electron microscope reveals an even more fascinating structure than can be seen by naked eye, hand lens, or microscope. This instrument enables us to study the structure of the cell itself and the way the cell-wall layers are built up. The cell wall consists of three layers: the middle lamella, a layer of relatively weak tissue between the cells that is made up largely of lignin; the primary cell wall, a relatively thin layer that is first laid down in cell formation; and the

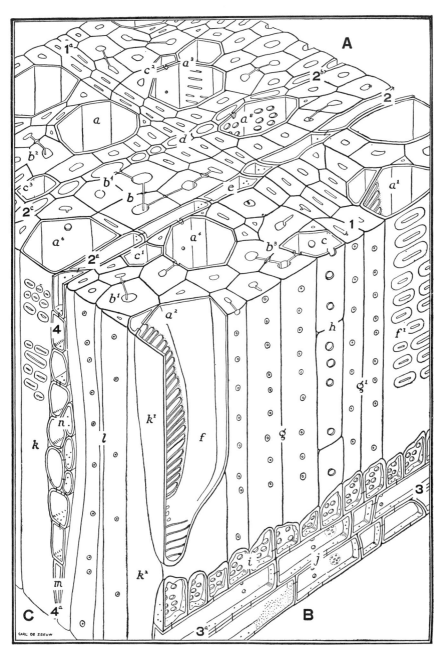

**Figure 15-2** Schematic drawing showing cellular structure of red gum. Face A, cross sections showing vessels a, fiber tracheids b. Face B, radial section showing ray 3 and wood parenchyma h. Face C, showing ray cells 4 and fiber tracheid 1. *(From Brown, Panshin, and Forsaith, "Textbook of Wood Technology," vol. 1.)*

secondary cell wall that is usually quite thick and forms the main supporting layer (Fig. 15-3). A warty layer is sometimes found on the inside of the secondary cell wall. The cell walls themselves are found to be made up of microfibrils that are wound around the cell in a spiral manner. These microfibrils may branch and rejoin or join other microfibrils. Hence, the cell itself is tube-shaped with walls composed of netted, spirally wound tissue of great complexity. The microfibrils are cemented together with lignin, thus making an amazingly complex, tough, and stiff system (Figs. 15-4 and 15-5). The microfibrils are made up of longitudinally oriented cellulose molecules that are too small to be revealed by the electron microscope.[1]

**The Physical Properties of Wood** The strength, hardness, toughness, shrinking and swelling, and other physical properties of wood

**Figure 15-3** Schematic drawing showing the structure of the woody fiber. *(Courtesy Modern Materials. Edited by Henry H. Hausner. Academic Press, Inc. 1958.)*

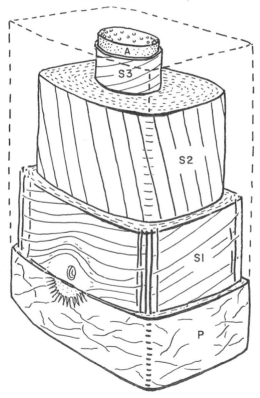

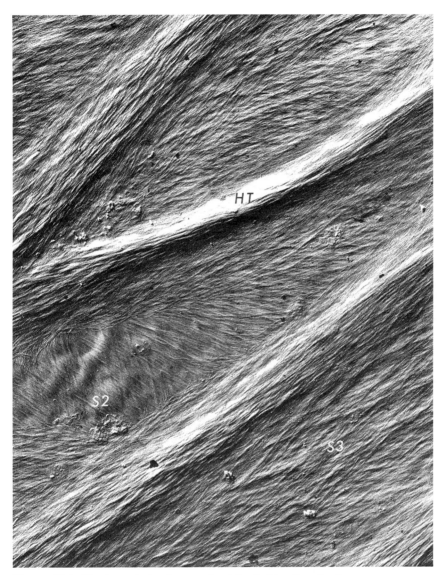

**Figure 15-4** Electron photomicrograph of tracheid cell wall of Douglas-fir. P-primary wall. S1, S2, and S3 are outer, middle, and inner layers of secondary cell wall. HT helical thickening of wall. *(Courtesy of Wilfred Côté and University of Washington Press.)*

depend to a considerable extent upon the type of wood and its microscopic and macroscopic structure. Balsa wood, for example, has thin-walled fibers, making it soft, light, and easily worked. Because of the

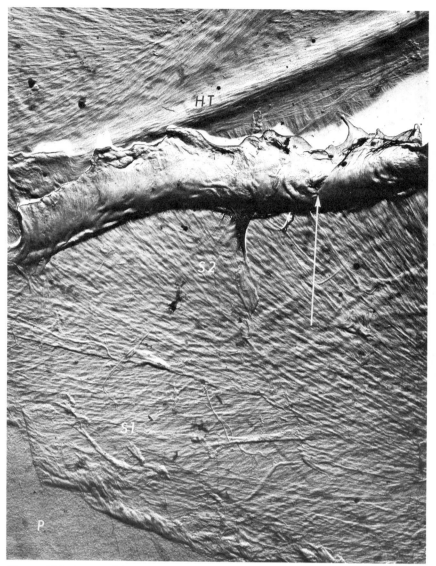

**Figure 15-5** Lumen lining of Douglas-fir tracheid showing helical thickenings and crisscross arrangement of microfibrils. *(Courtesy of Wilfred Côté and University of Washington Press.)*

lightness, stiffness, and high degree of porosity it is especially valuable for certain types of construction and insulation. Balsa wood is excellent as a core material in sandwich construction, as it holds glue well, is stiff, and is exceedingly light.

In contrast, hickory and oak are heavy. They have large pores, but they also have a high proportion of dense wood made up of heavy-walled fibers that make the wood hard and strong. Some kinds of wood are very straight-grained. White pine and redwood are good examples. Others may be spiral-grained or have interlocked grain as elm does. In the latter case, part of the woody elements may be laid on in a clockwise spiral direction and other layers in a counterclockwise direction. Such interlocked grained woods resemble plywood in that they are exceedingly difficult to split. A tough wood is one that will bend without breaking. Ash is noted for its toughness, and because of this it is used for making tennis rackets, snowshoes, handles, ball bats, and other articles requiring bending and shock resistance. Basswood, cedar, and redwood break quickly upon bending and hence may be called brittle woods. Woods that are strong in both compression and tension are those that are dense.

The primary woody substances, cellulose and lignin, are essentially the same for hardwoods and softwoods of all species. The major differences that occur in the chemical composition of woods are due to the gums, resins, minerals, and other material that may permeate the wood but do not enter directly into its cellular structure. Certain woods, such as aspen, basswood, and white pine, are relatively free from gums and resins. Longleaf and other southern pines, and particularly lignum vitae, are heavily impregnated with gums and resins that increase weight, add to stiffness and brittleness, and give them special properties. Lignum vitae is a standard wood for stuffing boxes in steamships.

Not only does wood vary with species of tree, but also within the individual tree and from one tree to another, depending upon growth conditions. Rapidly grown pine tends to be more brash or brittle than pine that has grown at a moderate rate and is therefore little suited for pattern and cabinetwork. Rapidly grown hickory, ash, and maple are harder than slowly grown wood of the same species and are consequently used where high strength and high resistance to shock are important. For a great many uses no specifications as to the number of rings per inch are required, but for special high-grade uses, such as for aircraft structural material, this becomes of major importance. A rough measure of the strength of wood can be obtained from its specific gravity.

The hardness of the wood influences its tendency to split upon nailing. Very hard woods, such as maple, ash, and hickory, are difficult to nail without splitting. Soft-textured woods, on the other hand, such as basswood, white pine, and aspen, are easily nailed and have only moderate tendency to split.[10]

All woods when dry have good heat- and electrical-insulating properties. Most gums and resins found in woods are themselves nonconductors of electricity. Some kinds of wood have a relatively high ash content.

These when wet will conduct electricity. For this reason their moisture content can be estimated by means of an electrical moisture meter that measures the resistance between two electrodes when these are pushed into the wood. The open pores and thin-walled tracheids that are common in wood give it excellent heat-insulating qualities.

A property of wood that causes considerable difficulty in use is its tendency to shrink and swell with change in moisture content. Water occurs in wood as free water in the lumina of the cells and as absorbed moisture within the cell walls. The free water can be evaporated from the wood without causing any pronounced shrinkage. The point at which all moisture has been withdrawn from the cellular spaces and yet cell walls are saturated is called the fiber saturation point. This occurs at about 25 to 30 percent moisture content. As the wood dries below the fiber saturation point, shrinkage occurs and continues to occur until the wood is thoroughly dried. Shrinkage across the grain exceeds many times that along the grain. Equilibrium moisture content is that at which the wood neither gains nor loses moisture when surrounded by air of a given relative humidity and temperature. Wood that is to be used for furniture must be dried to about a 6 percent equilibrium moisture content; that to be used for exterior walls, to about 12 percent. The tendency of wood to shrink or swell can be largely prevented by impregnating the wood with certain chemicals that will hold it in swollen condition. If chemicals having the capacity to swell wood are then polymerized, they form a plastic substance with the wood that maintains it at the swollen dimension.[5, 8]

The high affinity of cellulose for water not only causes wood to shrink and swell but also results in lowering the strength properties with increases in moisture content. The lignin fraction of wood is a plastic substance that can be softened at elevated temperature. Wood heated with steam, or even dry heated, to above 100°C can be bent into various shapes and will retain this bent form. Heating wood also reduces its stiffness. If wood is compressed or held in bent condition for a long time, it tends to hold its new shape because of lignin flow.[3]

## PROPERTIES IN RELATION TO USE

The wide use of wood is due largely to three factors—the intrinsic properties of wood as a material, its widespread availability, and its cheapness in relation to other products of comparable utility. These have made it the dominant material in America for one-family residences and other light construction. It is also the chief material used for railway ties, poles, piling, posts, concrete forms, and temporary structures. A recognition and understanding of use requirements in relation to wood properties can result in much economy in use.[9]

Since cellulose and lignin are essentially the same for all wood, it is the way in which these substances are distributed in the woody cells that largely determines wood properties. Because wood substance is laid down in the developing cell walls in response to local stress on the cells, wood formed in the main stem tends to be denser than that formed in the upper crown and in the roots. Also wood formed in the spring, when water and food requirements for developing leaves and new stem tissue are high, tends to have much thinner-walled cells with larger lumens than that formed in late summer. A youthful tree with plenty of root and crown space will form wide growth rings. After it becomes crowded above or below ground the rings become narrow. Wood properties within a tree, therefore, vary with location, with season of formation, and from year to year, depending on light, moisture, and nutrient supply.[2, 4] Growth rates also vary from species to species and within species with varieties in genetic makeup. The properties of wood profoundly affect its suitability for different uses. For some purposes almost any wood will do. For others only the most highly selected species and qualities are suitable.[3, 6, 7, 9, 10]

**Fuel Wood**   All dry, untreated wood will burn and give off heat. This is fortunate, for worldwide more wood is used for fuel than for any other purpose. The heating value varies directly with dry weight. A cord of light-structured wood such as aspen weighs but half as much and gives but half as much heat as a cord of shell bark hickory. Woods with high resin content such as southern pines are better heat producers per unit weight than white pine that is low in resin content. In the United States fuel wood is largely a luxury product, burned in fireplaces where most of the heat produced escapes up the chimney. A feature to be desired for fireplace wood is that it not rupture with explosive violence, throwing sparks over the living room carpet. Species differ widely in their tendency to throw sparks. Arbor vitae and locust behave as the wildest incendiary, whereas dry hard maple and birch are as docile as the family cat that sleeps by the hearth.[2]

**Structural Use**   Strength, stiffness, and ease of fabrication are the three most desirable qualities sought in wood for structural use. Hardness is also important for flooring. Poles and piling are used in the round, thereby taking advantage of the full strength of the tree stem. Sawn wood should retain its straightness, hence species that have a tendency to warp, twist, or bow are less desirable than those that hold their shape upon drying. Ease of nailing is a desirable trait in wood used for many construction purposes. Woods such as hickory that tend to split on nailing are less desirable than the conifers. This does not prevent using hard woods for posts, or for beams where nailing is not required.

For some construction uses, such as wall paneling and sheathing, almost any species except the extremely light-weight woods will be strong and stiff enough once they are fastened to studding or rafters. For insulation the very light woods, such as balsa, are preferred.

Durability is desired for wood used in contact with the ground, such as poles, piling, posts, and railway ties. Some woods have natural tannins, oils, and resins that make them resistant to wood-destroying organisms. Locust and cedar are favored where wood is to be used in the ground untreated. If the wood is to be treated with creosote or other preservatives, ease of treatment becomes more important than natural durability.[7, 8, 10]

**Esthetic Uses** For interior trim, furniture, exposed panels, floors, and many other such items, wood is selected on the basis of attractive grain pattern, color, and other decorative features. The most attractively patterned wood is usually cut into veneer so that it can be used to cover as much area as possible. Striking patterns can be obtained by cutting through crotches in roots and branches. For furniture legs and other structural members, strength as well as appearance must be considered. Walnut, maple, mahogany, teak, and many other species have wood of lustrous sheen, pleasing grain pattern, good dimensional stability, and strength that make them favored for furniture and other decorative use.

**Baskets and Veneer Containers** Baskets and many veneer containers need to be made of woods that will withstand bending and shaping without breaking. Ash, hickory, and beech are widely used. Elm veneer also has been used for wood-stiffened containers. By 1970 multiwalled corrugated board containers were replacing the veneer container for many uses.

**Athletic Equipment and Tool Handles** Wood for handles of striking tools and for hockey sticks, baseball bats, tennis and lacrosse rackets, as well as for many other types of athletic goods should be highly selected from the best quality of ash, hickory, and other species noted for high strength, toughness, and shock resistance.

**Musical Instruments** Among the most demanding uses of wood is that of the manufacture of harps, piano sounding boards and actions, violins and other stringed instruments, marimbas and xylophones, and the woodwind instruments. The quality of sound emitted depends upon both the absorptive and vibrating qualities of the wood used. Great care is taken in the selection of wood for musical instruments and top prices are paid. A special grade of spruce logs is even known as fiddle butts.

***Miscellaneous Uses*** Wooden pencils require a straight grain and a fine, uniform texture for ease of sharpening. For intricate wood-turning, even-textured woods such as birch and maple are favored over the ring porous hardwoods such as oak and hickory, even though these are used for handles. Ease of machining and painting are of prime importance for wood for toys. A reasonable amount of strength and toughness may also be needed if the toy is to get rough use. For artificial limbs a light, even-textured wood such as willow is preferred.

The above are but a few examples of selecting woods suitable for special uses.

## LITERATURE CITED

1 Côté, Wilfred A. (ed.). 1965. "Cellular Ultrastructure of Woody Plants," Syracuse University Press, Syracuse, N.Y. 603 pp.
2 Harlow, William M. 1970. "Inside Wood," American Forestry Association, Washington, D.C. 120 pp.
3 Kollmann, Franz F. P. and Wilfred A. Côté, Jr. 1968. "Principles of Wood Science and Technology," Springer Publishing Co., Inc., New York. 592 pp.
4 Larson, Philip R. 1969. Wood Formation and the Concept of Wood Quality, Yale University, School of Forestry Bulletin No. 74, New Haven, Conn. 54 pp.
5 Panshin, A. J., Carl deZeeuw, and H. D. Brown. 1964. "Textbook of Wood Technology, Vol. 1, Structure, Identification, Uses and Properties of the Commercial Woods of the United States," 2d ed., McGraw-Hill Book Company, New York. 643 pp.
6 Panshin, A. J., and Carl deZeeuw. 1970. "Textbook of Wood Technology, Vol. 1," 3d ed., McGraw-Hill Book Company, New York. 705 pp.
7 Panshin, A. J., E. S. Harrar, J. S. Bethel, and W. J. Baker. 1962. "Forest Products, Their Sources, Production and Utilization," 2d ed., McGraw-Hill Book Company, New York. 538 pp.
8 Stamm, Alfred J. 1964. "Wood and Cellulose Science," The Ronald Press Company, New York. 549 pp.
9 Tarkow, Harold. 1970. Properties and Behavior of Wood, *Journal of Forestry,* **68**:408–410.
10 U.S. Department of Agriculture, Forest Products Laboratory. 1940. "Wood Handbook," Washington. 326 pp.

Chapter 16

# Wood Processing and Secondary Products

Wood is used in three main forms—in round form, cut only to desired lengths; as sawn and mechanically shaped wood; and as wood pulp. Processing wood for pulp and use of wood pulp will be dealt with in the following chapter.

Because wood is laid down in response to stress in the tree stem, the full load-bearing strength and resistance to the shock of the pile-driver can best be realized by using posts, poles, and piling in round form. Timbers to be used in general construction must be joined to one another, a process difficult to perform with round timbers, to make tight walls and achieve strong joints. Wood shaped to uniform dimensions is much more suitable for most uses.

Wood may be shaped by hewing, sawing, planing, turning, and carving. Hand-hewing of timbers and railway ties, though widely practiced up to the 1920s, has disappeared from most industrial nations. It could still be found in use in remote regions of developing countries in the 1960s. Hewing is a slow, laborious process requiring skill to "hew to the line" literally and to shape parallel sides. It is also wasteful of wood. A sawmill performs this task with speed, ease, and precision.

## LUMBER MANUFACTURE

Lumber manufacture is the process by which round logs are converted into square-edged pieces of uniform width and thickness. These pieces, if 2 inches or less in thickness, are called lumber; if over 2 inches, planks, deals, or squared timber. The terms are used loosely. In simplest form lumber may be cut from logs using a whipsaw or sash saw pulled by men. In modern sawmills, many power-driven machines and processes are used to complete the conversion. It may take days to convert a large hardwood log to lumber by hand-sawing, but only minutes are required in a modern sawmill.

**The Sawmill** Lumber manufacture includes all operations that take place from the time a log is unloaded at the sawmill until the lumber, either seasoned or freshly cut, finished or rough, is loaded for transport to the purchaser.

At a modest-sized conventional sawmill of 1955, logs brought in by truck would be unloaded into the log pond whence a chain conveyor would bring them up to the sloping log deck. From the deck the logs would be rolled by hand or mechanically onto the saw carriage and clamped against "head blocks" with the bad face down. The headsaw, be it either circular or a band saw, would have swaged teeth (teeth broadened at the tips). As the log on the carriage moved past, the headsaw would remove first a slab and next one or more 1-inch boards. The head sawyer would recover as much lumber as possible in wide, high-grade boards. He did this by deciding which face to cut first, when to turn the log on the carriage, and what thickness to cut. The rough-edged boards were sent through parallel saws to cut square edges, trimmed to even lengths, and conveyed to the grading and sorting table. Slabs and edgings were cut up for fuel wood or sent along with the sawdust to a burner.

The 1955 sawmill cut some 6,000 to 12,000 board feet of hardwood lumber per day, or up to 18,000 feet of softwood. It employed 6 to 13 men, held 1 to 2 million board feet of lumber in the yard both for seasoning and in order to accumulate carload lots of a single species grade or dimension. The owner had some $40,000 or more invested in equipment and plant, and perhaps more in inventory of lumber. It differed little from the mill of a comparable size of the 1920s in design, equipment, and output per man-day of lumber (Fig. 16-1).

But portents of change were discernible. The millman's one great

**Figure 16-1** Ponderosa pine logs moving up the jack ladder to the sawmill, Graegle Lumber Company, Graegle, Calif. *(U.S. Forest Service.)*

problem had been to sell his low-grade lumber that accumulated in the yard. Meanwhile logs were getting smaller in size and lower in quality. Labor was getting scarce, more expensive, and less skilled. The market for low-grade hardwoods was picking up due to the growth of the pallet industry. And a market for slabs and edgings was developing, if they were debarked and reduced to pulpwood chips. Progressive operators were beginning to debark their logs and install pulpwood chippers. Some of the more thoughtful millmen, together with research men and machinery manufacturers, began asking why they should saw off slabs and edgings, thereby wasting the wood in sawdust that might be sold to the pulpmill as chips. Others had become familiar with the Finnish and Swedish sash gang saws that cut a narrow channel and reduced modest-sized logs to lumber in two passes through the saw. They asked themselves if it really paid to saw small-sized logs for grade by repeated passes through the head saw.

Acting on this thought, special sawmills were designed that could reduce the logs to lumber in one or two passes through a gang or tandem saw system. Such mills were used to convert small, short, or low-grade logs into such products as 2 × 4 studding and pallet lumber in which grade was of minor importance. Further refinements led to removing both slabs and edgings in the form of pulpwood chips in a single pass, and running the resulting "cant" through a resaw or a gang saw. Some mills do the complete operation in a single pass through a combined machine. Such machines, called high-speed chipping headrigs, were being installed by progressive mills throughout Canada and the United States in the late 1960s.[12]

The advantages of the new system are compelling. Less sawdust goes to the burner, so air pollution is reduced. The amount of pulpwood chips produced is increased. The labor and risk of injury in handling slabs and edgings is eliminated. Mill conveyors can be simplified. Production of lumber per man-hour rises sharply. Gross value of total products increases markedly. Less labor and fewer labor skills are required. Net income before taxes has shown increases from 4 up to 40 percent under good operating conditions.

The system does have its disadvantages. It does not work well on logs with large butt swell or crooks. Knots are not always chipped cleanly. And the percentage of high-grade lumber out-turn may be lowered. But the savings generally outweighed the disadvantages by far. Moreover the use of such machines is encouraging sawmill owners to buy lower-grade logs and to accept them in tree lengths on a weight or tree-length scale. Logs can then be cut to the most desirable length at the mill rather than at the stump or at the log landing in the forest. Material not suited for the sawmill can be diverted directly to a log chipper and the good sections can be run through the chipping headrig or, if they are of high quality, through a band mill or sent to a veneer mill. The point of control is thus moved closer to the point of final operation.

Even more startling developments are under way. At a mill in Sweden each log is sorted automatically into size classes and sent through appropriate debarkers. It then passes through a metal detector and on to an X-ray chamber that enables an operator to estimate quality and feed this information to a computer. The computer programs the log through the mill. The log proceeds automatically to a log-turning device where a second operator turns it for best sawing position. It is automatically scaled for diameter and length and this information is sent to the computer, which combines these figures with the other information already fed into it to program the log through the band saws, and the resulting lumber through automatic sorters and on to the automatic stackers.[2] Output is high and labor requirement low.

Stress grading of softwood lumber by use of a computer has been developed and used in the United States and Canada. Experiments were underway in 1970 to develop a method of sensing clear cuttings that could be obtained from hardwood lumber and so apply automatic grading.

Emphasis above has been placed on the task of sawing small-sized logs because millmen will be using these in increasing amounts in the future. The mills developed on the West Coast to handle the large logs available there are highly mechanized, turn out high volumes of high-grade products per man-hour, and hence are in a position to maintain their advantages without the pressure for output that the small-log mill operator faces.[10] The western mills are taking full advantage of the high-speed chipping headrigs described above for use on their small-sized logs.

**Lumber Seasoning** Freshly sawn lumber is a perishable commodity that must be properly seasoned to prevent degrade or decay. If sawn in the summer it may first be dipped in a solution to impede sap-staining organisms. It is then carefully piled, using "stickers" to separate the courses for free air circulation. The piles may be air-seasoned in the yard, or transferred to kilns in which air temperature, humidity, and movement are controlled. All lumber to be used in heated buildings should be kiln-dried to prevent further shrinkage. Special kiln schedules of temperature and humidity are maintained for different species and thicknesses of lumber. Kiln-drying takes from 1 to 4 weeks. Air-drying may take a minimum of 3 months of summer weather. Defects to be guarded against during kiln-drying include warping, checking, splitting, case-hardening, and internal collapse. Air-seasoning proceeds so slowly that if properly piled the lumber will season with little chance of defect, although it may still require some time in a kiln for final drying.

**Lumber Finishing** Most softwood lumber is finished at the sawmill by running it through a surfacer. These machines plane it to smooth, flat, parallel sides and edges. They may also form it into shiplap, tongue and groove, lap siding, drop siding, or many other decorative forms. Flooring is manufactured by ripping rough lumber to appropriate widths, then running it through a flooring machine that "matches" surfaces and ends by tongue and grooves. Various moldings and other small-sized products were formerly manufactured from slabs and edgings at the sawmill, but with the growth of chipping such material for the pulpmills this function is being transferred to planing mills as such.

**Lumber Handling**  Because lumber is both bulky and heavy, mechanical handling is essential to keep costs down. In the sawmill this is done largely by conveyors and rolls. In the yard, straddle trucks and lift trucks are used. These require good travel surfaces for efficient use. Automatic stackers as well as other transporting equipment are in use at large mills. Small-sized lumber handling equipment is used in retail outlets.

## COMPOSITE WOOD

**Veneer**  Wood that is cut in sheets of thicknesses of $1/4$ inch or less is known as veneer, whether it is to be used on the surface or as core for plywood. Use of veneer and plywood dates back at least to the third Egyptian Dynasty (2778–2723 B.C.) when a Pharaoh had 6-layered plywood used in the construction of a sarcophagus for his 10-year-old daughter.[9]

Fine veneers are used in furniture and office panels; common veneers are used in the manufacture of many small items that run into big-time business, items such as medical splints, tongue depressors, fruit containers, ice-cream holders, veneer baskets, and shipping containers, which come wholly or partially from veneer. The industry is characterized by small and large operations requiring varying degrees of technical skill, depending upon the quality of the product being sliced or sawed.

Veneer is manufactured by three separate processes and is called rotary veneer, sliced veneer, or sawed veneer. Rotary veneer is made from large logs cut into appropriate lengths corresponding to the width of the veneer desired. The bolts are soaked and steamed, then placed on a huge lathe. The veneer bolt turns against a heavy knife that extends the full width of the bolt.[10] The first few sheets that come off extend only partway around the bolt; but as the bolt becomes cylindrical, the veneer is unwound in a continuous sheet. It is clipped into appropriate lengths, and defective sections are discarded. After it passes through a dryer, it is ready for manufacture into plywood or other products. When the veneer is used for making baskets, tongue depressors, and novelties, the processing is carried to completion before the veneer is dried.

Sliced veneer is preferred to rotary cut for most exposed furniture surfaces. The sawed flitch or crotch is steamed, fastened in a heavy clamp, and moved in an oblique direction past a huge, sharp knife that cuts off the veneer in thin slices the length of the flitch being cut. Sliced veneer is kept in the order in which the sheets are removed to facilitate matching of grain. It is dried and shipped to furniture or other manufacturers. Skill is required in matching veneer to obtain the most beautiful patterns of grain.

Sawed veneer is cut from flitches by thin circular or band saws that remove the veneer in thicknesses of $1/8$ to $1/4$ inch. Veneer sawing is not common today.

*Plywood* Plywood consists of thin layers of wood glued together so that the grain of each successive layer is at an angle to the layer next to it. The finished product is strong in all directions, changes little in dimensions, and is superior to solid wood for many construction and allied purposes. The plywood industry is characterized by large corporations, and by highly technical problems of production. The industry is faced with the problem of maintaining production with a decreasing supply of large clear logs. Adjustments are being made so that small medium-grade logs may be used.

Veneer for plywood manufacture is usually rotary-cut unless the plywood is to be used for special furniture stock. The veneers are first separated into core stock and face stock. Core stock can be of less uniform texture and can include certain defects, provided they do not result in complete absence of woody material. Where knotholes exist, they must be patched.

Plywood is constructed of three, five, seven, or any other odd-numbered layers of veneer.[10] The two surface layers have the grain running in the same direction, but each successive layer of veneer is placed at right angles to the preceding. Crossbanded layers are coated top and bottom with glue. The laid-up panels are pressed at room or slightly elevated temperatures. When hot-presses are used, one or two panels are pressed between steam-heated platens.

Many adhesives used in modern plywood require heat for curing. Adhesives may be of the quick-setting, synthetic resins such as the phenol-formaldehydes, urea-formaldehydes, melamines and resorcinols; or they may be made of animal or vegetable glues of the cold-setting type. After the adhesives have cured, the plywood panels are taken from the presses, the edges are trimmed square, and the surfaces are sanded. The plywood is then ready to package for shipping to the consumer.

The veneer and plywood industries are among the most rapidly growing wood industries. Domestic production of softwood plywood, expressed as $^3/_8$-inch equivalent stock, has been by years as follows:[5]

| Year | Domestic production, billion sq ft |
|---|---|
| 1945 | 1.2 |
| 1950 | 2.7 |
| 1955 | 5.3 |
| 1960 | 7.8 |
| 1965 | 12.4 |
| 1969* | 14.0 |

*Preliminary estimate

**Laminated Wood**   Laminated wood consists of successive laminae glued together with the grain parallel. Large laminated beams, trusses, and arches make possible construction of huge, clear-floor-space buildings as wide as 300 feet.[6] Because of their beauty and ease of forming in any desired shape or design, laminated beams have been widely used for church construction and auditoriums. Laminated timbers may be small or huge.

Lamination permits fabrication from boards of timbers with greater strength ratios than solid wood members. It also permits wood to remain in the large-construction field despite scarcity of large-size timber.

**Particle Board**   Scraps from sawmills and secondary-wood processing as well as wood from small-sized trees may be converted into chips and flakes for manufacture into particle board. The wood particles are screened, coated with adhesive, and spread on plattens in layers with the fine material or flakes used on top and bottom and the rough chips in the center. The plattens are closed to compress the material into thicknesses of $^1/_2$ to 1 inch and heat is applied to cure the adhesive. Particle board can also be made in continuous sheets on steel belts that move over heated plattens, or by an extrusion process. The properties of the resulting product vary considerably with the precision of manufacturing and quality of the raw materials. Platten-formed board, because of random arrangement of particles, tends to be relatively stable against moisture changes. Extruded board is less so because of the tendency of the chips to be orientated at right angles to the long dimension of the board. In strength properties and screw-holding particle board is much inferior to plywood. It is suitable primarily for interior use as a core material for table tops, counters that are not subject to wetting, and other uses where a smooth, wide panel with only limited load-bearing ability is required. Though it was originally developed to use wood-waste of miscellaneous character, it has been found that a much better product can be made by using engineered chips—that is, those of uniform size and character. The process is well adapted to using small-sized material that might be removed in thinning young plantations.[1]

## WOOD TREATMENT

Wood may be treated with various chemicals to increase its resistance to decay-causing organisms, to render it fire-retardant, to densify it, to improve its adaptability for bending into various shapes, to harden it, and to improve its dimensional stability. The processes vary widely depending upon the objective of the treatment.[14]

**Decay Resistance** Wood on the moist forest floor decays rapidly. This is a useful and necessary means of disposing of dead plant material and returning its mineral constituents to the soil. Wood kept dry endures indefinitely. Carved wooden furniture and carts taken from the tomb of King Tutankhamen have shown little evidence of deterioration after more than 3,000 years.[3] It is therefore important to take special care to keep wooden structures dry. This means providing adequate ventilation to prevent accumulation of condensate from water vapor as well as protecting it from direct contact with water. Conversely wood kept continuously immersed in fresh water, as certain Roman and Viking ships have been, retains much of its strength for centuries. But wood that must be used in contact with moist soil, such as poles, posts, and railway ties, is subject to attack by fungi, bacteria, and insects, and piling in sea water is subject to attack by marine borers. For their work, such organisms require—in addition to a proper range of temperature and moisture content of the wood—oxygen, food materials, and certain mineral constituents and vitamins, notably thiamin. Thiamin can be destroyed by heating wood in ammonium hydroxide vapor. Decay resistance of heat-stabilized wood may be due in part to its lack of available thiamin.[14]

Man has long been aware of the losses caused by decay in wood and has taken such measures as he could to prevent them. These included using durable woods in contact with the soil, keeping wood dry, and timing cutting operations to avoid immediate attack.[9] The extent of damage done by decay organisms in the United States has been estimated at over $300 million annually. Such loss would be far greater were it not for preservative treatments of much of the wood that must be exposed to conditions favoring decay. Tests with railway ties, poles, and posts show that useful life may be increased from 5- to 10-fold by effective wood preservatives. A record for creosoted poles seems to have been set in Great Britain where 8,000 poles were still in service after 70 years.[7]

Railway ties were the first major product to be given preservative treatment on an extensive scale. This so prolonged the life of ties that annual replacements per mile have been reduced from an average of 249 in 1911 to 41 in 1965. Since 1958 poles have been the major item treated. Lumber has been treated in growing amounts since 1945.

To be effective a preservative must penetrate the wood and remain there. As wood tends to surface-check, if exposed under conditions causing frequent and rapid surface drying and wetting, the preservative should penetrate below such checks. Pressure treatment is necessary to achieve this. Coal-tar creosote is one of the most widely used and effective preservatives against all forms of wood-destroying organisms. Wood that must be painted can be treated with chlorinated phenols in oil solution, copper naphthenate, and various waterborne salts. A large number of chemicals are effective if properly applied and compatible with the service conditions to be encountered.[7] Wood can be made resistant to decay by treating it with certain nontoxic chemicals that reduce the hygroscopicity of wood.[14] Fiber-penetrating pheno-formaldehyde resin mix will both stabilize wood against moisture changes and preserve wood. Similar results may be obtained by acetylation of wood and other treatments. Chemical treatments to modify wood will be taken up in more detail in Chap. 17.

The economic advantages of treating wood against decay depend on the savings it effects in maintenance and replacement expense over the service life of the treated material, less the added initial cost of the treated product.

**Fire-retardant Treatments** Wood pressure treated with mono- and diammonium phosphate, phosphoric acid, and certain other chemicals either alone or preferably in mixture with one or more of the phosphates will not support combustion nor continue to glow once the source of heat is removed. Fire-retardant treated wood has been used in doors, interior trim, and floors of "fireproof" buildings and in places such as railway roundhouses where wood may be subject to ignition. It is also used in bridges, warehouses, mines, and many other places. Plywood also is treated with fire-retardants.[7]

**Miscellaneous Treatments of Wood** Wood may be treated with dyes to color it, with heat and pressure to stabilize and densify it, with resins that can be cured by heat to harden it and impart other useful properties. It can be treated with resins and compressed to form *compreg*. It may be treated with a chemical monomer such as ethylene glycol that can be polymerized inside the wood by gamma radiation or by chemical means to form a wood-plastic material that has many properties of wood and that requires only sanding and buffing to provide a finish. The entire treated member is alike from surface to center. Wood can be treated with steam, or more effectively with anhydrous or gaseous ammonia to render it flexible so that it can be bent into a variety of complex shapes which it will retain after the ammonia evaporates from the wood. The potentialities for industrial use of many of these newer treatments are still to be developed, but they offer exciting possibilities for adding many new useful products to the long list now manufactured from wood (Fig. 16-2).

## SECONDARY PROCESSING OF WOOD

Secondary processing of wood involves the use of lumber or wood in other forms to manufacture a great variety of products from matches to furniture, from candy-bar holders to guitars, and from baby cribs to burial caskets. For some products the manufacturing process is simple. Ice-cream holders are simply stamped out of veneer, tumbled in a cylinder with talc to smooth them, and packaged for shipment. Furniture, on the other hand, requires the entire gamut of wood manufacturing processes.

**Figure 16-2** Goblets, candle holder, and candy dish machined from wood-plastics material. *(State College of Forestry at Syracuse.)*

These include conditioning, machining, fastening, gluing, sanding, finishing, and packaging for shipment. Inspection along the line is needed for quality control.

**Conditioning Wood for Manufacture**  Wood must be properly conditioned for the process to which it is to be subjected. If it is to be bent, soaking and steaming or treatment with plasticizing chemicals is necessary. Most wood needs to be properly dried before it is machined for assembly. Each plant will have its own dry kiln for this purpose, and will follow appropriate schedules for the sizes of material it will be drying.

**Machining Wood**  Woodcutting operations include ripping and crosscutting, planing the surface with jointers and planers, molding the wood with variously shaped high-speed cutterheads, turning the wood on various types of lathes, boring it with bits, and carving it either by hand or with mechanical routers.

Each operation has its own technical problems. Rough lumber is resawed, ripped, and crosscut very much as these operations are carried out in the sawmill. To cut wood along curved lines, narrow-bladed band saws are used. For very intricate work the jigsaw is required.

Jointing, planing, molding, turning, and routing require careful workmanship and technical control to avoid splitting, roughened grain, and overheating of wood or cutter.

**Fastening of Wood**  The weakest parts of products manufactured of wood are likely to be the joints. Joints weakness is accentuated by the weakness of wood across the grain. Various mechanical devices have been constructed to facilitate and improve wood-fastening: wooden pegs or dowels; nails of various sorts, including rosin-coated, cement-coated, and roughened galvanized nails; corrugated metal fastenings; wood staples; screws, including lag screws; bolts; rods; and various forms of timber connectors. The effectiveness of all these mechanical fastenings varies with the way in which they are used and with the wood on which they are used.

**Gluing**  Use of modern adhesives is one of the most satisfactory means of fastening two pieces of wood together. A good glued joint is as strong as the wood itself. It avoids the removing or crushing of woody substance. Modern glues and adhesives are made from a great variety of raw material. Those that penetrate the cell walls form the best bonds. Some can be used on wood that is to be subject to repeated wetting and drying; others are moisture-resistant; some have little or no resistance to moisture. Some adhesives are subject to crazing or other forms of breakdown after prolonged use.

**Sanding and Finishing** Sanding is a relatively simple operation that requires but ordinary care for good results. It should be well done, because it is very difficult to cover up a rough surface by paints and varnishes.

Final finishing involves applying various seals, stains, lacquers, varnishes, paints, or other coatings to the wood surface. Many of these must be sanded or rubbed between coats and buffed after application of the final coating.

Modern wood-finishing is of two types: the penetrating oil seals, stains, varnishes, and resins; and those finishes applied to the surface of the wood, such as shellac, varnish, lacquer, paint, and related products. The penetrating seals, and particularly the penetration of wood with chemicals that form a hard resin upon curing, give a surface of maximum resistance to marring.

Few of the modern finishes applied to the surface of wood are as resistant to wear as the wood substance itself.

## TECHNOLOGY OF SECONDARY WOOD PROCESSING

**Design** Because wood is easy to machine and assemble into all sorts of shapes, variety in design of products is the rule. In fact furniture, in common with women's clothing, is subject to almost annual changes in styling. This places a considerable burden on the manufacturer as well as the furniture designer. To be functional, a well-designed article of wood should have grace and beauty, be easy to manufacture, and be durable in service. These objectives can best be achieved by a designer who understands the nature of wood and the modern technologies for modifying it to fit specific purposes.

Modern wood technology has opened a wide gap between what is known about wood and its treatment and what is practiced by wood-processing industries. This is especially the case in design and construction of furniture. In fact, many classical furniture designers, among them Hepplewhite, Sheraton, and Duncan Phyfe, designed pieces distinguished by grace and beauty and also gave great attention to the characteristics of wood so that their furniture gave high satisfaction in service. Unfortunately, few modern wood technologists are artists and few artists are wood technologists, although the need is great to combine the knowledge of the one with the artistic skill of the other.

A common failing in structures made of wood is weak joints. Wooden joints can be made as strong as the wood itself if they are properly designed and the highest quality of modern adhesives are used.

Second in importance to good joints is the allowance for changes in dimension across the grain with moisture changes. This is important in all wooden structures and especially so in those containing wide, solid wood panels.

The stiffness of a beam varies directly with the square of its depth. Therefore, wooden members that are to support weight, such as floors and heavy structures, need to be made deep in comparison to their breadth or thickness. Wooden structures should be designed for the type of loading they must support. Structures are strongest if the load is along the lengthwise dimension of the beam rather than at right angles to it. Other engineering principles that apply to all structural design should be followed. Trusses should be built up of triangular sections so that each section will be rigid. Those forms and shapes should be used that give the desired degree of stiffness or flexibility.

Among newer techniques are special chemical and other treatments of wood to make it dimensionally stable, to densify it where hardness is needed or heavy stresses are involved, to plasticize it temporarily for bending, and to form wood-polymer combinations of unique properties. Some of these will be dealt with below.

**Plant Layout** Since the basic steps in wood-processing are common to a great variety of products, it is not too difficult to draw up a plan for location of the several operations and for flow of materials through them. But because of frequent change in product design, and the general lack of interchangeable standard parts, a variety of machines, cutter heads and cutting tools, some of which must be custom-fashioned for a particular task, are commonly in use. This makes it important that both number of separate machining operations, as well as functions performed, be kept as flexible as possible in the flow of materials being processed.

**Programming Manufacture** If there is to be a smooth flow of products to the shipping room, all processes in their manufacture need to be programmed through the plant to make full use of manpower and equipment and avoid bottlenecks. This is especially important in processing large orders.

**Inspection and Quality Control** To avoid needless waste of time and material in assembly it is important that all parts be manufactured to precise dimensions. Removing samples for inspection at regular intervals will avoid having a large number of rejects. Careful inspection of the final product is also desirable to maintain the good name of the firm in the trade.

**Administrative Control** Management, supervision, warehousing, sales and service, budgeting, and planning are all a part of general administration in which the functions to be performed are analogous to that of counterparts in other lines of manufacturing.

## SECONDARY FOREST PRODUCTS

Lumber, pulpwood, and veneer account for over 90 percent of wood used for industrial purposes. Poles, posts, piling, and crossties cover the bulk of the remainder. There are a few specialty uses that are important. Wood shingles are still in use for siding and roof covering. Shingles of western red-cedar, redwood, cypress, and northern and southern white-cedar are all durable, light in weight, pleasing in appearance, relatively stable in dimensions, and have good insulating properties. Wood shingles of best quality have an average service life of 35 years on steep roofs. On sidewalls they may outlast the building. In appearance, long life, resistance to damage by hail, snow, and wind, insulating properties, and suitability for many types of climate, wood shingles surpass most other types of roof covering. The decline in use of shingles has been more from lack of suitable timber from which to make high-grade shingles, and cost of laying them, than from competition of other products.

When man had to move most of his containers of products by hand the wooden barrel that could be rolled was widely favored. It was strong for its weight and capacity, suitable for a wide variety of products, easily moved about, and readily disposed of when no longer needed. It has now largely been replaced by fiberboard cartons for dry materials and metal containers for liquids. Tight-cooperage is still used for aging wines and liquor, though white oak timber suitable for manufacture of tight-cooperage is exceedingly scarce. Wood-cooperage was also formerly used for water conduits and still has some appeal for this purpose. Wooden vats, pipes, and containers may be used for liquids that are corrosive of metal but have little effect on wood.

Until recent years excelsior was one of the major packing materials. Its lightness in weight, springy character whether wet or dry, compressibility, strength, mat-forming, and other desirable properties made it an ideal material for cushioning dishes and other fragile articles. A variety of other packing materials have now largely replaced it.

### Extractives and Derivatives

*Naval Stores* Turpentine and rosin are obtained by distilling the resinous exudate from tappings of longleaf and slash pine. Field operations consist of chipping through the bark of living trees, spraying the streak with sulfuric acid, hanging cups or bag containers to catch the flow of gum, and collecting the crude gum and transporting it to the turpentine distillery. Distilling must be carefully controlled to avoid degrading the product by burning. Because of the high labor requirement and the introduction of other solvents into the paint and varnish industry, the gum-turpentine industry has been declining since 1945 (Fig. 16-3).

Improved practices are being developed to reduce the number of chippings to three per year and to mechanize operations so as to increase output per man-day markedly.

Steam and solvent extraction of turpentine, rosin, and pine oil from chips of stumpwood of longleaf and slash pine has developed rapidly since 1937. Large plants with high technical control are used. Turpentine remains a preferred solvent for paint manufacture. Rosin is used for sizing paper, as a constituent of paint and varnish and in soap manufacture.

It is interesting to note that pitch distillation is an old practice. Makkonen[9] quotes the description given by Theophrastus of pitch-

**Figure 16-3** Selective cupping for naval stores, Olustee Experimental Forest. The first 10-year cycle is now in its fourth year. After 10 years of cupping, the worked trees will be harvested and a second cycle begun on the trees marked with an X. Unworked seed trees will still remain. *(U.S. Forest Service.)*

burning of pine as carried out in ancient Greece some 300 years before Christ. Makkonen observes that the practice as described was essentially identical to the practices in the Nordic countries up to the present century.

**Other Wood Extractives** Other extractive substances of wood are tannins, various phenolic constituents, and various dyes and coloring matters.[15]

The chief sources of natural tannins used in the United States are quebracho, imported from South America; chestnut and hemlock, from the United States; mangrove, from Borneo; and wattle from South Africa. Sumac leaves, scrub oaks, and the barks of red spruce and Douglas-fir can also be used as sources of tannin.[10, 14]

Various phenolic substances are found in wood, but these normally occur in low concentrations of 0.5 percent or less. The phenolics, nevertheless, have considerable influence on the durability of wood, for they are toxic to wood-destroying organisms.

Woods yield a number of natural coloring materials or dye substances. These belong to four major chemical groups: the quinones, hydroquinones, benzopyrans, and chromones.[13] Maclurin is a pale yellow dye material found in mulberry and in Osage orange. Walnut hulls contain a material, juglone, that produces a dark dye for wool and silk. Various tropical woods have been used as sources of dye materials, among them greenheart, woods of the Bignoniaceae, Caesalpinaceae, and the Lauraceae, *Haematoxylon,* sumac, and others. The chemistry of natural dyestuffs is complicated.

**Wood Distillation** When wood is heated with a limited amount of air, it breaks down into a number of products, among them methyl alcohol, acetic acid, various phenols that are found in wood tar, and other products. Charcoal is the end product from the wood itself. The gases formed upon heating wood are combustible and may be converted into liquid fuels.[4, 14]

The chief product of wood distillation, charcoal, must be made of wood. Charcoal is used in food refining, for broiling in restaurants, and for picnics and barbecues, and is the main cooking fuel of the tropics. Charcoal has limited use in the steel industry for manufacturing certain grades of high-carbon steel. Because it produces a reducing flame, charcoal blankets—fine charcoal treated with a fire-retardant—have been used by the steel industry to cover the molten metal while it is being poured. Activated carbon is made by treating fine charcoal with superheated steam. This has far greater adsorptive capacity than has ordinary charcoal and hence is widely used for gas masks, city water purification, and other purposes where high adsorptive properties are required. Charcoal is also used for curing tobacco.

The wood-distillation industry achieved substantial prosperity between 1910 and 1920 but declined again rapidly between 1920 and 1940. A sharp but short-lived revival occurred from 1940 to 1945, and an almost equally sharp decline followed. Both methyl alcohol and acetic acid can be recovered as a by-product of petroleum refining. A market for charcoal seems likely to continue indefinitely.

### Other Forest Products

**Maple Products** Maple products bring in a substantial income to farmers and others who engage in tapping the trees and evaporating the sap. Maple syrup is strictly a luxury commodity, selling for four or five times the price of cane-sugar syrups. Pure maple syrup is used to flavor other syrups and to flavor confections, ice cream, and other products; lower grades of syrup are used in curing tobacco.

**Christmas Trees and Greenery** It is estimated that approximately 28 million Christmas trees are used annually in the United States. Over 21 million of these are cut in the United States and more than 5 million are imported.

Many growers have found Christmas-tree plantations a profitable form of land use, yielding gross returns of $30 to over $100 per acre annually. Under sustained-yield production a grower could count on no more than 100 trees per acre annually. Favored species in the United States are balsam fir, Douglas-fir, black spruce, red-cedar, white spruce, red pine, and Scotch pine. Plantation-grown trees, because of their better shape and denser foliage, have increased greatly in popularity since 1950.

Holly for Christmas wreaths, mistletoe, laurel, and other evergreen material are valuable for decorative purposes. Various decorative materials are harvested from the forest at seasons of the year other than Christmas. Wood ferns are used in floral decorations. Ground pine and other club mosses are used both for Christmas wreaths and to form chains or ropes for weddings. Evergreen branches are extensively used for decorations throughout the winter, as are pussy willow, red osier dogwood, and other tree branches in early spring. Flowering dogwood is especially sought in late spring when the floral bracts are in full display, but its collection in many states is prohibited. Redbud, cherry, and peach boughs are also used during the season they are in flower.

**Tree Seed** The harvesting, extracting, and storage of tree seed is becoming an increasingly important business in the United States. Vast quantities of seed are used by state, federal and private nurseries. Seed for forest planting must be of high quality and must be collected from parent trees adapted to the climate in which the tree seedlings are to be set out. The U.S. Department of Agriculture makes it a policy to use only seed of known origin in all plantations on national forests and on other plantings the Department sponsors. Suppliers must provide adequate evidence to verify place and year of collection before seeds are purchased by public nurseries.[16]

**Fruits and Nuts** Many wild fruits and nuts enjoy good local and even regional markets. Black walnuts, hickory nuts, butternuts, pecans, hazelnuts, piñon pine, and various other edible nuts are collected and sold locally for human food. Beechnuts, acorns, and other nutritious fruits and nuts serve as food for wildlife and are often used by domestic animals, particularly swine. Locally, they form an important resource of the forest.

The forest also produces a variety of berries and fruits useful for human food. Many have a good local or regional market. Among these are blueberries, cranberries, persimmons, wild raspberries, strawberries, blackberries, and cherries. Little attempt is made to cultivate or control the harvest of products, except for cranberries and blueberries.

**Medicinal Herbs and Pharmaceuticals** In the past the forest has furnished a number of plants used for pharmaceutical purposes. Among these have been wild ginseng, sarsaparilla, yellowroot, witch hazel, bayberry, and a host of others. Medicinal drugs are now produced mainly by drug companies through synthetic processes or from cultivated plants.

Flavoring extracts may be obtained from the twigs and leaves of various forest plants. Oil of wintergreen is made from black birch twigs by steam distillation. Various other waxes, oils, and flavoring materials are obtained from the forest and enter into commercial use. Among them is the gum of the balsam, which is used as a cementing material in the optical industry and for the mounting of permanent microscope slides.

Tropical forests are by far the richest in plants containing useful alkaloids and other pharmaceuticals. Several species of plants produce strychnine, and the more useful of these species are now cultivated.

Cinchona bark produces the alkaloid quinine as well as closely related products that are useful in the prevention and cure of malaria and other fevers.

Rubber, chicle, dyestuffs, and tanning materials are still imported into the United States from tropical and subtropical regions.

Natural rubber is used for a great variety of products for which it is better suited than are modern synthetic rubbers.

## SIGNIFICANCE OF THE WOOD-PROCESSING INDUSTRIES

In 1963 the lumber and wood-products industries, exclusive of furniture and paper-products manufacturing, employed 563,000 workers, had payrolls of $2.34 billion, created $4.02 billion of value by manufacturing, and shipped products worth $9.20 billion. They thus accounted for more than

1 percent of the gross national product, a significant contribution to the national economy. By comparison the paper and allied industries employed 588 thousand men, had payrolls of $3.51 billion, created $7.40 billion of value by manufacturing, and shipped $16.36 billion worth of products.

The wood-processing industries were dominated by lumber manufacturing, their largest component. During the period from 1860 to 1957, lumber prices in terms of constant dollars rose 10-fold in contrast to a 3.5-fold increase in price of other building materials and a 1.7-fold increase in price for all commodities. Per capita consumption of lumber decreased from 320 board feet per year in 1900 to 210 board feet in 1957. Even when eliminating the effects of inflation on price increases, lumber prices in 1957 were 5-fold those of 1870. From such behavior of price and use trends, Potter and Christy[11] conclude that lumber displays the attributes of a resource of increasing scarcity. In terms of the quality of accessible stands of sawtimber, studies by the United States Forest Service bear this out.

In contrast to the sharp advance in lumber prices, prices of pulpwood at the mill in noninflated dollars in 1953 were no higher than in 1900. During this period the use of pulpwood per capita increased 6-fold.[11] So the lumber industry following 1900 lost markets to competing industries and encountered a growing scarcity of high-quality accessible timber to process. Pulpwood supply during the same period remained abundant due to new pulping techniques that make it possible to use hardwoods and pines. The fact that relative prices held within the same general range permitted paper use to grow.

Since 1948 the price and use trends upon which Potter and Christy postulated increasing scarcity of lumber seem to have changed. Standing-timber volumes ceased to decline and so did use of lumber. Also the price trend in relation to other construction materials has reversed to favor growing use of lumber.

During the 20-year period from 1948 to 1968, the price of lumber relative to the general price level increased by 10 percent, that of clay products by 41 percent, of cement by 47 percent, and of steel by 80 percent. The relative price of plywood declined 45 percent. As a result of lumber prices leveling off and plywood prices declining, use of both expanded.[8]

With all the advances in modern technology—among which the lumber industry has been classed as a laggard—wood-frame construction has remained the dominant pattern for detached single-family homes and other light construction from colonial times to the late decades of the twentieth century. It has been estimated that in 1970, 85 to 90 percent of owner-built homes being constructed in the North and the West of the United States were of wood frame.

There are many reasons for builders of their own homes to select wood. Its flexibility in construction makes it adaptable both to the needs of the family and to the site. Its durability, insulating properties, and many other qualities promote gracious living. Its competitive-price basis adds to its popularity—even though the owner of a custom-built home cannot benefit from economies of scale in manufacture.

Factory-built homes have long been available but still are not popular with the man who wants to build a house of his own. Certain parts, however, were being factory-built in increasing volume in the late 1960s. Among them were doors and windows prehung in their respective frames and modular parts. But erecting the wooden frame of a house, all things considered, can be done about as cheaply on the site as it can be done in a factory. Factory construction requires large work areas for construction and temporary storage, special trucks and permission to move the bulky products on streets and highways, and cranes to set the house walls on the foundation at the site. Unless a large number of houses of similar design are to be erected in the same vicinity, the economy of scale is negligible. Many homes are erected on hilly ground where heavy equipment cannot readily be used. On such ground, building plans must fit the site for owner satisfaction.

Two trends in residential construction were in evidence in 1970: the increase in high-rise apartment housing, and the increase in mobile houses. Both had significance for the lumber and wood-products industries. The average amount of lumber and plywood required for a single-family dwelling in 1970 was about 18,000 board feet equivalent, for multifamily units 6,000 board feet equivalent, and for mobile homes 3,400 board feet equivalent.[8] It is also evident to those familiar with apartment house and mobile-home living that such dwellers, due to space limitations, tend to accumulate less furniture than occupants of conventional single-family dwellings.

The significance of the lumber and wood-processing industries to the people of America, therefore, extends far beyond the number of men employed, wages paid, values created, and products shipped. They play a key role in national housing. It has been declared by Congress that the United States needs to build 2.6 million new dwelling units per year for the decade 1970–1980. To do this will place a heavy burden on the wood-processing industries as well as on the nation's forests. The cost will be high. But the cost of not building them may be higher in terms of discontent, human unrest, disunity, and frustration of people's creativity.

The wood-processing industries, therefore, create a social value that needs to be weighed alongside the dollar value of their shipments.

Worldwide the use of lumber has been growing much as it did in the United States from colonial times up to 1907. This growing use seems likely to continue for a long time as people in developing nations strive for better housing. Trade in wood products has also been growing since 1945, both among nations with advanced economies and in exports of wood from developing nations to industrialized nations. This has made possible an overall rise in the output of forest industries in developing lands and has led to considerable investment of outside capital in new plants. Increased use of timber has also stimulated the development of forestry in many such countries.

It is a source of satisfaction to those interested in the use of wood that markets lost to substitute materials do not always stay lost. This is especially true of those in which the intrinsic properties of wood are better suited to the intended use than substitute material.

Both nationwide and worldwide, during the remainder of the twentieth century the outlook is for a significant upward trend in use of lumber, panel board, and other products made of wood. The new machinery that has been developed for wood-processing, and the new treatments of wood still to be commercialized, should bring new products to the market at attractive prices. Lumber seems likely to hold its prominent place in the housing, general light-construction, furniture, fixture-manufacturing, and associated fields. Many research findings have high potential for increased use of wood in both natural and modified forms. Because wood is such a versatile and easily worked material, it is well suited to new uses, especially those that require originality in design and innovation in processing.

**LITERATURE CITED**

1 Akers, L. E. 1966. "Particle Board and Hardboard," Pergamon Press, London. 172 pp.
2 Anonymous. 1968. Computerized Lumber Mill Grades by X-Ray, Scales Electronically, Wood and Wood Products. 73(2):28–31.
3 Desroches-Noblecourt, Christiane. 1963. "Tutankhamen, Life and Death of a Pharaoh," New York Graphic Society, Greenwich, Conn. 312 pp.
4 Gillet, Alfred. 1949. The Future of Wood Pyrolysis, pp. 45-51, *Fourth Meeting of the Technical Committee on Wood Chemistry,* Food and Agriculture Organization of the United Nations, Rome.
5 Hair, Dwight, and Alice H. Ulrich. 1970. The Demand and Price Situation for

Forest Products 1969–70, U.S. Department of Agriculture, Forest Service, Miscellaneous Publication No. 1165, Washington. 77 pp.
6  Harlow, William M. 1970. "Inside Wood," The American Forestry Association, Washington. 120 pp.
7  Hunt, George M., and George A. Garratt. 1967. "Wood Preservation," 3d edition, McGraw-Hill Book Company, New York. 433 pp.
8  Josephson, H. R. 1969. Substitution—A Problem for Wood? [Talk presented before the Western Forestry and Conservation Association, Spokane, Wash., December 1969] U.S. Department of Agriculture, Forest Service. 11 pp.
9  Makkonen, Olli. 1969. Ancient Forestry, An Historical Study, Part II. The Procurement and Trade of Forest Products, *Acta Forestalia Fennica*, **95**:1–46.
10  Panshin, A. J., E. S. Harrar, J. S. Bethel, and W. J. Baker. 1962. "Forest Products, Their Sources, Production and Utilization," 2d edition, McGraw-Hill Book Company, New York. 538 pp.
11  Potter, Neal, and Francis T. Christy, Jr. 1962. "Trends in Natural Resource Commodities," The Johns Hopkins Press, Baltimore. 584 pp.
12  *Proceedings—High-Speed Headrig Conference*, November 12–15, 1968, State University College of Forestry at Syracuse University, N.Y. 103 pp.
13  Shriner, Ralph. 1948. A Review of Natural Coloring Matters in Wood, in West, Clarence J. (ed.), "Nature of the Chemical Components of Wood," TAPPI Monograph Series 6, New York, pp. 182–219.
14  Stamm, Alfred J. 1964. "Wood and Cellulose Science," The Ronald Press Company, New York. 549 pp.
15  West, Clarence J. (ed.). 1948. Nature of the Chemical Components of Wood, TAPPI Monograph Series 6, New York. 234 pp.
16  U.S. Department of Agriculture, Forest Service. 1948. Woody Plant Seed Manual, Miscellaneous Publication 654, Washington. 713 pp.

Chapter 17

# Wood Chemistry

In the foregoing chapters, we saw that wood has a marvelously complex physical structure ideally suited to supporting a tree crown. We saw also that the same complex structure makes it highly useful for constructing buildings, packing crates, and many other products requiring a strong, relatively light, easily worked material. Its chemical nature is also fascinating. Wood is built up from two abundant and simple chemicals—water and carbon dioxide—and eventually it returns to these. Both synthesis and decomposition of wood are chemical processes readily comprehended in general outline yet consisting of profoundly complex intermediary steps.

In the sections to follow, the chemical composition of wood will be explored as it relates to wood properties and behavior and how wood is transformed by chemical processes to give rise to a whole family of products indispensable in modern society.

## CHEMICAL COMPONENTS OF WOOD

Chemically, wood contains three types of substances: those that remain after combustion—wood ash; those that can be removed by steam or neutral solvents—wood extractives; and those that are insoluble in neutral solvents or are nonvolatile with steam, or both—the wood or cell-wall substance.[11, 12]

The ash content of wood is due largely to the deposition of chemicals in the tree stem from the sap stream. Some of these chemicals enter into the live tissue in the wood rays, cambium, and bark. Consequently wood ash contains all those elements that are essential for plant growth except carbon, hydrogen, oxygen, nitrogen, and sulfur, that are lost during combustion. The more common components of ash are calcium and potassium. Pioneers in America used wood ash as a source of potash for soap manufacture. The chemicals found in ash add little to the strength of wood but those dissolved in the moisture content of wood account for its electrical conductivity. A high silica content occurs in some woods, notably in a number of tropical species. This is objectionable for it rapidly dulls machining tools.[8]

The substances that can be removed by neutral solvents or steam are classed as extractives or extraneous substances. They include tannins, waxes, oils, resins, gums, sugars, pigments, and other substances. Some of these may be present in wood in sufficient amounts to add to its stiffness and strength, and others affect its decay-resistance. In the living tree they may seal over wounds, repel insect attacks, and serve other functions.

The major components of wood, however, are those that are insoluble and nonvolatile, though combustible—the wood substance or cell-wall substances. These substances are made up of lignin and the polysaccharide system termed holocellulose, which is in turn made up of cellulose and the alkali-soluble hemicelluloses together with some pectic materials. The hemicelluloses may be broken down by weak acids. About 50 to 60 percent of the cell wall of wood is composed of cellulose, and about 20 to 30 percent of hemicelluloses. The remainder is lignin. In wood the individual fibers are held together with an amorphous material in the middle lamellae, which is primarily lignin together with some hemicellulose and perhaps some cellulose. The lignin not only cements the cells and microfibrils together but provides rigidity. It largely accounts for the compression strength of wood. If a tree stem contained no lignin it would be prostrate. Both lignin and hemicellulose are found together with cellulose in all portions of cell walls throughout the woody stem.

As both cellulose and lignin are inert to ordinary chemicals, wood is not easily corroded in the air or in salt water. Cellulose and lignin both, however, can be decomposed to yield foodstuffs that will support the life of fungi and various other wood-destroying organisms. For this reason wood decays.

Both cellulose and lignin belong to the family of chemical compounds known as polymers. A polymer is made up of a number of simple units chemically linked to form long chains or networks.

## THE CHEMICAL NATURE OF CELLULOSE

The cotton fiber is an example of almost pure cellulose. Cellulose is a chemical compound of high molecular weight made up of units of anhydroglucose, that is, glucose from which a molecule of water has been subtracted. The empirical formula for cellulose is written as $(C_6H_{10}O_5)_n$, where $n$ stands for the number of anhydroglucose units in the cellulose molecule. In naturally occurring pure cellulose, such as the fibers of flax, $n$, the degree of polymerization, may be in the range of 9,500 to 11,000. Corresponding values for other celluloses are: cotton fibers, 8,000 to 9,000; wood cellulose, 5,000 to 8,000; wood pulp for paper, 500 to 1,500; and rayon-tire yarn, 490. Cellulose in solution may have as low values as 50.[9] The degree of polymerization in hemicelluloses is in the order of 50 to 300.[12] In general the longer the cellulose chain the stronger the fiber formed from it. For this reason the highest grades of papers are made of cotton and linen rags.

Cellulose can be arbitrarily divided into three groups—alpha-, beta-, and gamma-cellulose. Alpha-cellulose has the longest molecular chains and gamma the shortest. Cotton has about 98 percent alpha-cellulose. Wood pulp has a much lower alpha-cellulose content and is higher in beta-cellulose and gamma-cellulose. The cellulose molecule is large enough to be classed with colloids. Cellulose can be quantitatively hydrolyzed to yield glucose.[12] This reaction is performed by bacteria in the stomachs of ruminants.

In a highly organized parallel condition, cellulose molecules form microcrystals, whose existence can be proved by x-rays. It is the linear nature of the cellulose molecules, together with their packed parallel organizations, that gives cellulose fibers their high tensile strength.

Cellulose from all plants has the same chemical composition. Chemically, cellulose is closely related to starch and has the same empirical formula. In starch, the cellular units are thought to be linked to form a spirally coiled chain. In cellulose, they are linked to form a linear chain. This small difference in structure is responsible for vast differences in physical properties and chemical behavior.

It is possible by appropriate treatment of wood to dissolve away the lignin and leave the total polysacharide fraction called holocellulose. Woody plants contain 70 to 80 percent holocellulose. Cellulose is separated from most other constituents of wood by the industrial chemical-pulping processes.

**Properties and Derivatives of Cellulose** Cellulose has high affinity for water. It is one of the most powerful dehydrating agents. Dry, fibrous materials are good electrical insulators, but it is almost impossible to keep woody material dry. The electrical conductivity of wood increases over 100,000-fold in the range from 7 to 30 percent moisture.

Wood pulp or cellulose fiber may be mechanically "hydrated." This is one of the actions that takes place when wood pulp is refined. The fibers hold more water because of the increased area of the finely broken fibers and shortened cellulose chains. The pulp becomes gelatinous after prolonged refining.

Cellulose can be dispersed or dissolved by appropriate alkali treatment and can be recoagulated by acidification. This property is made use of in the manufacture of rayon and cellophane. Cellulose can also be treated with nitric and sulfuric acids to form nitrocellulose known as guncotton that has high-explosive properties.[12] Nitrocellulose containing less nitrogen than guncotton is used for lacquers and plastics. Cellulose forms esters with acetic acid. This combination is known as cellulose acetate, which, being thermoplastic, may be softened by heat.[7] Various plastic materials are made of cellulose acetate, including such items as rayon and films for photographic work. Ethyl and methyl ethers can also be made from cellulose. Plasticized ethyl cellulose may be used as an insulating coating for wire and may be extruded in strips for the manufacturing of materials for furniture seats. It is flexible, tough, and dimensionally stable; has low moisture-absorption; can be easily fabricated; and takes printing.

The affinity of cellulose for water, due to the formation of weak hydrogen bridges, is useful for many products, notably surgical dressings and underclothing. In wood, it gives rise to dimensional instability that is undesirable. Chemists have recently been able to graft various chemicals with polymeric properties onto the cellulose molecule. The result can then be polymerized to form graft polymers. It is possible in industrial processing to impregnate wood with certain polymeric chemicals and by use of gamma radiation or appropriate chemical means polymerize these within the wood to form a wood-plastic material. Some of these have high dimensional stability and are also resistant to decay organisms. In addition they have many of the desirable properties of both wood and plastic materials. Such treatment can serve as a finish that is uniform throughout the treated wood (Fig. 17-1).[4]

## LIGNIN

Lignin, in contrast to cellulose, is not a linear molecule. Its basic repeating units are $C_9H_9O_5 \cdot OCH_3$. Judging from molecular-weight determinations

**Figure 17-1** Ornamental screen built of wood that was shaped while softened by ammonia, and folding chair constructed of wood-polyethylene glycol plastic. Note wood-plastic hinge. *(State University College of Forestry at Syracuse.)*

by solution processes, its degree of polymerization is in the order of 16 to 18. The degree of polymerization of lignin in situ in the wood may be higher than this. In fact, lignin may not be a discrete chemical compound but may be a mixture of two or more compounds.[9]

When lignin is chemically broken down by various methods, it yields a number of compounds that are believed to be present in modified form in the lignin molecule. Among these are vanillin, used in flavorings, perfumes and pharmaceuticals.[1] Vanillin may be manufactured from lignosulfonic acid formed during the sulfite cooking process of making wood pulp.

The structure of lignin is known to be very complex. The lignin of conifers differs from that of hardwoods in the higher methoxyl content of aromatic rings in hardwoods.[1,12] A property of lignin highly undesirable for paper manufacture is its tendency to give color reactions with various chemicals. Lignin in itself has plastic properties, as will be discussed below. It may ultimately prove to be a valuable agent in the wood-plastics field. Lignin, in common with coal and other products, can be hydrogenated—that is, treated with hydrogen under elevated temperatures and pressures to yield various petroleum-type derivatives. These include products resembling gasolines and lubricating oils.

Both cellulose and lignin are decomposed by biological action. Lignin, being the more resistant, tends to accumulate as a constituent of soil humus.

## MAKING PAPER PULP FROM WOOD

The Chinese made paper in A.D. 105, using bast fibers from mulberry bark, old linen, and fish nets. The Arabs learned paper-making from the Chinese and improved and perfected the process. They used papyrus fibers in place of mulberry bark. It was not until early in the eleventh century that paper was first made in Europe (in Spain) and not until 1690 that the first mill was established in America. During most of this period, paper was made from linen rags. Because of the scarcity of raw material the production of paper did not keep pace with the demand.

Paper was not successfully made from wood until the middle of the nineteenth century. Limited amounts of paper are still made from rags, straw, cane bagasse, bamboo, and marsh grasses. These materials, though important locally, taken together contribute only a small percentage to the total paper manufactured. Wood is now the predominant paper-making fiber.

The two major operations in making paper products are pulp production and paper and paperboard manufacture.

Pulp is manufactured by treating wood or other fibrous raw material in such a way that the fibers are separated into a pulpy mass. This can be done by either chemical or mechanical means or by a combination of both. [6, 10]

A number of different pulping processes have been developed. The relative amounts of different pulps consumed in the United States are set forth by years in Table 17-1 with projections to the years 1975 and 1985.[3]

In all pulping processes except those for pulps to be made into coarse grades of building boards or floor coverings, clean, bark-free wood is required.

**Groundwood Pulp** Groundwood pulp is produced by pressing billets of wood against the face of a large rotating grindstone upon which water is played to keep the stone cooled to about 160°F and to carry away the fiber-bearing slurry. The axes of the billets during grinding are parallel to that of the rotating stone. The pulp is screened to remove knots and lumps of wood, thickened by removing water, and then sent to a paper mill to be made into paper.

Table 17-1  Apparent Pulp Fiber Consumption by Type 1920–1985

| Year | Total | Virgin wood pulps | | | | | | Other | | |
|---|---|---|---|---|---|---|---|---|---|---|
| | | Dissolving and alpha | Sulfite | Sulfate | Soda | Groundwood | Semichemical | Whole wood Fiber | Nonwood pulps | Waste paper |
| | | | | | | Million tons | | | | |
| 1920 | 4.7 | | | 0.4 | | 1.8 | | | 0.7 | 1.9 |
| 1930 | 6.4 | | 2.6 | 1.4 | 0.5 | 1.9 | | | 1.4 | 3.8 |
| 1940 | 9.7 | 0.3 | 2.7 | 3.9 | 0.5 | 1.8 | 0.2 | 0.3 | 1.0 | 4.7 |
| 1950 | 17.1 | 0.7 | 3.2 | 8.4 | 0.6 | 2.5 | 0.7 | 1.1 | 1.4 | 8.0 |
| 1960 | 26.6 | 1.0 | 3.1 | 15.2 | 0.5 | 3.6 | 2.0 | 1.3 | 1.0 | 9.0 |
| 1966 | 37.4 | 1.3 | 3.3 | 23.7 | 0.2 | 4.3 | 3.2 | 1.5 | 1.0 | 9.9 |
| | | | | | | Projections | | | | |
| 1975 | 57.5 | 1.7 | 3.5 | 38.0 | 0.2 | 5.7 | 6.4 | 2.0 | 1.0 | 13.3 |
| 1985 | 86.4 | 2.2 | 3.8 | 59.0 | 0.2 | 7.7 | 10.9 | 2.6 | 1.0 | 17.1 |

*Source:* Hair, Dwight. 1967. Use of Regression Equations for Projecting Trends in Demand for Paper and Board. U.S. Department of Agriculture, Forest Service, Forest Resource Report No. 18.

Groundwood is a low-grade, low-cost pulp, which is suitable only for the cheaper grades of paper and paperboard. The pulp consists largely of broken fibers, fiber fragments, and fiber bundles. Groundwood, which has low strength, must generally be blended with a stronger pulp. Newsprint, for example, is made from 85 percent groundwood and 15 percent stronger pulp. Papers made from groundwood are not permanent. Their chief advantage is low cost, which is derived from the high yield (92 percent) of fiber from the original wood. Recent improvements in making "bleached" groundwood have added to the number of grades of paper in which this pulp can be used. Total production of groundwood pulps in the United States in 1966 was some 3.5 million tons, of which 2.4 million was manufactured into newsprint. In addition to local production the United States imported 7 million tons of newsprint, mostly from Canada.[3]

The principal item of processing cost is power. The power requirement for grinding spruce and fir is between 50 and 60 horsepower days per ton (895 to 1,074 kilowatt-hours). Only softwoods such as spruce and fir and soft-textured hardwoods such as aspen can be successfully pulped by the ordinary grinding process.

A mechanical pulp that closely resembles groundwood in properties and use is made by defibrating untreated chips in a disk refining mill (Fig. 17-2). Again, power consumption is a major item in total cost. Spruce, fir, hemlock, and Douglas-fir may be used.

**Sulfite Process** To produce a high-grade paper pulp from wood, lignin must be dissolved by chemical means. All commercial chemical treatments in use in 1971 result in removal of lignin. In the process they remove hemicelluloses that would otherwise contribute bulk and strength to the paper. They also cause cellulose degradation and loss. Roughly half the wood weight is lost in the process of removing lignin.

The chemical acid-sulfite pulping process consists of cooking wood chips in a digester under pressure with a liquor made up of a solution of calcium bisulfite and sulfurous acid that contains about 6 to 7 percent sulfur dioxide. During the cooking the sulfur dioxide reacts with the lignin to form ligno-sulfonic acids. After cooking the chips are blown into a large tank where they disintegrate. The spent cooking liquor, with the dissolved lignin, is washed from the pulp fibers and discharged. Various useful products, among them alcohol, fodder yeast, and other lignin derivatives can be recovered from the waste liquor.[6, 9]

The sulfite process produces a high-quality pulp. In the unbleached grades it is used in newsprint, catalog, glassine, paperboard, and wrapping papers. After bleaching with hypochlorite it is suitable for white bond and ledger papers. The species used are the long-fibered coniferous woods such as spruce, hemlock, and fir. Hardwoods may also be used.

**Figure 17-2** The two faces of an opened disk refiner for processing wood pulp and softened chips.

**Soda Process** The soda process is an alkaline chemical process that uses sodium hydroxide as the cooking chemical, which is made at the mill by reacting sodium carbonate with lime. The process permits recovery and reuse of the cooking chemicals by concentrating the liquor by evaporation. The residue of dissolved lignin and dissolved organic compounds may then be burned and the alkali recovered in the ash as sodium carbonate. The yield of pulp after bleaching is about 43 percent of the original wood weight. The pulp is low in strength and is used in combination with stronger pulps to manufacture book and magazine papers. The use of the soda process in the United States has not increased significantly since 1930. No new mills have been built in recent years.

**Sulfate Process** The cooking liquor in the sulfate process consists of a 5-percent solution of alkali of which about 75 percent is sodium hydroxide and 25 percent is sodium sulfide. Enough liquor is added to the digester to equal 20 percent of the weight of wood.

The chemicals are recovered from the waste liquor by evaporating most of the water, adding sodium sulfate to replace lost alkali, and burning the cake that contains inflammable lignin and saponified resins.

The heat produced is sufficient to evaporate the water and to supply steam for the digesters. The pulp yield varies from 45 to 55 percent of the weight of the wood.

The sulfate process produces a strong pulp in a short cooking cycle. The process is well suited to the pulping of pine and is widely used in the southern states. The unbleached grades, or "kraft" pulp, are used in wrapping papers, corrugated board, fiber boxes, and paper bags. The pulp is rather difficult to bleach, but improved bleaching processes have made available white grades suitable for writing and other high-grade papers. Almost any species can be pulped by the sulfate process.

The sulfate has been the fastest-growing pulping process (Table 17-1).

**Neutral Sulfite Process** The neutral sulfite process is known as the semichemical or chemicomechanical process. This less-acid cook results in increasing the yield of the unbleached pulp to about 70 percent. Following bleaching, a yield of 55 percent is normal. This is some 5 to 10 percent higher than that from the acid-sulfite process. The cooking, being less drastic, softens but does not cause the chips to disintegrate. They must be run through a disk-refiner to reduce them to a pulp. The higher pulp yields possible over those of the acid cook, as well as the reduced stream pollution, have led to widespread adoption of the process. It has the further advantage that pulps of a wide range of properties may be made by varying the cooking conditions and times, the refining of the chips in the disk mill, and the bleaching process.

**Cold-Soda Process** The cold-soda process was developed for simple small-sized operations. The chips are steeped in caustic soda and then refined in a disk mill. A high-pressure digester is not required. Hardwood chips are used. The pulp can be used in place of groundwood for container board or it can be bleached for lighter-colored papers such as newsprint and groundwood specialties. The process has been little used in the United States but may have value in developing nations where small scale operations may be practical.

The chemigroundwood process[5] involves cooking the wood in billet form in neutral sulfite or other liquors, then defibrating the billet on a pulpwood stone. The softening that occurs during the cooking process makes it possible to grind hardwood billets. The power requirement is less than half that required for grinding untreated fresh spruce wood. The quality of the resulting pulp is intermediate between that of groundwood and that of chemical pulp. In yield, chemigroundwood approaches groundwood pulp. The process is in use in Europe and at the Great Northern Paper Company in Maine for making newsprint.

The neutral-sulfite, cold-soda, and chemigroundwood pulps are intermediate grades used for paperboard and other products requiring low-cost pulps. The pulps can be bleached and refined for use in high-grade papers. Bleaching with chlorine dioxide decreases the yield from around 70 percent for the unbleached semichemical to 55 percent for bleached grades. Bleached neutral-sulfite pulp is suitable for glassine paper, magazine stock, and related grades. A problem is to get good fiber separation without lowering the rate at which water drains from the pulp suspension on the paper machine. Considerable power is used by the disk mill in refining the partially cooked chips.

A disadvantage of chemical pulping processes has been the pollution caused by venting digester fumes to the air and discharging spent cooking liquors into streams. Acrid, malodorous digester fumes can be detected as far as 20 miles downwind from the pulp plant. Spent cooking liquors contain pulping chemicals and dissolved lignin, hemicelluloses, and degraded cellulose that create a biological oxygen demand harmful to fish and other aquatic life. This harm is augmented by a discharge of white water from the paper machine, which bears broken fibers and other particles that pass through the wire cloth on which the paper is formed.

The pulp and paper industry has long been aware of these pollution problems and has spent much effort and money to overcome them. A high degree of success has been attained. Use of the milder neutral sulfite and other semichemical processes reduces both fumes and cellulose degradation. Recovery of chemicals from spent liquors, such as is practiced in the soda and sulfate processes, can be extended to the sulfite process by using magnesium and ammonium-base chemicals instead of calcium. By using a continuous-flow digester, pressures can be controlled. Fumes from a continuous digester and from a multieffect evaporator of spent liquor may be vented through the recovery boiler where they are burned. Such treatment virtually eliminates objectionable odors.

In modern plants mill water can be used three to four times before discharge, thereby reducing total processing-water requirements by one-third or more. Waste water can be given primary treatment for removing settleable solids and secondary treatment for biologic oxidation of suspended organic material. Effluent from such waste-treatment plants is essentially free from substances harmful to aquatic life (Figs. 17-3 and 17-4).

Unfortunately it is infeasible to install continuous digesters in old mills or to build elaborate effluent-treatment plants for mills of low output and limited future competitive life.

**Whole Wood Fiber** Special pulping processes reduce wood to the fibrous state with little or no chemical. One of the most important of these

**Figure 17-3** Primary waste-water clarifiers remove setteable solids at the Ticonderoga, N.Y., mill. The tall structure in the right background is the continuous digester. *(International Paper Company.)*

is the Masonite process, in which wood chips are heated for a very short time in a special "gun" with steam at a pressure of 1,000 pounds per square inch. The chips are then discharged through a quick-opening valve where the explosive action of the entrapped steam reduces the softened chips to a fibrous state. The fibers can then be formed into a loose, porous insulating material or pressed into hardboard.

Another whole-fiber pulp is made by the Asplund process. Wood chips are forced into a cylindrical reaction chamber by means of a screw-feed. Chemicals and steam are added and, after being retained up to 20 minutes, the softened chips are fed into a disk refiner. The yield is 90 percent. The pulp is used for coarse building boards. In 1966, 2.6 million tons of pulp were made by these methods.

## TECHNOLOGY OF PAPER MANUFACTURE

Paper manufacture is a separate operation from pulp manufacture. Pulp received at the paper mill is mixed with water and subjected to refining in special beater or refining machines. During refining the pulp is further subjected to a rubbing or crushing action that "hydrates" and fibrillates the individual fibers. Without this treatment it would be impossible to make strong paper.

To the refined pulp may be added rosin soap sizing to prevent absorption of ink and water, dyestuffs to color paper, urea- or melamine-formaldehyde resins to impart wet strength, and clay to improve opacity and smoothness.

The thoroughly beaten pulp mixture is diluted to form a suspension of about 1 percent fiber and pumped to the head box of the paper machine. The modern paper machine is one of the largest and most expensive individual machines used in any industry. The Fourdrinier paper machine consists of a fine endless wire cloth on which the dilute fiber suspension is flowed (Fig. 17-5). The water drains through the "wire" by gravity and suction as the fibers are felted and interlocked with each

**Figure 17-4** The 14-acre aeration lagoon is shown in the right foreground. Suspended organic material of mill waste water is oxidized here, sent through a second clarifier and foam trap before finally being discharged into Lake Champlain, shown on the left. The pulp and paper plant is in the left background. *(International Paper Company.)*

**Figure 17-5** The machine room of the Ticonderoga, N.Y., pulp and paper mill, completed in 1971. The machine on the left, one of the largest in use at the time, is 500 feet long, has a forming-wire screen 306 inches wide, and operates at 2,000 feet per minute. From extreme left to right on this machine can be seen the head box from which the pulp suspension flows, the forming wire opposite the man on the platform, and the press rolls. Beyond the press rolls are the enclosed drying rolls and, at the far end, the takeup rolls on which the paper is wound. The two machines produce some 200,000 tons of fine business and printing papers annually. *(International Paper Company.)*

other on it. The wet fibrous mat is picked up by a special felt belt and carried first through a series of presses that compact the paper and press out the water, and then over a series of steam-heated dryers that lower the moisture content of the paper to 5 to 7 percent. When it reaches the end of the machine the paper is passed through a series of highly polished metal rolls, called a calender stack, where the paper is given a final polish. Since all these operations are carried out in continuous sequence, precise synchronization of the various machine sections is necessary. Fourdrinier machines operate at speeds of 200 to over 3,000 feet per minute. A paper mill may operate anywhere from 1 to 10 machines.

Cylinder machines differ from Fourdrinier machines in that the paper or paperboard is built up in plies formed on hollow wire-covered cylinders revolving in vats containing pulp stocks. The multiple layered mat is then pressed and dried in the same manner as is paper made on a Fourdrinier machine. The particular advantage of cylinder machines is that different stock can be used in the various plies.

The inverformer permits manufacturing laminated paper on the Fourdrinier machine. After leaving the paper machine, the paper may be given one or more special treatments in a converting plant. It may be run through a supercalender, embossed, surface-sized with glue or starch, or coated with pigment and adhesive.

**Paper Coating** The printing characteristics of paper can be greatly improved by coating it with clay and other materials to brighten the color and to fill in the rough spots between fibers. Coated paper is widely used where fine detail in printing is desired. As the technology of coating improves, making the process applicable to lower grades of paper, the use increases rapidly. Production of coated book paper in the United States increased almost threefold between 1950 and 1966, whereas the production of noncoated book paper during the same period increased by only 49 percent.[3]

## IMPORTANCE OF THE PULP AND PAPER INDUSTRY

The largest wood-chemical industry by far is the pulp and paper industry. In 1970 this industry used about 45 percent of all commercial wood processed in the United States. The pulp and paper industry is about tenth largest among manufacturing groups in the United States. The importance of the paper industry in the national welfare is even greater because of the essential part that paper plays in every phase of modern life.

The paper industry has shown remarkable growth, particularly in view of the ancient origin of the industry. The widespread use and increasing consumption can be attributed to the low cost and versatility of paper. No low-cost substitute exists for paper as a wrapping, packaging, printing, and writing material.

Paper consumption moves up and down with the general business cycle, but the long-term trend in paper consumption has been steadily up. In 1966 the estimated total consumption of paper and paperboard in the United States was 52 million tons. This would amount to an apparent per capita consumption of 531 pounds for the year.[3] Increase in per capita consumption coupled with a rising population spells long-term growth in paper consumption (Table 17-2).

In 1960 the pulp and paper industry reported 447,000 wage earners receiving $2,230 million in wages. The total value of sales was $23.5

Table 17-2 U.S. Consumption of Paper and Paperboard by Grades 1920–1985

| Year | Total paper and board | All paper | Newsprint | Groundwood | Book paper | Fine paper | Sanitary and tissue | Coarse and industrial | Construction paper | Board |
|---|---|---|---|---|---|---|---|---|---|---|
| | | | | | Million tons | | | | | |
| 1920 | 7.7 | 5.4 | 2.2 | 0.2 | 0.9 | 0.4 | 0.2 | 1.2 | 0.4 | 2.3 |
| 1930 | 12.3 | 8.4 | 3.5 | 0.2 | 1.4 | 0.7 | 0.4 | 1.8 | 0.5 | 3.9 |
| 1940 | 16.8 | 10.6 | 3.7 | 0.6 | 1.6 | 0.7 | 0.7 | 2.6 | 0.7 | 6.2 |
| 1950 | 29.1 | 16.8 | 5.9 | 0.7 | 2.6 | 1.2 | 1.4 | 3.7 | 1.4 | 12.3 |
| 1960 | 39.3 | 22.1 | 7.4 | 0.9 | 3.8 | 1.7 | 2.2 | 4.7 | 1.4 | 17.2 |
| 1966 | 52.3 | 28.4 | 9.1 | 1.1 | 5.5 | 2.6 | 3.0 | 5.6 | 1.5 | 23.9 |
| | | | | | Projections | | | | | |
| 1975 | 72.1 | 37.7 | 11.0 | 1.3 | 7.8 | 3.7 | 4.7 | 7.4 | 1.8 | 34.4 |
| 1985 | 101.5 | 51.7 | 14.3 | 1.5 | 11.4 | 5.6 | 7.1 | 9.8 | 2.0 | 49.8 |

*Source*: Hair, Dwight. 1967. Use of Regression Equations for Projecting Trends in Demand for Paper and Board. U.S. Department of Agriculture, Forest Service, Forest Resource Report No. 18.

billion. Production in 1966 was 47 million tons, a record high to that date.

The paper industry requires heavy investment and large-scale operations to function efficiently.

The Ticonderoga, N.Y., mill completed in 1971 and shown in Figures 17-3, 17-4, and 17-5 has a daily capacity of 700 tons of paper and cost, with its $5 million air- and water-pollution control systems, a total of $76 million. It employed 1,000 people in the mill and provided the equivalent of 350 full-time jobs to men engaged in harvesting and delivering the 700 cords of wood it processed daily.

The paper industry ranks highest among all industries in the capital invested per worker. The industry is normally highly competitive. As a consequence, in terms of all commodity prices, the consumer paid but little more for paper in 1966 than he did in 1926. In terms of 1959–1969 dollars the relative price of all paper in 1926 was 90.3, and in 1966, 101.3. Meanwhile quality had been substantially improved. During the same period, annual per capita consumption of paper exclusive of paperboard rose from 136 pounds to 289 pounds.[3]

Growth of the pulp and paper industry in Canada, Europe, Australia, and Japan shows similar upward trends to that in the United States. In many developing countries, notably Chile, the growth of the industry has been even more striking. Need for books, magazines, newspapers, wrapping and packaging materials, expands rapidly with increase in gross national product and with population.

The basic essentials for a pulp and paper industry are an adequate supply of wood and chemicals, an ample supply of good water, power, or fuel to produce it, a labor supply that can be trained to carry out the complex techniques involved in paper making, good transportation facilities, and of course plenty of capital and competent technical and administrative talent. Plantations of rapidly growing hardwoods and conifers that can be harvested on a short-rotation basis may be established to supply the wood needed.

## CELLULOSE AND PLASTICS INDUSTRIES

Industries that manufacture rayon, cellulose, plastics, and cellulose derivatives are closely related to the pulp and paper industries. Much of the alpha-cellulose used for dissolving purposes is produced by companies that also manufacture paper, and closely allied techniques are used.

The plastics industries also use wood in other forms. Wood-flour and finely ground charcoal are used as extenders in various plastics. Wood fiber or pulp may similarly be used, though the plastic mixture may require preforming because fiber interferes with plastic flow. Lignin is used to a limited extent as an extender and also as a plasticizing agent. Many companies in the wood-fiber and lignin-plastic fields entered these fields as a sideline to paper or pulp-product manufacture.

An important and growing field is that of paper-plastic combinations. These may be prepared by immersing paper in a bath of plastic material, then curing it in single or laminated sheets. A second method is to add the plastic material as a water-soluble resin to the paper stock before running it on the paper machine. A third possibility, still to be developed, is to treat the wood fibers themselves with a chemical, such as ammonia, that softens the fibers so that they may be pressed into molds, as any other plastic material. If perfected, this method offers a new source of basic raw material for a whole series of plastic uses in which the toughness of cellulose offers promising technical properties.[2]

## CHEMICAL TREATMENT OF WOOD

Wood itself may be given various chemical treatments to modify its properties for specific uses. It may be softened by steaming so that it can be bent, as for bent-wood chairs. It can be heated when moist and compressed to form *staypak,* a densified and comparatively dimensionally stable product. It can be impregnated with soluble resins that, when cured, impart hardness and produce the decay-resistant, dimensionally stable *impreg.* If it is compressed before the resin is cured, *compreg* results, which has high density and stability. These dimension-stabilizing processes cause the product to become more brittle than the original wood. Treatment of the wood to replace the hydroxyl groups with acetyl groups reduces swelling upon wetting by 70 percent, with little change in other properties. Wood and plywood so treated show little checking or darkening when exposed to the weather. Paper may also be improved in stability by acetylation.[9]

The newer treatment is to impregnate wood with polymeric chemicals and polymerize these within the wood itself with gamma radiation or chemical means. This process, already referred to, leads to considerable stabilization as well as imparting other desirable properties. The properties vary with the polymeric chemical used. Methyl methacrylate-wood polymer has much of the toughness of wood and affords a good finish but does not prevent moisture-induced dimensional changes. T-butyl styrene, on the other hand, gives good dimensional stability (Fig. 17-1).[4]

Wood treated with polyethylene glycol so that the polymer forms 16 percent of the weight of the treated wood is resistant to checking and splitting.[9] Parts of a swollen pine log, buried for some 31,000 years, were air-dried to a sound condition after being treated with polyethylene glycol, whereas similar untreated specimens were reduced to splinters. Various other treatments offer promise of inhibiting wood from swelling and shrinking. In addition, most of these treatments afford a high degree of protection against wood-destroying organisms, as the wood must be moist for their enzymes to attack it.

Treatment of wood with ammonia in liquid or vapor stage has been mentioned as a means of plasticizing the wood. The ammonia appears to soften the lignin and to permit the cellulose crystals to slip over one another. Wood so treated may be bent into far more intricate shapes than steamed wood and appears to retain its new shape better after wetting. Perhaps a better way to state it is that ammonia treatment seems to relax the internal stress caused by bending better than does steam (Fig. 17-6). Ammonia-treated wood of both hard maple and sugar pine has been sliced with a knife both along the grain and at right angles to it to produce perfectly smooth surfaces. Also, veneer has been cut from bolts treated with ammonia. The lathe checks were reduced in both number and depth by the treatment. Chips treated with ammonia can be reduced to pulp in an attrition mill. The possibilities of industrial application of the process appear to be both wide and revolutionary in the wood-processing, wood-plastic, wood-pulping, art, and ornamental fields.[2]

**Figure 17-6** An artist engaged in sculpture with ammonia-plasticized wood. *(State College of Forestry at Syracuse.)*

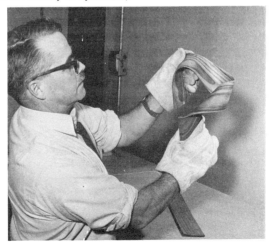

## LITERATURE CITED

1. Brauns, F. E. 1948. The Proven Chemistry of Lignin, in West, Clarence J. (ed.). "Nature of the Chemical Components of Wood," TAPPI Monograph Series 6, New York, pp. 108–132.
2. Davidson, R. W., and W. G. Baumgardt. 1970. Plasticizing Wood with Ammonia—A Progress Report, *Forest Products Journal,* **20**(3):19–24.
3. Hair, Dwight. 1967. Use of Regression Equations for Projecting Trends in Demand for Paper and Board, U.S. Department of Agriculture, Forest Service, Forest Resource Report No. 18, Washington. 178 pp.
4. Langwig, J. E., J. A. Meyer, and R. W. Davidson. 1969. New Monomers Used in Making Wood-Plastics, *Forest Products Journal,* **19**:57–61.
5. Libby, C. Earl, and Frederic W. O'Neil. 1950. The Manufacture of Chemigroundwood Pulp from Hardwoods, New York State College of Forestry, Syracuse, N.Y. 42 pp.
6. Pearl, Irwin A. 1965. Silvichemicals, Products of the Forest, *Journal of Forestry,* **63**:163–167.
7. Plunguian, Mark. 1943. "Cellulose Chemistry," Chemical Publishing Company, Inc., New York. 96 pp.
8. *Proceedings: Conference on Tropical Hardwoods,* August 18–21, 1969, State University College of Forestry at Syracuse University, Syracuse, N.Y.
9. Stamm, Alfred J. 1964. "Wood and Cellulose Science," The Ronald Press Company, New York. 549 pp.
10. Stephenson, J. Newell (ed.). 1950. "Pulp and Paper Manufacture, Vol. 1, Preparation and Treatment of Wood Pulp," McGraw-Hill Book Company, New York. 1,043 pp.
11. West, Clarence J. (ed.). 1948. Nature of the Chemical Components of Wood, TAPPI Monograph Series 6, New York. 234 pp.
12. Wise, Louis E., and Edwin C. Jahn. 1952. "Wood Chemistry," 2d ed., Reinhold Publishing Corporation, New York, vol.1. 688 pp.

Chapter 18

# Forest Products in World and National Economy*

In the Ituri Forest of Africa, in the upper reaches of the Orinoco and Amazon River systems, in Borneo, in New Guinea, and in a few other places live primitive forest-dwelling people. They use wood for the shafts of weapons and tools, for fuel, rafts, and bridges, and for erecting rude shelters. They meet their physical needs from the soil, waters, trees, food plants, insects, animal life, and fish of the forest, and from the maize, tubers, and other plants they can grow on temporary clearings. Theirs is a forest-dependent existence. Still, such people make little impress on the forest and use little wood per capita.

*The basic data, projections, and conclusions in this chapter, with respect to countries other than the United States, are based upon publications of the UN Food and Agriculture Organization, Department of Forestry, especially on "Wood: World Trends and Prospects," published in *Unasylva,* **20**:1–134, 1966, and the Forestry Section of the Indicative World Plan for Agriculture, still in preliminary draft in 1970. Data and conclusions on United States forestry are from the U.S. Forest Service publication, "Timber Trends in the United States," Forest Resource Report No. 17, 1965, and from preliminary releases of a study in progress (1971) of the United States timber resource situation. For English and metric equivalents of units used in this chapter, see Chap. 11.

## TRENDS IN USE OF WOOD

As man's control over his food supply increases, he begins to live in organized societies. His use of wood then increases. First he uses wood in round form, then as split posts, rails, and planks, then as hewn timbers, and ultimately as sawn timber. With modern technology wood is fashioned into veneer, plywood, particle board, fiberboard, paper, paperboard, and cellulose products. Industrial requirements cause wood use to increase enormously over that of primitive man, but stone, glass, clay products, and metals also come into use. Thus we have the paradox that as man's reliance on the forest decreases, his use of its products increases.

It is of little avail to consider total timber resources on a worldwide basis, for the vast timber lands of the Amazon, for example, seem destined to remain but little used for two or more decades to come. Rather the timber potential must be weighed in terms of the expected industrial development that makes it feasible to harvest and process such timber for local and world markets.

Certain conclusions emerge from a study of trends in worldwide use of wood. In general as national per capita income increases the following trends occur:

1  Total wood use per capita increases.
2  Use of wood for fuel declines.
3  Use of wood in round form and as sawn wood tends to reach a plateau at the national level with an actual decline in per capita use occurring.
4  Use of wood in panel boards (plywood, particle board, block board and fiberboard) increases markedly.
5  Use of paper and paper board increases markedly.
6  Within the several classes of products the relative importance shifts. For example, use of $1/4$-inch plywood panels increases more slowly than does the use of $3/4$-inch plywood and block-board panels. Newsprint use increases less rapidly than does paperboard for packaging.
7  When a new, highly versatile, and useful product, such as particle board, comes on the market its use grows with great rapidity, often at the expense of other materials such as lumber,
8  Vastly increased use of wood products, especially of panel boards and paper, is badly needed in developing countries. Still the increase in wood use between 1960 and 1985 was expected to be due largely to increased use in North America, Europe, and Japan, the nations that already have the highest level of use.
9  In general the greatest increases in wood use will occur where the greatest increases in per capita income occur.
10  The rate of increase in product per capita in the developed

countries is expected to decline while that in the developing countries is expected to rise over the closing decades of the twentieth century.

## GROWTH OF POPULATION AND ECONOMIC OUTPUT

Over the 25-year period ending in 1963, world population increased at an average rate of about 1.5 percent per year, which resulted in a 50-percent increase for the period. As of 1970 the rates of population increase in Latin America and Africa are enough to double their populations within 25 to 30 years, whereas it would take almost a century to double the population of Europe at its 1970 rate of increase. In 1950 the developed regions of the world—Europe, North America, Japan, and Oceania—had one-third of the world's total population. The developing regions—Latin America, Asia, and Africa—had two-thirds.[13] By the year 2000 the developed regions are expected to have little more than one-fifth of the total world population (Table 18-1). Population projection, though far from exact, is still one of the more precise of economic projections.

The total product output of regions with developed market economies in 1960 was $920 billion; that of developing market economies was but $169 billion. So 84 percent of the world's product output was produced and consumed by some 32 percent of the world's people.

Average annual per capita income in 1961 was $88 in Asia, $135 in Africa, and $295 in Latin America. This contrasted with $1,095 in western Europe and $2,620 in the United States.

Table 18-1 World Population Growth 1920–1967 with Projections to 1975 and 2000

| Year | Developed regions[1] In millions of people | Developed regions[1] In percent | Developing regions[2] In millions of people | Developing regions[2] In percent |
|---|---|---|---|---|
| 1920 | 662 | 36 | 1201 | 64 |
| 1930 | 742 | 36 | 1328 | 64 |
| 1940 | 801 | 35 | 1494 | 65 |
| 1950 | 834 | 33 | 1683 | 67 |
| 1960 | 947 | 32 | 2043 | 68 |
| 1967 | 1036 | 30 | 2384 | 70 |
| 1975 | 1097 | 28 | 2810 | 72 |
| 2000 | 1388 | 23 | 4577 | 77 |

[1]Europe, U.S.S.R., North America, Japan, Oceania.
[2]Latin America, Africa, Asia.
*Source:* Wood: World Trends and Prospects, 1966, *Unasylva,* **20:**6.

## THE INDICATIVE WORLD PLAN FOR AGRICULTURE AND FORESTRY

Developing nations have both burgeoning populations and growing expectations of their people. They seek and badly need increased per capita income. To acquaint them with how they might bring this about, around 1966 the UN Food and Agriculture Organization (UN-FAO) began to develop indicative world plans for the several regions of the world. The indicative world plan (IWP) projected population and potential developments from the base years 1960 to 1962 to the years 1975 and 1985.

It was clearly recognized that although agriculture and forest products are both basic to development of an industrial economy, their production and use cannot proceed rapidly without developing the economic infrastructure to support them. Transportation, communications, banking, and distributors of machinery and equipment are other important elements of economic infrastructure in a self-energizing development. So, of course, are incentives for people with the capital and managerial skills to devote their energies to developing these elements. As of 1970 the indicative world plans for agriculture and forestry were still in review draft. They do, however, outline what seems to be possible to accomplish in the several nations of the world, what such development might cost, and what trained manpower will be required to carry the plans to completion. It remains for leaders in the individual nations to build popular support for such development and to take the actions necessary to obtain financing. The IWP has two basic virtues: It recognizes the limitations imposed on each of the nations by the present state of their economic and industrial development, and hence it does not hold forth the promise of unrealizable goals. It is at the same time optimistic, however, for most of the developing nations are rich in resources that can be drawn upon to expand their output. Due attention is also given to the price relationships that are likely to exist and to the expected changes in commodity use that occur with increasing per capita income. Plans for most developing nations include a sharp increase in timber harvest, export, and use, together with forestry programs to assure a continuous supply.

## THE PLACE OF WOOD PRODUCTS IN WORLD ECONOMY

In a primitive economy, obtaining wood for fuel and shelter takes second place only to obtaining food. As technology develops, other products come into use. By the year 1961 metal products made up about 30 percent of worldwide values added by manufacture. The respective importance of the various categories of manufacturing as measured by value added and men employed is shown in Table 18-2. Wood products accounted for 6

Table 18-2  Relative Importance of Various Categories of Manufacturing Activity in the World for 1961

| Category | Percent of total value added | Percent of total employment |
|---|---|---|
| Metal products | 30 | 27 |
| Coal and petroleum products | 18 | 10 |
| Food, beverages, and tobacco | 11 | 14 |
| Textiles, clothing, and leather products | 10 | 20 |
| Metal mining and basic metals | 9 | 6 |
| Wood products, furniture, paper, and paper products | 6 | 9 |
| Nonmetallic minerals | 5 | 6 |
| Electricity and gas | 5 | 2 |
| Other | 6 | 6 |
| Total | 100 | 100 |

Source: Wood: World Trends and Prospects, 1966, *Unasylva*, 20:10.

percent of the value added and 9 percent of the employment. Eleven years earlier, wood products accounted for 7 percent of value added. The decline to 6 percent in 1961 was not due to less wood use, but rather to a slower rate of increase in wood use than in the other categories of manufacturing.[13]

The production for the base period with projections to the year 1975 are shown for four categories of wood products in Table 18-3. It will be noted that round-wood and fuel-wood use is projected to increase by 8 percent; sawn wood use by 23 percent; pulp and paper products by 109 percent; and wood-based panels by 149 percent. By 1968 wood-based panel production in developing regions seemed to be well on its way to the 1975 goal shown in Table 18-3. In Europe, North America, and the U.S.S.R., on the other hand, production was not up to the expected schedule.[14]

Most developing nations have a surplus of labor and a dearth of capital for investment. Wood-based industries appeal because compared with many others they are labor-intensive. To produce 6 percent of the product they employ 9 percent of the workers (Table 18-2). Developing nations are generally obliged to export raw materials, such as sawlogs and veneer logs, together with agricultural products and other labor-intensive products, to earn the money with which to purchase the machinery

Table 18-3 Annual Consumption by Regions of Various Wood Products in 1960–1962 and Projected Consumption in 1975

| Region | Sawn wood million m³ | | Wood-based panels million m³ | | | Round wood and fuel million m³ | | Pulp and paper products* million M tons | |
|---|---|---|---|---|---|---|---|---|---|
| | 1960–1962 | 1975 | 1960–1962 | 1969† | 1975 | 1960–1962 | 1975 | 1960–1962 | 1975 |
| Europe | 78 | 87 | 8.4 | 12.7 | 22.6 | 144 | 98 | 22.9 | 50.6 |
| U.S.S.R. | 100 | 111 | 2.2 | 4.2 | 14.4 | 168 | 140 | 3.5 | 15.0 |
| North America | 94 | 108 | 16.3 | 19.2 | 26.0 | 65 | 45 | 37.4 | 56.4 |
| Latin America | 12 | 25 | 0.5 | 1.5 | 1.9 | 201 | 232 | 2.7 | 6.9 |
| Africa | 4 | 7 | 0.4 | 0.5 | 1.1 | 196 | 262 | 0.9 | 2.5 |
| Near East | 1 | 2 | 0.2 | | 0.6 | 12 | 17 | 0.2 | 0.6 |
| Far East | 40 | 59 | 2.0 | 6.9 | 6.1 | 366 | 452 | 6.5 | 19.8 |
| Pacific | 6 | 8 | 0.3 | 0.6 | 0.6 | 9 | 8 | 1.1 | 2.2 |
| Mainland China | 11 | 20 | 0.2 | -- | 2.5 | 115 | 130 | 2.4 | 8.0 |
| World total | 346 | 427 | 30.5 | 45.9 | 75.8 | 1,276 | 1,384 | 77.6 | 162.0 |

Source: Wood: World Trends and Prospects, 1966, Unasylve, 20:42–45.
*1 metric ton = 1.102 avoirdupois tons.
†1 m³ = 35.3 cu ft.

essential for industrial development. It is the capital-intensive industries that have done the most to increase the output per man-day of labor and hence to increase the per capita income in industrial nations.

## USE OF WOOD PRODUCTS PER CAPITA

In contrast to food, use of wood products per capita varies over a wide range between developed and developing nations, and within each group according to the availability of local supplies and industries to process them (Table 18-4). Wood use is low in India and China where wood is scarce. Wood use for fuel declines in developed nations as more convenient sources of energy come into use, such as coal, petroleum products, gas, and electricity. To some extent use of lumber per capita also declines as steel, concrete, and other construction products become available. Plywood panels have been used for a variety of purposes for which lumber formerly was used. Paper and paperboard use appear to continue to rise with increasing per capita income because of their utility for printed communications and for packaging. No substitute for printing paper exists nor at present is there a cheap substitute for the paperboard carton for packaging many products.

## WORLD TIMBER RESOURCE OUTLOOK

The production and consumption of total industrial wood by regions is set forth in Table 18-5 together with projected consumption for the year 1975. Of the major regions, only North America and the U.S.S.R. produced substantial exportable surpluses in 1961. Europe and the Far East had the largest deficits. Use of all industrial wood was expected to increase by 42 percent during the 14-year period. About 78 percent of the industrial

Table 18-4 Consumption of Wood Products per 1,000 Capita in Six Regions in 1960–1962

| Product and unit | Northern Europe | Southern Europe | Latin America | Pacific | Africa | Near East |
|---|---|---|---|---|---|---|
| Sawn wood $m^3$ | 492 | 55 | 52 | 420 | 15 | 20 |
| Wood-based panel products $m^3$ | 56 | 4 | 3 | 25 | 1 | 2 |
| Paper products M tons | 104 | 10 | 12 | 78 | 3 | 3 |
| Round wood and fuel wood $m^3$ | 984 | 299 | 12 | 570 | 28 | 96 |

Source: Wood: World Trends and Prospects, 1966, *Unasylva*, **20**:42–45.

Table 18-5 Outlook for Industrial Wood Balance, 1975 (Million Cubic Meter of Round Wood)

| | 1961 | | | 1975 |
|---|---|---|---|---|
| Region | Production | Consumption | Surplus or deficit | Projected consumption |
| Europe | 238 | 259 | −21 | 376 |
| U.S.S.R. | 259 | 243 | +16 | 305 |
| North America | 339 | 320 | +19 | 420 |
| Latin America | 39 | 40 | −1 | 76 |
| Africa | 26 | 25 | +1 | 36 |
| Near East | 6 | 9 | −3 | 14 |
| Far East | 94 | 104 | −10 | 180 |
| Pacific | 16 | 18 | −2 | 26 |
| China (mainland) | 34 | 34 | -- | 62 |
| World total | 1,055 | 1,054 | -- | 1,495 |

Source: Wood: World Trends and Prospects, 1966, Unasylva, 20:117.

wood was consumed in Europe, the U.S.S.R., and North America. Japan alone accounted for an additional 9 percent. Though substantial percentage increases in use of industrial wood are expected to occur in developing regions by 1975, the consumption of developed regions will still account for 85 percent of the total. It is to these regions that major concern should be directed.

Europe, which had a timber deficit of 21 million cubic meters in 1961, is expected to have a deficit of 79 million cubic meters in 1975. The chief suppliers of Europe's deficit in 1961 were Canada and the U.S.S.R. It is expected that these same two nations will also supply most of the 1975 deficit except for tropical hardwoods that will come from West Africa and Asia. It is known that the allowable cut for both Canada and the U.S.S.R. exceeds their local use plus export. European nations can therefore expect that their needs to and beyond 1975 can be met by imports from these and other regions.

The situation in the United States will be dealt with in detail at the end of this chapter. In general it will be seen that its needs can be met from its own supplies plus some imports from Canada and Latin America. The other major consumer, Japan, will continue to look to the Pacific area, North America, and possibly the U.S.S.R. for imports.

Among the developing nations individual countries will face shortages but most of these can be supplied by other nations within the region. Such nations will, however, remain importers of pulp and paper and to some extent of wood-based panel products. The main sources of pulp and paper for the world trade are expected to remain northern Europe, Canada, and the United States, as they were in 1970. In fact, the immense outlay of capital required to build a pulp and paper mill that can compete with world standards of price and quality is such as to make it unlikely that any developing nation, with the possible exception of Chile, can expect to compete in more than regional markets until 1985. This statement should be qualified by recognizing that such competition might be possible for mills built and operated by strong Japanese, European, or North American corporations engaged in world trade. But the growth in demand by the developing nations themselves for paper products is expected to be so high as to require the entire output of such mills as they may build. That developing nations can enter the regional market is well borne out by the case of Chile, which has recently built up a respectable pulp and paper industry supplied by plantation-grown timber. She was already exporting to her Latin American neighbors in 1968 and is expected to continue to do so. Brazil, Argentina, Uruguay, Mexico, Venezuela, Turkey, Pakistan, India, the Philippines, and other developing nations were engaged in pulp and paper manufacture during the period from 1960 to 1970.

**The Indian Subcontinent** Two developing regions deserve special mention: the Indian subcontinent (India, Pakistan, and Ceylon), and mainland China. In 1970 they accounted for 19 and 21 percent respectively of the world's population yet occupied but 3 and 7 percent of the world's land area. Both have high population pressures on the land. Both have large areas of mountain lands that receive high precipitation. Both have much arid land. Both have current shortages of wood and paper. Both are making important progress toward industrial development, and both have a forest potential that could contribute substantially more than at present to improved living standards.

Pakistan was in the unfortunate position of being divided into two widely separated land areas that had populations of different ethnic makeup.* East Pakistan was a low-lying, well-watered tropical region with fertile soils and very dense population. Its 2.2 million hectares of natural and man-made forests were highly productive, making it essentially self-sufficient in timber and able to help supply West Pakistan's deficit. But it is dependent on outside sources for coniferous wood. West Pakistan has some 920,000 hectares of coniferous forests and 1,675 thousand hectares of miscellaneous hardwood forests. Its local timber harvest in 1965 covered barely half its needs. It depended both on East Pakistan and imports from other countries to make up the deficit. West Pakistan has substantial areas of mountainous land that receive enough

*Data on Pakistan forests are based on the nation as it existed in 1969.

precipitation to support coniferous forests. Forest cover is badly needed on these mountains to prevent serious erosion and silting of power and irrigation reservoirs. But heavy grazing of domestic animals together with clearing for food crops impedes both natural reseeding and plantation development.*

Pakistan's wood needs by the year 1985 are projected to be 28 million cubic meters. To produce this from its 5 million hectares of forest land would require the rather high average annual increment of 5.6 cubic meters per hectare. It is questionable if the nation will meet such a goal, even with the high potential productivity of East Pakistan forests. Continued imports seem likely to be needed (Fig. 18-1).

India has a somewhat more favorable situation in that mountain lands supporting forests are more widely dispersed than in Pakistan. But it also has most of its coniferous forests in the Himalayas where transportation is difficult. India's total forest area is 69 million hectares. To this should be added a substantial area of plantations both to stabilize watersheds in its mountainous areas and to augment wood supply. An annual planting program of 200,000 hectares per year is suggested by FAO. Industrial

---

*National Forestry Research and Training Programme Pakistan Final Report, 1970, UN Development Programme, Food and Agriculture Organization of the United Nations, Rome. 145 pp.

**Figure 18-1** Impressive headquarters of Pakistan Forest Institute, Peshawar, Pakistan.

wood consumption is projected to approach some 26,500 cubic meters by 1985. Fuel-wood requirements are expected to be 141 million cubic meters. The total requirement of some 167 million cubic meters would, if met from local forests, mean an annual production of 2.4 cubic meters per hectare per year from India's forest land. This would seem to be possible of realization.

**The People's Republic of China** In 1963 China was reported to have 96 million hectares of forest land that supports a growing stock of 7,500 million cubic meters of timber volume.[10] To this was being added annually a substantial area of forest plantations variously reported to be in the order of 1 to 2 million hectares. Industrial timber production was reported as 35 million cubic meters in 1958 and 39 million in 1960. With a net population growth of almost 2 percent per year, by 1990 total industrial-wood requirements would be 50 million cubic meters and by 2000, 62 million cubic meters at the 1963 rate of per capita consumption of 0.056 cubic meters per year. Planned available supply for 1990, assumed to be equal to consumption, was 300 million cubic meters or sixfold the above projection. Whatever projection is accepted, it is apparent that China will need substantially more wood in the future than she was using in 1958.

China's land-use classes are roughly as follows:

| Land-use classes | Area, in million ha |
| --- | --- |
| Cultivated land | 106 |
| Forest land | 96 |
| Grazing, barren, mountainous, and eroding land | 758 |
| Total | 960 |

There is plenty of land that would seem to be available for forest planting, much of which once supported tree growth. If only as much as 100 million hectares could be brought up to an average annual increment of 3 cubic meters per hectare, a level that would be considered but average in Europe, the 1990 needs could be met. Otherwise China will be obliged to import substantial quantities of wood or to lower her consumption goal. If she becomes able to obtain sufficient foreign exchange to do so, she could become an important wood-importing nation between 1975 and 2000. By the year 2000, yield from plantations and improvement of natural stands may suffice to support an annual consumption of 300 million cubic meters or more.

## MEETING WORLD DEMAND

Converting the data in Table 18-3 to the round wood equivalent, the total wood use for the base period 1960–1962 was 2,132 million cubic meters. The projected use for the year 1975 would be 2,686 million cubic meters. Will the world's forests be able to sustain such an annual harvest? A precise answer cannot be given as of 1972, for at this time but 55 percent of the world's forests had been surveyed and inventoried. The growing stock on the inventoried forests was reported to be 237 billion cubic meters. Conifers made up 48 percent and hardwoods 52 percent of inventoried volumes. The forests of Europe, the U.S.S.R., the United States, Canada, and Japan contain 55 percent of the total inventoried volume. The net annual growth increment of these forests is reported to be 1,900 million cubic meters. This alone would sustain 70 percent of the projected 1975 demand. The potential increment from the other two-thirds of the world forest area should be adequate to sustain the remaining 30 percent of demand. It seems likely that the even much higher requirements projected for the year 1985 also can be met without serious depletion of growing stock provided the forestry measures set forth for each nation in the indicative world plan are activated. This cannot be done without some increase in price of timber on the world market. The remote forests of Siberia and many tropical forests can be exploited only if made accessible by roads. These require heavy expenditures for construction and maintenance. Management practices themselves will add to costs. Plantation establishment and other forest-productive efforts will have to be financed to produce the timber the world will need. As of 1971 it was too early to determine whether the needed forestry measures will be taken by the many nations involved. The trend does seem to be favorable and the world timber reserve still remains large. It is well to bear in mind though that restraints on timber cutting are likely to increase also. This may complicate the task.

The uninventoried forests are mainly of tropical hardwoods, few species of which have yet become significant items in the world timber trade. Interest in these woods is growing among timber-importing nations and corporations. In the 1960s explorations by foresters were in progress in New Guinea, Borneo, the Amazon, and Orinoco regions.[9] Other regions with vast timber reserves are to be found in Malaysia and the coastal rain forests of Venezuela, French Guiana, Surinam, Guyana, Colombia, and Ecuador. Most of these are reasonably accessible. The forests of Malaysia and Indonesia, which are relatively accessible and in which timber-needy Japan is interested, seem likely to be developed soon.[12]

**Increasing Importance of Man-made Forests** Forests planted by man have long been the main source of supply of wood products for most of the countries of Europe. New Zealand, South Africa, and Chile harvest most of their wood supply from plantations. Since World War II, planting of forests has been greatly extended in the United States, U.S.S.R., Argentina, Brazil, Spain, other Mediterranean countries, Taiwan, and the People's Republic of China. Pulp and paper and mining companies in southern Brazil have shifted rapidly from native growth to planted forests for their wood supply, rather than attempt exploitation of Amazon forests. Since 1950, forest planting in the People's Republic of China has been on a heroic scale. Reports on achievement have been conflicting. Survival is known to have been low in early plantations, as they were given no protection or care. Since 1960, practice is reported to have improved markedly. Cutting in remote forests recently opened to exploitation is budgeted at a rate presumed to provide a continuous supply until the plantations reach harvestable sizes.[10]

Growth rates in forest plantations of many countries are high and rotations short. Compared with the average growth for all commercial land of the United States of 2.4 cubic meters per hectare per year, European plantations grow about 4 cubic meters per hectare per year, teak in Honduras up to 14 cubic meters, eucalyptus in Brazil up to 15 cubic meters, and plantations in the Philippines up to 30 cubic meters. Hybrid poplars, various species of eucalyptus, Monterey and Mexican pines, Cryptomeria of Japan, several conifers native to China, and various tropical leguminous trees have been used extensively, as well as the highly valued teak, rosewood, mahogany, and dipterocarps. Rotations are as short as 7 to 8 years for pulpwood, 12 years for small sawlogs, and 20 years for veneer logs. By contrast, rotations in natural forests generally exceed 50 years. Forest plantations are as a rule managed intensively.

Much of the planting in China, Uruguay, Pakistan, the U.S.S.R., and the Mediterranean region has been for the dual purpose of timber production and protection of soil against water and wind erosion. Such planting must often be done on lands subject to annual or periodic drought. Prescription of species and methods for such planting have been extensively outlined by Goor and Barney.[2]

## NONTIMBER USE OF FORESTS

Forests used for parks, for soil and watershed protection, on steep slopes, for campgrounds, for natural and wilderness areas, and as reservations for primitive people are usually closed to commercial timber operations, though some of the wood may be used locally.

The losses of forest to suburban development, transportation, and industrial parks have been more than offset in recent years by the contraction of agriculture in industrial nations. In the United States some 13.2 million hectares were retired from agriculture for forests between 1943 and 1963. In Europe it was estimated that 3.5 million hectares had been released from agriculture and made available for forest use by the year 1961. Another 5 million hectares may be given up by 1980. This is not a recent land-use change, for one can often find evidences of past agricultural use of lands in Europe that now bear mature timber stands. In the northeastern United States, stone fences that were constructed with great expense of time and labor—both to protect the cultivated fields and to remove the stones from them—now crisscross timber stands of 100 years of age or more.

## OUTLOOK FOR WORLD TRADE

The four great timber-deficit regions are Europe, Japan, the Middle East, and China. Of these Europe and Japan are the major importers. Their needs to import seem likely to increase substantially by 1985. The Middle East's needs will also increase substantially as the general level of income rises. China is trying desperately to increase her own forest resources and to conserve and reprocess as much wood and paper products as possible. If she finds means of gaining sufficient foreign exchange to do so it seems probable that she will also import considerable wood products until her own forests and plantations can meet domestic wood needs. This seems unlikely to occur before 1985.

The major need of all four importing regions is for coniferous timber. Until well after 1985 this must come from Canada and the U.S.S.R. Eventually some may come from plantations of fast-growing pines, should extensive areas be established in West Africa and the Southeast Asian areas.

Opportunities for developing countries to export high-quality hardwoods to Japan, Europe, and the United States seem certain to grow. A large task involved is to increase the number of species acceptable in world trade. Importing nations are themselves interested in furthering such trade and have done extensive testing of tropical woods. Still, the number of species is so great and species dispersal in the tropical forest is so wide that extensive exportation seems unlikely until local markets are able to absorb the major volume of the less desirable species (Fig. 18-2).

Most developing nations have great need to expand their own consumption of wood products to improve housing, commercial buildings, schools, and transportation. They need especially to expand their

**Figure 18-2** Experimental plantation of teak and other tropical hardwoods in Honduras. The tree shown is 24 years old and about 30 inches in diameter. United Fruit Company lands near Lancetilla.

use of paper products. The key to expanded exports of forest products for most nations is increased use within the country. This will encourage the development of lumber, plywood, particle board, and fiberboard mills that can place their products on the world market at competitive prices.

An interesting case of how this can work out is that of a lumber company in the Philippines. To use its waste slabs and edgings from the sawmill, it built a fiberboard mill that turned out an oil-tempered hardboard. This found ready acceptance on the world market. Soon local people also became aware of the merits of this product for exterior wall construction. Within 7 years' time exports dropped from 70 percent to 10 percent of output while volume of production increased by some threefold. The company was soon processing not only its own sawmill waste, together with all it could procure from nearby mills, but was also sending plantation thinnings to the hardboard mill. Thus, in an effort to make maximum use of its timber resource, the company made a substantial contribution to domestic housing, continued to be a substantial earner of foreign exchange, and strengthened its reputation for high-quality products both at home and abroad. The company was able to make the key investment in the fiberboard mill as a result of its earnings from export of lumber and high-quality logs.

## TIMBER TRENDS AND OUTLOOK IN THE UNITED STATES

With the arrival of colonists in America, a long period of clearing land for agriculture set in. It reached its height in New England by the 1880s, in New York by 1900, and in the Lake states and the South by the 1920s and 1930s. Growing public concern over the disappearance of forests led the U.S. Forest Service to accept as one of its responsibilities the making of periodic reviews of timber resources, the current and likely future demands to be made upon them, and the programs needed to assure that these demands will be met. The last general survey was undertaken for the base year 1968 with projections by decades to the year 2000.

**The General Situation as of 1968** The original forest area of the territory now included in the 50 states was some 942 million acres. In addition, there were some 93 million acres of open woodland where trees were too sparse to form a canopy. Even with the losses to agriculture, urban development, and transportation there remained in forest in 1968 a total of 762 million acres or almost one-third of the nation's area. Of this, 510 million acres were considered capable of producing commercial timber.[7, 16] The remaining 251 million acres of noncommercial forest land included land in parks, wilderness areas, other areas reserved from timber harvesting, inaccessible lands, and those incapable of producing timber of merchantable sizes and quality. The largest of these areas was 113 million acres in interior Alaska.

From 1910 to 1968 some 85 million acres were given up by agriculture, most of which returned to forest. During the same period 35 million acres of forests were taken up by agriculture, suburban development, and transportation. Further losses to wilderness use, agriculture, and suburban use can be anticipated.

When Columbus landed in America the volume of timber standing on the territory now included in the United States was estimated to be 5,200 billion board feet. By 1910 it had been reduced to one-half that number. The decline in sawtimber volume has apparently continued to 1968, when only 2,490 billion board feet remained. The reserves of high-quality sawtimber were still being drawn down. The annual increment due to growth has advanced substantially from the 32 billion board feet level of 1938 to 54 billion in 1967 (Table 18-6). The total growth in cubic feet, which includes trees below sawtimber size, reached a balance by 1944 and has exceeded the cut since then.

This rapid increase in timber growth has been due to small timber growing to sawtimber sizes as a result of applied forestry measures. The trend of increasing annual increments is expected to continue to 1985 and possibly beyond then. As of 1967 the average annual increment per acre was but 105 board feet, or 34 cubic feet. The best forests of the South and the West Coast were producing almost five fold this annual volume per acre. Additional land, if intensively managed, could be made to produce 200 board feet per acre or more.

From the above it might be concluded that as of 1971 the forest situation in the United States was well in hand, and that it need not become a question of public concern until 1985 or later. Unhappily such optimism is not justified. The quality of standing timber has declined due to the cut being concentrated on the larger sizes and the better quality of trees, as well as on the preferred species and the most accessible land. The remaining reserves of quality sawtimber are concentrated in the uncut lands of the West. In Pennsylvania, for instance, timber-users were complaining in 1970 of the growing shortage of merchantable sawtimber even though U.S. Forest Service reports showed it to be increasing. This anomaly stems from two causes: the fact that the Forest Service lists trees 9 inches and larger as sawtimber whereas millmen are interested mainly in trees 14 inches and larger, and the fact that the Forest Service

Table 18-6 Growth and Cut of Sawtimber and of All Growing Stock for Selected Years [15,16,7]

| Year | Sawtimber in billion board feet | | All growing stock in billion cubic feet | |
|---|---|---|---|---|
| | Growth | Cut | Growth | Cut |
| 1938 | 32.0 | 42.4 | 11.3 | -- |
| 1944 | 43.4 | 49.7 | 12.5 | 11.5 |
| 1952 | 47.3 | 48.8 | 13.3 | 11.5 |
| 1962 | 54.8 | 48.4 | 15.9 | 11.5 |
| 1967 | 53.7 | 57.5 | 17.3 | 13.3 |

counts all accessible private timber as available for cutting irrespective of whether or not the owner is willing to permit his timber to be harvested. Fragmentation of ownership mentioned earlier is also reducing the availability to the commercial logger of much of the timber standing on the ground. These were problems of 1971. The outlook beyond 1975 is even less reassuring.

**Projections of Future United States Timber Requirements and Supply** The U.S. Forest Service has accepted a low Bureau of Census projection of population for the year 2000 of 280 million people. The gross national product in constant dollars is projected by 2000 to triple its 1970 level. The Forest Service expects a continued decline in use of industrial raw material per dollar of gross national product, but even so the projected timber use for 2000 is more than double that of 1970.

Residential construction constitutes the heaviest use of lumber, plywood, and other wood-based panels. The annual number of housing units to be constructed is projected to increase from 1.5 million in 1969 to 2.8 million in 2000. The Forest Service estimate appears modest in view of the goal set by the United States Congress in 1968 to build 26 million housing units between 1969 and 1979. The use of lumber for construction, for farm use, for railways, for mining, and for manufacturing, have all been separately projected from past trends to the year 2000. The total annual demand is expected to approach some 138 billion board feet by the year 2000.

Instead of declining, the per capita use of wood-based panel products is expected to continue to increase in the future as they can economically be substituted for lumber for many structural purposes. Total demand for wood-based panels could triple between 1968 and 2000.

The demand for paper and paperboard is also expected to increase, both in per capita use and in total use from 42.4 million tons to 115 million—an increase of 2.7-fold.

U.S. Forest Service estimates of 1970 indicated a rise by 1980 of domestic and export demand for sawlogs from 46 to 64 billion board feet, of veneer logs from 9 to 15 billion board feet, and of pulpwood from 81 to 121 million cords.[5]

To dramatize the timber situation the Forest Service expressed it in these terms. Between 1970 and 2000 the nation will be obliged to build enough houses, schools, stores, and factories to establish 100 cities the size of Indianapolis, a city of 750,000 people. It will also be necessary to rebuild the central core of most existing cities and to replace much of our industrial and commercial plants. It is estimated that by the year 2000 such an undertaking will require that the nation double its 1970 consumption of sawlogs, triple its consumption of veneer logs, and quadruple its consumption of pulpwood.[5]

**Meeting Future Needs** Four major lines of action lie open to meet the nation's future needs: increase imports, reduce use, reduce waste and recycle used material, and increase output of the nation's forests.

*Increasing Imports* The United States is both an importer and exporter of logs, lumber, plywood, and paper products. Both imports and exports are expected to grow. The major supplier of softwood lumber and of paper products to the United States is Canada. The major supplier of hardwood, plywood, and veneer are tropical countries and those manufacturing these products from tropical species. The United States has exported veneer logs to Japan, and to European nations. This trade also is expected to increase. Imports of hardwood logs, lumber, and plywood in 1968 totaled 275 million cubic feet.[3] Asia has been the major supplier, accounting in 1968 for 71 percent of the total. The outlook for continued import of hardwood products appeared bright as of 1971. The outlook for softwood imports from Canada also seemed good. Canada's allowable annual cut of softwood timber for lumber and paper pulp has been substantially above the annual harvest up to the year 1970.

In 1968 the United States imported 345 million board feet of hardwood lumber, 3,841 million square feet of hardwood plywood, and 2,179 million square feet of hardwood veneer.[3] In 1969 total United States timber products imports were 2.5 billion cubic feet. This was approximately 19 percent of total consumption. Timber products export was equivalent to 1.2 billion cubic feet during the same year. Assuming that past trends in both imports and exports of timber products continue, the net imports above exports seem unlikely to increase to any significant extent.[5]

*Reduce Use* Some spokesmen deeply concerned about our environment have proposed that the nation reduce its drain on its timber resources by curtailing consumption. This might be done. The United States is one of the highest users of wood per capita of any nation. To reduce the consumption of lumber, wood-panel boards, and veneer would require a corresponding increase in use of other products for housing, light construction, packaging, and other purposes. Use of such materials has been growing, in part because of the increased cost of lumber in relation to that of commodities in general, though as noted earlier this trend may be changing.[6]

But there is another point for those concerned about environment to consider. Timber is produced with no impairment of environment and is processed and used with less use of power and emission of pollutants than any other products of comparable utility. If environmental protection is the aim, wood should be used in preference to materials that require high consumption of power and emit pollutants in manufacture and processing. Also, many substitutes are not as disposable as is wood.

**Use of Wood Residues** Residues from logging and from primary and secondary processing of wood were estimated by the U.S. Forest Service in 1968 to be the equivalent of 4.4 billion cubic feet. Logging accounted for the largest amount, 2.0 billion cubic feet, and sawmills and veneer plants for 0.9 billion cubic feet. Secondary processing produces 0.7 billion cubic feet of residues annually.[8]

Vast steps have been taken to capture and use for paper manufacture the slabs and edgings of sawmills and the scraps of veneer mills. With the introduction of chipper headrigs on sawmills, the material removed to square the log for lumber is converted directly to chips for the pulp mill. Also logs and boards too low in quality for sale as lumber can be converted into pulpwood chips. Tree-length logging in the eastern United States now results in using for pulp manufacture the tops of hardwoods as well as conifers, up to a minimum $3^{1}/_{2}$-inch-top diameter. The widespread use of hardwood pulping processes, together with improved pulping techniques for conifers, make possible very close utilization of the timber that is commercially logged for veneer, lumber, and pulpwood. Possibilities still exist to reduce the waste of wood in secondary processing plants, but these are not large consumers. Waste in construction has been greatly reduced by manufacturing lumber to the sizes that are used so that little sawing and fitting are needed on the construction job. There seems to be but limited opportunity to save timber by reducing waste in manufacturing in the forest, sawmill, veneer plant, or pulp mill, where integrated use is practiced. It must be recognized that many sawmills are too far from pulp mills to justify converting residues to chips. Also, in 1971 the material in partially rotten or broken logs, heavy hardwood and conifer knots, bark, and related material had little value for other than fuel if, in fact, it was useful for that. Limited use was possible for cattle bedding, poultry litter, and soil amendment.

**Recycling Wood and Paper** Recycling of lumber was practiced little in 1971 in America. The cost of dismantling a building to salvage the lumber has exceeded the cost of new lumber. Presumably, if the used lumber were freed of paint and nails, it could be chipped and made into pulp. Some weathered lumber is used for decorative purposes. In wood-hungry China, Japan, and even Europe, reuse of lumber does take place.

Pallets are ordinarily used until they break or become too weak for reuse. By this time there is little in them to salvage except for possible use for pulp manufacture.

Much recycling takes place in the normal manufacture of paper. Paper broke, that which is torn or creased in changing reels, in starting the machine, and in rethreading after a break occurs on the machine, is immediately recycled along with that removed in trimming. White water, water that flows through the paper machine wire and that contains broken fibers and other fine material, is recycled to capture as much of the material as possible.

The largest opportunity for recycling, however, occurs in the use of wastepaper and paperboard. Here the major problem is in collecting and sorting the material. Already a technique has been perfected for deinking newsprint. The material known as chipboard, a board made mostly from waste newspapers, has long been a familiar product for cheap cartons and tablet backs. The U.S. Forest Service Products Laboratory has made a low-grade news sheet from fiber salvaged from the solid-waste disposal plant of Madison, Wisconsin. In the United States only 18 percent of wastepaper and board was reused to manufacture pulp in 1969, as opposed to some 50 percent in Japan and Western Europe.

Other possibilities exist for savings in use of wood through increased efficiency in construction, better design of homes and wooden furniture, and the chemical treatment of wood for bending, improving properties, and making it decay-resistant. The U.S. Forest Service estimates that by 1980 it may become possible to increase the amount of wood potential by 4.7 billion cubic feet by use of residues, increased efficiency in use and reuse of paper, fiberboard, and other wood debris. Much technical, economic, and social research remains to be carried out before such a wood saving becomes possible.

**Increase Output of the Nation's Forests** If the trend of improved management practices on the nation's forests continues at the rate that prevailed during the 1950s and 1960s, during the 1970s, and on to the year 2000, the annual supply through growth will be overtaken by timber harvesting probably around the year 1990, and thereafter the growth rate will decline due to lowered growing stock in the sizes suitable for commercial use. If management improvement is accelerated at a sufficiently rapid rate, the growth rate could keep ahead of cutting well into the twenty-first century.[4] Projections made by the U.S. Forest Service for the year 1980 illustrate the potentialities. With continuation of management trends of the 1960s and at 1968 prices, softwood timber supplies could be increased from 47.5 billion board feet to 55.3 during the period from 1968 to 1980.[17] At accelerated management trends, the supply by 1980 could be 63.5 billion board feet. Similar increases by accelerated management trends could be expected for hardwood timber supplies. Such a trend would go far toward meeting the needs of the year 2000 and beyond (Fig. 18-3).

**The South's Third Forest**  An example of the type of planning and application of forestry measures that will be required to meet timber needs of the year 2000 is that done by foresters in the South.[11] The South's third forest is the one that will supply the timber cut for the year 2000 and beyond, and which was being and will be established during the period 1955 to 1985. The growth and cut for recent years and the anticipated growth and cut for the year 2000 is as follows, in billion cubic feet:

**Figure 18-3** Hybrid of limber pine and Himalayan white pine outgrows the native limber pine in the row at the left. Institute of Forest Genetics. *(U.S. Forest Service.)*

| Year | Growth | Cut |
|---|---|---|
| 1962 | 7.5 | 4.2 |
| 1968 | 8.2 | 5.7 |
| 2000 | 13.0 | 13.0 |
| Biological potential | 15.7 | |

To attain an annual growth increment of 13.0 billion cubic feet in the South by 2000, it will be necessary to increase the annual growth per acre on forest industry and national forest land from an average of 56 to 96 cubic feet, and to increase growth on other private lands from less than 40 cubic feet to 53 cubic feet.

The forest-improvement measures that need to be taken include planting some 10 million acres mostly to pine, converting 20 million acres of low-quality hardwoods to pine and timber-stand improvement on 90 million acres. Protection against fire, insects, and diseases must also be improved. The most important and most difficult part of the task will be to increase the growth rate on the 72 million acres in nonindustrial private lands. The program is an ambitious one, most of which will have to be carried out between 1970 and 1985 if the year 2000 goal is to be achieved. It will be no mean task to attain an average growth rate over the whole of the South equivalent to 83 percent of the growth potential of the land.[11]

## CANADA'S FORESTS

Canada has some 1,090 million acres of forest land. In 1966 it was harvesting from this about 3.2 billion cubic feet. It exported 4.5 billion board feet of lumber, 2.5 million tons of wood pulp, and 6.5 tons of paper. The United States was one of the major importers of Canadian forest products. It is generally agreed that Canada could cut between 7.5 and 8.2 billion cubic feet annually without overcutting its forests. But Canada's population is growing at a faster rate per capita than is that of the United States. By the year 2000 it is expected that the demands on her forests will be for an annual cut of from 8.2 to 10.9 billion cubic feet. To sustain such an annual wood harvest would require that the 1966 growth rate of her forests be increased by $1/3$, or from an average of 18.4 to 24.1 cubic feet per acre per year.[1] Since much of Canada's forests lie in northern latitudes where growth rate is low, it may not be easy to attain this average.

Still, it is clear that Canada can indeed supply some of the long-fibered pulp and softwood lumber that the United States will need by the year 2000. The total imports from Canada by 2000 will likely make up scarcely more than 10 to 20 percent of total United States consumption. Nations other than the United States will be seeking to import forest products from Canada, even as they do today.

## NORTH AMERICA'S ROLE IN THE FUTURE WORLD TIMBER MARKET

From the economist's viewpoint there are two serious weaknesses in the world and United States' outlooks on the future timber situation. Although the UN-FAO attempts to present information on a broad regional basis, the country-by-country presentations are based mainly on proposals toward increasing national self-sufficiency in timber. The U.S. Forest Service forecasts, on the other hand, are based almost entirely on meeting our national needs rather than our national needs in relation to that of the world timber trade. On looking at this question, Zivnuska[18] considers it provincial not to include Canada along with the United States in forecasts of future demands and supplies. He notes that Canada furnishes 88 percent of United States wood imports. Already wood-using firms operate in both nations. The United States uses 50 percent more wood from Canada than is consumed within Canada. When the forest resources of both countries are considered together, a more realistic economic picture emerges of a wood-surplus region that can export more wood than it imports. With North America as the unit for projection, Zivnuska states that the wood requirements can readily be met in the year 2000 with a substantial concurrent increase in export to Europe, Japan, and other wood-deficit regions. He believes that this can be done without exceeding Canada's allowable cut or causing a significant decline in the United States' timber inventory.[18]

The second weakness in forecasts is the lack of economic data adequate to treat demand and supply in terms of price. As the price for stumpage increases, timber now deemed inaccessible may become operable. Also, due to changes in ownership, much now considered to be merchantable may actually be economically unavailable if holdings are too small to be logged profitably. Zivnuska also raises the question of whether the United States forest industries will, in fact, through research, new-product development, and aggressive marketing, be able to hold their traditional markets against the plastics, steel, and other industries' vigorous push into the housing, furniture, and construction fields.

Studies made in 1969 show that steel can already be used more economically than wood for roof trusses, studding, and other framing members. The steel industry is actively promoting steel-framed houses as opposed to wood.

On the other hand, the U.S. Forest Service in cooperation with the wood industries is making studies to keep wood in the construction field. A new type of factory-built house has been designed, using treated wood foundations. It can be erected and enclosed on the site in 2 days. So possibilities for imaginative innovation in wood use also abound.

## SUMMARY AND OUTLOOK

Forests occupy more than one-fourth the land area of the earth and one-third that of the United States. They exert a major influence on environment. Their chief commercial product is wood.

Nationally and worldwide the current trend and future outlook is for sharply increased use of wood products.

Per capita use of wood varies widely with its availability, the presence of wood-processing plants, and the price of the product. Among nations it increases rapidly with increase in per capita income.

The prospects are bright for augmenting substantially the per capita income and foreign trade of developing countries through developing their forests and forest-based industries.

Logging, lumber manufacturing, and much secondary wood-processing are labor-intensive rather than capital-intensive. Exports of logs, lumber, veneer, and processed wood offer opportunities for developing nations to obtain foreign exchange for investing in paper-manufacturing plants and others that are capital-intensive.

Worldwide, wood-based industries in 1961 accounted for 6 percent of the total value created by manufacturing and employed 9 percent of the workers engaged in manufacturing.

With rising per capita income, use of fuel wood tends to decline, of lumber to level off, but of wood-based panel products and paper to increase sharply.

Europe, except for Sweden, Finland, and European U.S.S.R., is expected to remain a wood-importing region with growing demands for the indefinite future. The exports from Sweden and Finland will be consumed mainly by the other European nations.

The Mediterranean and Middle East region, including Pakistan and India, can scarcely become self-sufficient. Even the best-situated nations seem likely to remain wood importers. Imports will likely increase in the future as the income of these nations rises. Considerable forest planting for soil and watershed protection is expected, which will supply wood for local use.

The U.S.S.R. is expected to become increasingly important as an exporter of coniferous timber as the forests of Siberia are opened up through railway and highway construction.

China seems likely to remain a wood-deficit nation for a long time to come. It is expected that what wood she will be able to harvest from her residual forests and from plantations will be almost all consumed within the nation. China, together with Mexico, is expected to become an important source of genetic stock of pines and other valuable conifers for planting in many parts of the world. The extensive planting within the country will furnish evidence of the versatility of the various species for use in other lands.

For the foreseeable future, Japan appears able to hold her position as one of the most important among wood-importing nations. Her merchant fleet, many good seaports, and well-developed industrial and mercantile companies, as well as her experience in forestry and timber trade, seem to assure her a good future as a wood importer, processor, and merchandiser.

The Southeast Asian region seems likely to hold its position as an important exporter of teak, dipterocarps, and other valuable tropical woods. With the prospect of opening the forests of Malaysia and Indonesia to extensive timber use, the timber-export potential of the region will be enormously increased. Since much of the land is relatively accessible to the seacoast, early development seems likely.

African forests have been but little developed for timber export. Development has begun in West Africa where good timber can be harvested along the navigable rivers. These forests will likely play an important role in timber trade in the future, but much of their timber will be required within Africa itself. East Africa and southern Africa have some important plantations but little natural forests of commercial quality. These regions are likely to continue to be modest importers of timber.

Australia can probably become largely self-sufficient from natural and plantation-grown timber, as New Zealand now is. She also has large areas of eucalyptus forests that supply both wood for national needs and useful species for planting elsewhere in the world.

Canada and the U.S.S.R. seem likely to remain the two main exporters of coniferous timber for an indefinite period.

The United States is expected to continue to be an importer of softwood from Canada and of tropical hardwoods from Latin America. She will probably be self-sufficient for the bulk of her timber needs until 1985 and perhaps much longer.

Mexico will probably be largely able to meet her timber needs from her own forests.

Central America and the Caribbean have considerable areas of valuable forests but also have needs for additional timber from the world market.

South America has the world's largest reserves of hardwood forests that may some day play an important role in the world timber trade. The accessible rain forests near the north and west coasts seem likely to furnish considerable wood for export. So also do strategic locations along the navigable sections of the Amazon and a few other major rivers. Plantation-grown wood seems likely to become a major source of supply for southern Brazil, Uruguay, Argentina, and Chile.

The projected trends in future wood use are strongly upward both for the United States and for the world as a whole, with the rate of increase for the developing nations higher than that for the developed nations.

Wood resources seem ample to meet effective demand up to the year 1985 and beyond for the world as a whole as well as for the United States. Considerable rise in price may be necessary to effect adequate harvesting, processing, and distribution of wood products.

The above suggests two viewpoints from which to look at the future of commercial wood use. The first, and pessimistic one, is of increasing costs, declining markets, increasing economic nationalism with its restrictions on international trade, and consequently a gloomy future for wood industries and the people who use their products. The second, and optimistic viewpoint, looks at the opportunities for (1) increasing the productivity of forest land; (2) closer utilization of the wood that is grown; (3) more economical harvesting, processing, and distribution; (4) improvement of product quality; and (5) vigorous market promotion of wood-based products. It also looks to research to discover new uses, to develop new processing techniques, and to reduce timber losses due to disease, decay, insect damage, and poor design in use. And most of all it looks to a growing market for decades to come. Such a positive and optimistic outlook for the future would appear to be more in keeping with the aspirations of man and with orderly progress toward a more peaceful world than the pessimistic and somewhat miserly approach.

Accepting the positive outlook, and for developing nations an annual increase in per capita gross national product of some 3 to 6 percent, the world's foresters and forest industries will have a huge task to perform. This will embrace every facet of forest establishment, care, and use, and of timber harvesting, processing, and distribution. It will also embrace the additional products and services of forests such as forage, water, wildlife resources, environment protection, public amenities, and recreation.

New developments and improvements of methodology have been growing rapidly in the wood industries during the past two decades. Much of this has already spread to developing countries. There seems to be realism, therefore, in expecting forests and forestry to play a significant and growing role in future world economy.

The outlook, therefore, is challenging and need not be defeatist.

## LITERATURE CITED

1. Besley, Lowell. 1967. Canada's Pulp and Paper Industry, *Journal of Forestry,* **65**:696–707.
2. Goor, A. Y., and C. W. Barney. 1968. "Forest Tree Planting in Arid Zones," The Ronald Press Company, New York. 409 pp.
3. Hair, Dwight. 1969. Current and Prospective Trends in Imports of Hardwood Timber Products into the United States, U.S. Department of Agriculture, Forest Service, May 15, 1969. 29 pp.
4. Hair, Dwight. 1970. The Prospective Demand and Supply Situation for Timber and the Role of the National Forests in Increasing Timber Supplies, U.S. Department of Agriculture, Forest Service, November 1970. 8 pp.
5. Hair, Dwight, and H. O. Fleischer. 1970. Meeting Growing Demands for Wood Products, U.S. Department of Agriculture, Forest Service, November 1970. 22 pp.
6. Josephson, H. R. 1969. Substitution—A Problem for Wood?, U.S. Department of Agriculture, Forest Service, December 3, 1969. 11 pp.
7. Josephson, H. R., and Dwight Hair. 1970. The Timber Resource Situation in the United States, Review Draft, U.S. Department of Agriculture, Forest Service. 29 pp.
8. Lassen, L. E., and Dwight Hair. 1970. Potential Gains in Wood Supply through Improved Technology, *Journal of Forestry,* **68**:404–407.
9. *Proceedings—Conference on Tropical Hardwoods,* Syracuse, N.Y., August 18–21, 1969, State University College of Forestry at Syracuse University, Syracuse, N.Y. Various paging.
10. Richardson, S. D. 1966. "Forestry in Communist China," The Johns Hopkins Press, Baltimore. 237 pp.
11. Southern Forest Resource Analysis Committee. 1969. "The South's Third Forest," Forest Farmers Association, Atlanta, Ga. 111 pp.
12. Stadelman, Russell C. 1966. "Forests of Southeast Asia," Russell C. Stadelman, Memphis, Tenn. 245 pp.
13. United Nations Food and Agriculture Organization Secretariat. 1966. Wood: World Trends and Prospects, *Unasylva,* **20**:1–136.
14. United Nations Food and Agriculture Organization Secretariat. 1968. Commodity Report. Plywood, Particle Board and Fibreboard—An FAO World Survey of Production Capacity for 1968, *Unasylva,* **22**(4):24–46.
15. U.S. Department of Agriculture, Forest Service. 1958. "Timber Resources for America's Future," Forest Resource Report No. 14, U.S. Government Printing Office, Washington. 713 pp.
16. U.S. Department of Agriculture, Forest Service. 1965. "Timber Trends in the United States," Forest Resource Report No. 17, U.S. Government Printing Office, Washington. 235 pp.
17. U.S. Department of Agriculture, Forest Service. 1969. Possibilities for Meeting Future Demands for Softwood Timber in the United States, Prepared for Working Group of the Cabinet Task Force on Lumber, Revised September 29, 1969. 25 pp.
18. Zivnuska, John A. 1967. "U.S. Timber Resources in a World Economy," The Johns Hopkins Press, Baltimore. 125 pp.

# Part Six

# Forestry and Society

Chapter 19

# Social Benefits of Forestry

A mission of the forestry profession is to prevent and to ameliorate human impoverishment in material goods and spirit brought about through misuse of forest land. In 1927, a property of 2,200 acres was acquired by a college as a demonstration forest. At the time of purchase the property included 11 farms, of which only 3 were occupied. The owners of these were dependent on outside work for income. Most of the timber had been cut off. The college employed a forester to plant the open lands, thin the young stands, and convert the cut trees into lumber, which was sold locally. Cutting was concentrated on the poorer-quality trees and the white pines infected with blister-rust fungus. It was kept in line with growth of the old trees. After 20 years of forest management the property was providing full-time work for 9 men, a level it has since maintained for more than 2 decades. The timber volume, instead of decreasing, was increasing. Land that had been a social liability became a social asset through application of the forester's skill.[6]

The Bartrams and other early travelers along the east coastal plain and piedmont regions of the southern United States remarked on the splendid pine forests that this land supported.[14] Yet rapid exploitation of this land began even in colonial times, especially along the rivers and other accessible areas. Rice and even sericulture were introduced, along with conventional crops. After the pineries of the Lake states were cut out, the lumbermen moved to the south and began cutting the southern pines. As timber was cut off, farmers moved in to plant cotton, tobacco, and other crops. Under such culture the sandy soils of the coastal plain were soon leached of their fertility, and the piedmont soils lost their fertility by erosion. Field after field was abandoned. Pines reseeded these open areas and grew up into the South's second forest. This forest, with the lumber, plywood, and pulp and paper industries it supported, brought new wealth to a region from which the primary wealth had been removed by logging and by cotton-growing that depleted the soils.

In some respects the second forest has been even more productive than the first. In 1935, the second forest contained 120 billion cubic feet of wood. Between 1935 and 1969 the forest provided 169 billion cubic feet to supply the forest industries, yet still it had standing 150 billion cubic feet at the end of the period. This activity added $10 billion to the economy of the South and $4 billion to that of the North.[10, 15] The trend of wood use in the South is sharply upward. If the forest industries are to continue to expand to meet anticipated demands, by the year 2000 the South's third forest must be 2.3 times as productive as the second forest was in 1969. A committee of the South's leading foresters have drafted recommendations that, if adopted, would hopefully enable the third forest to meet the year 2000 goal. Steps were taken in 1970 to put these recommendations into effect.

The forests of Spain and other Mediterranean countries were cut off centuries ago, and the land has been overgrazed and badly eroded since that time. About 1945 the Spanish government embarked on a great land-restoration program involving erosion, torrent and flood control, livestock and pasture improvement, water impoundment for irrigation and power, forest planting, and development of new agricultural and forest industries. The rehabilitation task itself provided employment and income to the impoverished local people. The program, begun at the end of World War II, was well advanced by 1970, with prosperity slowly returning to regions depleted of resources for centuries. Illustrative of new values created is that of one 20-year-old pine plantation. As the pines grew up, volunteer mushrooms came in. The harvest of these in one year produced revenue equal to half the planting costs. Both pines and mushrooms were still growing.

Much land in the states of Minas Gerais, São Paulo, and Parana,

Brazil, were cleared of forests for coffee growing. Disease and soil depletion through erosion so reduced coffee yields that landowners turned to sugar cane growing and eventually to cattle grazing. Both caused even further soil depletion. Mining and pulp and paper companies have bought up some of such depleted lands and planted them to trees which grow remarkably well. One tract of 150,000 hectares of cutover and grazed land was bought up by a paper company that planted 50,000 hectares to eucalyptus. This could be harvested on a 7-year rotation for pulpwood. Within a few years the 150,000 hectares that supported but 100 families in cattle grazing was supporting some 6,000 families in timber growing and paper manufacture.

These are but a few examples of the social benefits of forestry.

## THE NATURE OF SOCIAL BENEFITS

Social benefits of forests include soil and watershed protection, amelioration of floods, favorable modification of local climate, fostering of useful wildlife, public recreation, raw material for industry, maintenance of reserve timber for national emergencies, a source of livelihood for local workers, environmental protection and improvement, and esthetic enjoyment. It is the sum of such social benefits that justifies public expenditures to support forestry on public and private lands. Most of these benefits have already been dealt with in earlier chapters.

**Reserve for National Emergencies** The people of the United States have always used natural resources in great abundance when at war and their use of wood has been no exception. Yet when it is realized how much labor must go into the harvesting, processing, and use of wood for any purpose, it becomes apparent how vital it is to national security. Lumber for cantonments, for pallets and packaging for overseas shipment, for construction of bases, warehouses, and docks, and for repair of bomb damage is required in huge amounts. In 1944 the lumber to package war material for overseas shipment alone required 16 billion board feet or half of all lumber produced that year. In the fighting area, lumber was needed in generous amounts for bridges, road repair, supply dumps, hospitals, and temporary shelters.

**Employment for People** During the Great Depression of the 1930s any resource that could provide work was looked upon with high favor. Public leaders found in our undeveloped forests an almost inexhaustible supply of useful work that would add to the future value of the resource. Building needed roads and telephone lines, planting trees, and improving cutover lands, figuratively soaked up man-days of labor. Needed work on public forests alone had by no means been exhausted when the programs terminated after a total of 9 years of operation.

Of far greater importance than emergency programs has been the year-in, year-out employment in protecting forests and in harvesting and processing products (Fig. 19-1).

## FORESTRY AS A FORM OF LAND USE

The examples cited at the beginning of the chapter afford some concept of the big role forests can play in rejuvenating resource-depleted regions. But to gain a more perceptive view of some of the physical and economic forces at work requires study in greater depth.

The older sections of our country that have gone through a high stage of industrial development well illustrate the trends. It is here that wealth is abundant, markets for labor and products highly developed, and

**Figure 19-1** An integrated hardwood sawmill, Connor Lumber and Land Company, Laona, Wis. The company operates a sawmill, dry kilns, flooring mill, dimension mill, furniture factory, and wood-flour plant. Operations are supported by 45,000 acres of industrial forest land. *(American Forest Institute.)*

evolution in land use well advanced. It is in some of these states also that progressive steps have been taken to change the use of land to fit modern economic conditions. Such a case is Broome County, New York.

**Forests of Broome County** Broome County, New York, is one of the more prosperous counties of the Empire State. The valley in which the Triple Cities—Endicott, Johnson City, and Binghamton—are located has been called the Valley of Opportunity. It has been the valley of opportunity for many people, for in these three cities are located some of the most prosperous and rapidly growing industries of the United States.

Outside of the cities, opportunities in Broome County have been far more circumscribed. Farmlands have been abandoned and have reverted to forest. During a 30-year period agricultural land has decreased by approximately one-half while forest area has doubled.[4, 5] The trend has been as shown below:

| Year | Agricultural land, in thousands of acres | Forest land, in thousands of acres |
|---|---|---|
| 1938 | 303 | 121 |
| 1948 | 258 | 156 |
| 1968 | 152 | 241 |

Two questions arise: Why was this land cleared for agriculture in the first place? What happened to drive or lure the farmers off the land?

The early settlers who came to New York found the uplands much the easiest to clear for cultivation; the land was high, away from the "miasma" and mosquitoes that infested the lowlands, and natural drainage existed. But erosion washed away much of the fertile topsoil. The advent of machine tillage and other advances in scientific farming added to production of hill lands, but it added the same percentage of increase to fertile, easily tilled valley farms. The income-spread between the good and the poor farms expanded enormously, and hill farmers were forced out of competition. What once seemed a logical land-use pattern had been completely reversed.

In Broome County, one-third of the farmland was classified as submarginal in 1938, and by 1950 one-half was so classified.[8] Farmers' sons, realizing the poor living to be obtained from the land, either found better farms or sought employment in the cities. Farmers themselves often followed their sons to city jobs. Some retained their residence on the land but allowed the pastures and fields to lie idle. These gradually reverted to forests. Many urban workers themselves purchased submarginal farms for a rural residence and they also allowed the land to grow up to trees. Some even hastened the process by planting conifers.

The expansion of the three cities and the outlying villages also took land from agriculture, though this was of far less significance to the area than that which reverted to forest.

Of the land given up by agriculture during the 30-year period from 1938 to 1968, some 36,000 acres remained nonstocked with trees in 1968. Changes between 1948 and 1968 were:

|  | 1948[4] | 1968[5] |
|---|---|---|
| Seedlings and saplings (thousand acres) | 33 | 111 |
| Plantations (thousand acres) | 2 | 18 |
| Sawtimber volume (million bd ft) | 235 | 272 |
| 20-year timber growth (million bd ft) |  | 239 |
| Presumed 20-year timber harvest (million bd ft) |  | 202 |

In Broome County, during 20 years the people have harvested almost as much timber as they had standing at the beginning and yet they have increased their total growing stock by 16 percent. Total annual growth increment can be expected to increase substantially during the following years as seedling and sapling stands increase to pole and sawtimber sizes. Even so, in 1968 the average growth rate per acre remained less than a quarter what it may be expected to become when all forests are fully stocked and a well-managed distribution of age-classes is attained.

Fortunately for the people of Broome County, their main economic dependence is neither on agriculture nor on forests, but on industry. This makes their expanding forest area all the more useful for its contribution to environmental improvement. Instead of derelict farm buildings punctuating a desolate landscape of barren, abandoned fields, neat rural residences now look out over forest-covered hills. Recreational use of this land is increasing rapidly as individual home owners, resort operators, and associations take over the land.

It is fortunate also that alert citizens' groups are taking advantage of economic trends to enhance the county environment.

**The Forest Lands of the Lake States** The three states of Michigan, Wisconsin, and Minnesota loomed large in timber production at the turn of the century and in the 2 following decades. Immense stands of white pine and red pine covered almost one-third the forest area. Valuable hardwoods were plentiful in Wisconsin and the upper peninsula of Michigan. These lands, for the most part, were cut with no thought for future productivity of the land. Fires burned through the slashings. What had been superb forests were converted into brush fields, into grassland, or into stands of aspen, jack pine, and paper birch that had negligible commercial value.

Minnesota and Michigan were covered three times by forest surveys, and Wisconsin twice, Adjusting results to the base years 1936, 1956, and 1966 makes examination of trends possible.[1, 2, 3, 11, 12] The total forest land decreased in each of the 3 states during the 30-year period, for the 3 together by some 6 percent. Timber volumes by contrast showed increases over 1936 by some 30 percent in 1956 and 70 percent in 1966. This increase was due to timber growth and to regeneration of nonstocked forest land. The 30-year period saw a decline in volume of softwood and a large increase in hardwood. This reflected the heavier softwood cutting due to the better market that prevailed for them. There was also a decline in the volume of trees in the large sawtimber sizes, as these are the ones most wanted by the mills.

The trends in increased timber-growth rates and in volume of standing timber that prevailed over the 30-year period from 1936 to 1966, if continued, even at a somewhat reduced rate for the period from 1966 to 1986, would result in almost trebling the growth rate of 1936 and doubling that of 1956. A substantial program of forest-planting, gradual conversion of brush land to timber species, and growth of seedlings and saplings into pole sizes, have all been responsible for this more-than-doubling of the growth rate. If the expected growth rate is attained by the year 1985, it still will but average some 100 board feet per acre per year. Eventually it could approach twice this value.

On one specific area, the Chippewa National Forest, by 1952 the timber-growth rate had increased far beyond the three-state average. This area deserves special consideration. Between 1902 and 1925 the timber, which belonged to the Chippewa Indians, was cut off and the cutover land was placed under the administration of the U.S. Forest Service. During the heyday of harvesting the old-growth timber, the village of Cass Lake, which served as the headquarters for large mills and for the U.S. Forest Supervisor, went through a tremendous boom development. Stores, hotels, churches, and saloons prospered—as did the sawmills. Many dwellings were erected. But when the big mills shut down and the lumbermen moved out in 1925, the village of Cass Lake suffered a disastrous economic decline. Saloons, stores, and hotels closed. Owners who once could place $30,000 orders were reduced to penury. Houses were no longer painted or kept in repair. Churches were barely able to keep open. The village relied heavily on the limited payrolls of the National Forest headquarters and a local Indian office. During the depression years the emergency work-relief activities were the main source of income.

Meanwhile, the forest administration began seeking markets for the despised aspen and jack pine and for small-sized red and white pines. Production of piling, poles, mine timbers, box lumber, and fence posts was added to the sale of sawtimber and pulpwood. Applied research greatly increased the effectiveness of forestry measures. A cooperative of timber operators provided finances for men with little or no capital. As a result of all these measures, employment spread, stumpage prices increased, and prosperity gradually returned. As local people gained confidence in the stability of their livelihood, civic pride grew. The community changed from a debilitated lumber camp to an orderly, growing village of confident, loyal citizens.

**Western North America** Guthrie and Armstrong studied the forest industries of British Columbia and the 12 western states of Alaska, Arizona, California, Colorado, Idaho, Montana, Nevada, New Mexico, Oregon, Utah, Washington, and Wyoming.[7]

The population of this region was 27 million in 1958. The forest-products industries accounted for 13 percent of the value of industrial production in this region and for 17 percent of the total employment. In 1957 the region contained more than 3 trillion board feet of sawtimber, over half the sawtimber volume in North America.

Guthrie and Armstrong sought to determine whether this region could expect to maintain its position in lumber, pulp and paper, and veneer and plywood production in the years ahead, looked at from two viewpoints: (1) could these industries maintain their competitive position in the eastern and the world market; and (2) would the forest resources of the region be sufficient to permit this continued high rate of use without serious depletion of the timber supply? Their conclusions were positive on both questions. They felt that because of high quality and rapid growth of its timber and, particularly, because of the reserve old-growth timber available and the efficiency of logging, sawmill operations, pulp and paper plants, and veneer and plywood manufacturing, their products could remain competitive in the eastern market. They recognized, however, that there would have to be a substantial shift in the species of timber used for the three products with an expansion in the use of the true firs, spruces, and lodgepole pine. This would necessitate a shifting of location by individual industries that would not have sufficient resources near their present mill sites. Also it would be necessary to use lower-quality logs for veneer. Continued timber cutting would result in some reduction in the reserve supply of old-growth timber. They recognized further that the success of the industry in adjusting to these changes would depend to some extent upon the policies that the industries themselves adopted and the policies that the governments adopted in managing public forest lands. For the region as a whole, 83 percent of the commercial forest land and about 81 percent of the total timber volume were under the jurisdiction of forestry agencies of the United States government and the Provincial Government of British Columbia. These agencies practice sustained-yield management, interpreting this to mean production at the highest practical rate that the land can indefinitely support.

## WORLDWIDE EXAMPLES OF GOOD AND POOR LAND USE

The Danes have converted the heath on Jutland into good farms and forests. Through use of forest plantations, shelterbelts, proper farming practices, patience, and hard work, they have built up a prosperous community, supporting as many as 25,000 people in a region that once supported only scattered villages and a few nomadic shepherds. The Swedes have been particularly successful in building up a high standard of living through proper care of their forests, fields, pasturelands, and mineral resources.

Norwegians inhabit a most inhospitable land. The western coast is outstanding for its beauty and ruggedness, but for this very reason it has a minimum of level land on which agriculture and forestry can be practiced. Through fishing and through diligent cultivation of forests and crops on the small patches of moderately sloping land along the fjords, however, the people maintain a good standard of living, educate their children, and live in harmony with their environment. Switzerland has Europe's most rugged topography, but the Swiss practice intensive dairying, farming, and forestry.

Italy, Greece, Albania, much of North Africa, and the Asia Minor region stand in sharp contrast to the good land-use practices just described. Mountain slopes have been denuded of timber, erosion is severe, and human poverty and misery have become widespread throughout a region that once was the garden spot of the world (Fig. 19-2). Yet here, too, there has been a turn for the better. The Italians are making important conifer plantations in the Apennines, depending meanwhile on rapidly growing hybrid poplars and eucalyptus planted on agricultural soil to tide them over. Spain, France, Greece, Turkey, Israel, and the countries of North Africa are engaged in a vast Mediterranean land-rehabilitation program involving forest planting, pastureland improvement, and irrigated cropland.

On the Island of Formosa, Chinese and the native Taiwanese have

**Figure 19-2** Remnants of impressive ancient cities testify to a rich past on land now impoverished by centuries of overuse. Greek columns at Aspendos in a desolate Anatolian landscape.

planted more than 500,000 acres of steep, eroding land to forests. In this way they are helping to tame their flash floods that wash out rice paddies and leave streams strewn with boulders. It has taken hard, dogged work. Every acre that can be held in agriculture is being saved and supported by tree planting on the steep, surrounding slopes. Land is precious on this densely populated island, and the hardworking people are striving heroically to save and expand it.[13]

The Philippine Islands have some of the best tropical hardwood forests to be found anywhere in the world. Huge stands running 80 percent or more to red and white luan, the Philippine mahogany of commerce, cover extensive areas in Mindanao and other southern islands. These trees may rise to a height of 80 feet or more clear of limbs, with diameters up to 50 or 60 inches. The quality of the timber is generally sound. Following partial cutting, the stands can grow at a rate of 1,000 board feet per acre annually; and under favorable circumstances, twice this fast. The Philippine government owns most of the forest land. Under its Bureau of Forestry, standing timber is being sold to various operating companies. The Bureau requires that all timber operators harvest on a partial-cutting basis, leaving smaller-sized trees for future cut.

As of 1967, Philippine forests covered 14.6 million hectares. The local forest industries employed 160,000 men who produced the major lumber, hardboard, and veneer products consumed by the Republic. In addition, these industries exported $182 million worth of forest products that accounted for one-fourth of the foreign exchange earnings for the year. Overall, forest industries ranked second among the three major producing segments of Philippine economy. A planting program of 26,000 hectares per year was restoring abandoned cultivated land to commercial forests. The reserve of standing timber was estimated to be 1,667 million cubic meters—enough to support cutting at the 1966 legal rate of 8 million cubic meters for 208 years. Given reasonable care, the forests could easily sustain even a much higher rate of cutting indefinitely. Other tropical nations have looked upon the Philippine timber exploitation with favor and plan similar operations themselves.

But as mentioned in Chap. 2, the picture is not all bright. Squatters, shifting cultivators, and others were estimated to be illegally destroying 15.3 million cubic meters annually on the island of Mindanao alone. Little of the timber killed is used by the kaingineros, as the shifting cultivators are called (Fig. 2-3).

The Philippines is by no means the only nation whose forests are threatened by population pressure. A Pakistan district forester, gazing out from a forested ridge at the recently cleared patches along sharply dissected ravines, felt impelled to express his concern in these words, "Pakistan forests are but tiny islands in a sea of humanity that beats incessantly on their shore lines."

In the Far East Asia region alone some 24.5 million men get their livelihood mainly from practicing shifting cultivation. As much as 8.5 million hectares are cleared annually and almost as many are abandoned to scrub growth. The total area subjected to shifting cultivation in this region is estimated to be 103 million hectares. In Africa south of the Sahara, 100 million hectares of high forest are reported to have been destroyed mainly by shifting cultivation.

The immediate reaction of any forester first confronted with the destruction of forests by shifting cultivation is that the practice must be stopped forthwith. Further acquaintance with it leads him to conclude that the only way in which it can be stopped is by providing these people with alternate sources of livelihood that are more attractive than the grueling work of annually clearing new lands.

Is there no good word to be spoken for the shifting cultivator? In terms of world timber supply the practice is probably still of only local importance. Shifting cultivation in many areas is far less destructive to soil than is continuous cultivation on steep slopes. Many abandoned clearings do return to valuable pioneer-forest types in the tropics, as they did in the southern United States. And whatever may be said about the shifting cultivator, he is after all a rugged, self-reliant pioneer seeking merely to provide for himself and his family.

Moderate grazing may also be practiced in the forest of some regions without doing significant damage. Heavy grazing, on the other hand, leads to browsing of young growth, compaction of the soil, and concomitant high runoff and accelerated erosion. Eventually overgrazing may convert a forest to brushland.

The People's Republic of China suffers from the greatest wood deficit of any major nation. Almost every scrap of wood and paper is salvaged for reuse. Yet plenty of land exists that is unsuitable for tilled crops and on which forests can be grown. About 1950, a program of land restoration through water conservation and forest planting was begun on a heroic scale. Old-growth timber is being opened to use by extending transportation lines. It is hoped that yield from plantations will become available to offset planned increased harvesting of existing supplies.[9]

## THE INDICATIVE WORLD PLAN

Extending the social benefits of good land use to developing nations throughout the world is the major mission of the United Nations Food and Agriculture Organization, including its Department of Forestry and Forest Industries. It is also a major objective of bilateral aid programs sponsored by the United States, many European nations, and Japan. Forestry has been included in many such programs. Results in individual countries have been impressive. But very much more can be achieved and is in the process of being brought to fruition. The FAO indicative world plan for agriculture, including forestry, points to proven measures that would increase social benefits by 50 to 100 percent through forest and land-use improvements. These are such as could be brought about within a 20-year period. Activation of the proposals in the indicative world plan is a question for each nation to decide for itself.

What FAO attempts to do through the plan is to acquaint each nation with the potential contribution its forests might make to the general welfare if practices well tested in other lands are carried out.

## BROAD SOCIAL BENEFITS

Added to the benefits inherent in commodity values based on forest products are social values of many intangible kinds that are dealt with in

Chaps. 9, 10 and 20. But how much social benefit does accrue to the thousands of people who each year spend leisure days in Zurich's Sihlwald, in England's New Forest, or in the small but heavily used forests of the Netherlands? What are the benefits to Boy Scout troops who go overnight-camping winter and summer in county and state forests? What sends bird watchers and flower lovers into the forests in spring, summer, winter, and fall? What sends millions to national parks and forests each summer? Such social benefits have been mentioned frequently and are but small examples of the benefits of life-sustaining forest ecosystems.

Such benefits cannot be quantified in tangible expressions such as monetary value, visitor days, or vacation expenditures. Even less is it possible to put a tangible value on a quality environment that enables a man to lead a life that affords time and stimulus for creative thought and harmonious living. Forests help to provide all such benefits, together with those that contribute to life's physical necessities.

**LITERATURE CITED**

1  Chase, Clarence D., Ray E. Pfeifer, and John S. Spencer, Jr. 1970. The Growing Timber Resource of Michigan, 1966, North Central Forest Experiment Station, St. Paul, Minn. 62 pp. U.S. Department of Agriculture Forest Service Resource Bulletin NC-9.
2  Cunningham, R. N., and Forest Survey Staff. 1950. Forest Resources of the Lake States Region, Forest Resource Report 1, U.S. Department of Agriculture, Washington. 57 pp.
3  Cunningham, R. N., and Forest Survey Staff. 1956. Lake States Timber Resources, U.S. Department of Agriculture, Lake States Forest Experiment Station Paper 37, St. Paul, Minn. 31 pp.
4  Davis, James E., Miles J. Ferree, and Neil J. Stout. 1953. Broome County Forest Data, State University College of Forestry at Syracuse University, Bulletin No. 27. 63 pp.
5  Ferguson, Roland H., and Carl E. Mayer. 1970. The Timber Resources of New York, Northeast Forest Experiment Station, Upper Darby, Pa. 193 pp. U.S. Department of Agriculture Forest Service Resource Bulletin NE-20.
6  Foster, Clifford H., and Burt P. Kirkland. 1949. Results of Twenty Years of Intensive Forest Management, The Charles Lathrop Pack Forestry Foundation, Washington, D.C. 36 pp.
7  Guthrie, John A., and George R. Armstrong. 1961. "Western Forest Industry: An Economic Outlook," The Johns Hopkins Press, Baltimore. 324 pp.
8  Pasto, Jerome K., and Howard E. Conklin. 1949. An Economic Classification of Rural Land, Broome County, N.Y., Cornell Economic Land Classification Leaflet 1, Cornell University Press, Ithaca, N.Y.
9  Richardson, S. D. 1966. "Forestry in Communist China," The Johns Hopkins Press, Baltimore. 237 pp.

10  Southern Forest Resource Analysis Committee. 1969. The South's Third Forest, How It Can Meet Future Demands, Forest Farmers Association, Atlanta, Ga. 111 pp.
11  Stone, Robert N. 1966. A Third Look at Minnesota's Timber, North Central Forest Experiment Station, St. Paul, Minn. 64 pp. U.S. Forest Service Resource Bulletin NC-1.
12  Stone, Robert N., and Harry W. Thorne. 1961. Wisconsin's Forest Resources, Station Paper No. 90, Lake States Forest Experiment Station, St. Paul, Minn. 52 pp.
13  Tao, Yu Tien. 1960. Forestry of Taiwan, Taiwan Forestry Bureau, Forestry Information Bulletin No. 21, Taiwan, China. 47 pp.
14  Watson, Winslow C. (ed.). 1968. "Men and Times of the Revolution," 2d ed., Crown Point Press, Inc., Elizabethtown, N.Y. 557 pp.
15  Wheeler, Philip R. 1970. The South's Third Forest, *Journal of Forestry*, **68**:142–146.

Chapter 20

# Forests and Outdoor Recreation

Outdoor recreation activities may be thought of as forming a continuum from tending the flowers in a window box to penetrating the most remote wilderness. They include hunting, fishing, and gathering nuts, mushrooms, and berries, which date back to the time when these were necessary food-acquiring activities; outdoor games and sports; hiking, skiing, swimming, canoeing, motor boating, sailing, camping, bird watching, mountaineering, rock climbing, snowmobiling, and many other activities. Participants in each of these activities look to forests and other open lands for opportunity to engage in the sport currently popular. To accommodate all these demands, which seem to be increasing at an exponential rate, is a task of growing complexity for public and private land administrators.

A major task is providing access and facilities so that recreationists in large numbers may visit a forest or other outdoor attraction to enjoy an outing. The time has long passed when a simple sign by the roadside, labeled "spring," sufficed for camp and picnic grounds. One by one were added fireplaces, pit toilets, a protected and safe water supply, picnic tables, garbage cans, tent sites and platforms, flush toilets, showers and laundry facilities, and with the arrival of campers and trailers, water, sewage, and electric connections. At lakes, bathing beaches were developed with elaborate bathhouses, also often with nearby boat-launching sites and marinas. Graded trails led to points of interest, including rail-guarded overlooks. Museums were sometimes built. Generous parking areas had to be provided, some with attendants on duty during busy days (Fig. 20-1).

To plan and construct such facilities on national and state forests requires expert knowledge far beyond that required to lay out a camp for loggers or fire fighters. A camp site to take care of 500 modern campers and trailers must be planned to provide many of the facilities that would be found in a modern village accommodating as many families. It includes developing a safe water system, sewage collection and treatment, solid-waste disposal, a transportation network, electric lines and outlets, as well as bathing facilities, marinas, and trails to points of interest. Facilities of such a nature require the care of a permanent maintenance and custodial force.

Most recreation sites on public and private forests are planned with facilities that lie somewhere between the most primitive, which are recognized today as suitable only in large wilderness areas, and the highly sophisticated camps provided in national parks, some national and state forests, and in a few privately developed camp sites. Planning the

**Figure 20-1** Evaluating the qualities of a wild and scenic river. *(U.S. Department of Interior, Bureau of Outdoor Recreation.)*

elaborate camp sites is clearly the work for a landscape architect, while building simple picnic grounds can be done by foresters. The points to be borne in mind are: use soon tends to overrun simple layouts; no sharp line can be drawn between what the forester should do unaided and what falls clearly within the sphere of the landscape architect; that facilities for public use also form a continuum varying from the primitive to elaborate.

One facet of the task is to provide outlets for a multitude of people to engage in diverse and newly popular outdoor activities. Another is to provide the wide range of facilities and conveniences the participants will require. A third and most important task is to harmonize the whole and fit it to the needs of all classes of people. Foresters must inevitably participate in all three tasks. They need to recognize the several approaches available and the limitations involved. Park developers have been obliged to provide for heavy concentrations of people much longer than foresters have and have crystallized their thinking more completely.

In this chapter we shall first consider use of recreation areas, then the park approach to development and how it can be applied on public and private forests, and finally some of the questions that lie ahead concerning how to provide an integrated system of forests and parks for amenities and outdoor recreation.

## PUBLIC RECREATION LANDS AND THEIR USE

With the discovery in our country of great natural wonders, including hot springs, waterfalls, geyser basins, big trees, and rugged mountain scenery, came recognition of need for national parks. The thought of such areas falling into the hands of commercial exploiters was distasteful to high-minded citizens, who desired that this wonderland heritage be preserved for public enjoyment. Areas of historical or lesser scenic value were set aside for national monuments and state parks. Still later has come the demand for national recreation areas and for public fishing and shooting grounds. Parks and recreation areas are administered by schools, villages, counties, cities, states, and the federal government.

The extent of large outdoor recreational areas and their uses is shown in Table 20-1.

Between 1956 and 1970 the public outdoor recreation areas as a whole increased from 2,687 to 3,894 separate units, counting the 85 wilderness areas as separate units from the national forests and the wildlife refuge—one only—in which they occurred. State parks alone increased from 2,100 to 3,202. A corresponding increase in area also occurred. National parks increased in area by 10 percent, state parks by 45 percent, national wildlife refuges by 68 percent, and national recreation

Table 20-1 Large Outdoor Recreation Areas and Their Uses, 1970

| Type of area | Number | Area, in thousands of acres | Number of visits, in millions |
|---|---|---|---|
| National parks[1] | 35 | 14,464 | 45.9 |
| National monuments[1] | 85 | 10,223 | 16.0 |
| National recreation areas[1] | 13 | 3,809 | 11.5 |
| National forests[2] | 154 | 219,826 | 172.6 |
| Wilderness areas[3] | 85 | 10,258[6] | 5.8 |
| Wildlife refuges[4] | 320 | 29,000 | 18.0 |
| State parks[5] | 3,202 | 7,352 | 354.8 |
| Total | 3,894 | 284,674 | 624.6 |

Sources: [1]National Park Service. [2]Forest Service (visitor days). [3]Wilderness Society and Forest Service (visitor days). [4]Bureau of Sport Fisheries and Wildlife. [5]Bureau of Outdoor Recreation (data as of 1967). [6]9,925 thousand acres included in national forests. Remainder in wildlife refuges.

areas by 86 percent. The number of visits more than doubled, increasing from 311 million to 625 million. In 1956 state parks had 201 million visits, federal areas 110 million. In 1970 state park visits accounted for 355 of the 625 million visits. In terms of accommodating people, state parks exceed national areas by over 100 percent, though their total area is but 2.6 percent of the national areas. Expressed differently, the state parks accommodated 48 visits per acre of park land, the federal areas accommodated but one visit per acre. This reflects both the nearness of state parks to population centers and also their wide dispersion. It is also a good argument for increasing their numbers and total area.

**Planning Outdoor Recreation** A survey of recreational needs and anticipated use is the first step in planning an outdoor recreation system. This requires determining the population, its density, growth rate, occupation groups, ethnic makeup, average financial income, and time available for recreation. Use of an outdoor recreational site probably depends upon its distance from the user, its attractiveness, how well it fits the user's desires, and the general recreational environment in which it is located.[7] From such consideration the user population to be expected can be estimated. It should be kept in mind that people tend to form groups having a common recreational interest, be it snowmobiling, skiing, sailing, or folk festivals.

Having sized up the population and the recreational needs of the various subgroups within it, the recreation planner must next survey existing facilities. In appraising these, he will want to give consideration to their present use, the effectiveness with which they serve the people who now frequent them, and whether or not improvements might be made that would enable them to serve a greater number of people or serve the present number more effectively. Many parks are overused. Only by decreasing use can these be restored to a standard that will bring to the users the degree of satisfaction desired from outdoor recreation.

A park planner must think of the population anticipated 15 years in the future. He must study the surrounding territory with an eye toward possible park locations. It is far better to acquire areas well in advance of their need than to wait until private development has increased property values and destroyed many of the recreational features of an area. The park planner will wish to consider a number of locations and a variety of types of parks which, taken together with existing facilities, will provide a park system tailored to meet the needs of the metropolitan region concerned. He will then establish priorities for acquisition and development, estimate the overall cost for each unit, and present his findings to the public park boards for their consideration and action.

**Minimizing Conflicting Use** The park planner clearly recognizes conflicting uses and will seek to keep them segregated. On the same road or trail, motoring conflicts with cycling, horseback riding with hiking, snowmobiling with skiing. On water bodies, fast power boats with water skiers will conflict with fishing, canoeing, sailing, and swimming. Intensive use destroys quietude for bird watching. Use by organized groups may conflict with family use. And each age, ethnic, or special-interest group wishes to carry out its form of recreation without being disturbed by others.

The skillful park planner will try, insofar as he is able, to provide facilities for mass recreation, such as picnic areas, swimming areas, and boating areas, and, at the same time, will try to reserve a portion of his park in as near a natural condition as possible. Here the bird lovers and the seekers for wild flowers can wander off on the little-frequented trails, gaining the satisfaction of adventure and exploration, as well as an opportunity to become acquainted with the natural forest environment where it has been little disturbed by man. The truest test of successful park planning is to provide for active and passive recreation, for group activities and family activities, and for the pleasures of solitude.

The seasonal needs of the population must also be considered. More and more people are seeking year-round outdoor activities. Skiing, snowmobiling, snowshoeing, and hiking across the snow-covered land bring as much enjoyment as do summer sports.

**Planning Individual Parks** Every improvement in a public park area needs to be carefully designed and located. First, consideration must be given to overall planning so that traffic in the park flows with a minimum of congestion and interruption. The major recreational features of the park are first located. Road system, parking area, and trail system must be provided so that people may move readily from their cars into the park to major points of interest. The trails should radiate from the necessary centers of congestion, such as parking areas, bathhouses, museums, and other features. Appropriate signs must be erected to guide park-users to the various facilities. These may be simple lettered signs or pictorial signs; but, in any case, they must carry their message clearly so that the visitor is not confused or misdirected. Parking areas, picnicking areas, picnic tables and benches, drinking water, fireplaces, shelters, bathhouses, toilets—all the essentials for the comfort and enjoyment of the people must be properly located, attractively designed, and integrated with other features.

**Building the Park** Park construction requires engineering skill in laying out roads, trails, facilities, and buildings, and in their construction. Roadways and trails should be naturalized insofar as possible. Concrete and asphalt are to be avoided. Gravel and grass walkways and trails should be used wherever these will stand up under the traffic to which they will be subjected. Buildings and facilities must be serviceable, simple yet rugged in design, and easy to service and keep in order. Above all, they must be adequate for their purpose.

**Maintaining the Park** Park maintenance may involve simple custodianship of the facilities and protection of the park area against fire and misuse. Use of any park area by the public introduces a major disturbing factor that tends to destroy many of the features most desired in a park. Constant trampling wears out grass, compacts soil, leads to erosion, tends to kill trees and other vegetation, invites attack of trees by insects and fungi. Constant vigilance is required to direct public use so as to minimize such damage. All parks require maintenance. In rare cases this may be provided by the users themselves if they are well disciplined and truly appreciate the facilities placed at their command. More often a park custodian is employed to keep the park in a neat and orderly condition and to supervise the use of the park by the public.

In addition to simple custodianship, there is regulation of traffic, assignment of facilities, and the appointment of guards to direct the public in the use of the park area. Where swimming and boating facilities are provided, lifeguards are needed. The park manager will have guards wear uniforms or carry appropriate badges so that the public will know that they are in charge and will cooperate with them. Collecting entrance fees provides an opportunity to point out special features or attractions and to build an understanding between the park authority and the using public.

***Promoting Understanding*** The most intensive type of park management is that which involves recreational and educational programs, guided tours, museums, nature trails, and other special educational activities (Fig. 20-2). Restored battlefields, old forts, and other historic places may enable the imaginative citizen to relive in his mind the events of history. Here the utmost scholarship and subtlety are required if such an educational program is to be effective. The public will be educated only insofar as the displays provide a person the imaginative thrill that comes from projecting himself into historic situations.

The national parks offer conducted tours and nature talks that add greatly to the enjoyment and understanding of park visitors. A properly planned park system for a metropolitan area will normally include parks and historic sites that have educational value (Fig. 20-3).

**Park Administration** Strictly speaking, a forester is concerned only with large natural park areas, such as national and state parks; yet in these as well as in large city and county parks, such as those of Westchester County, New York, or the city of Denver's mountain parks, the activities of the park planner and park superintendent merge with those of a city-park manager and the national-park superintendent. All have many

**Figure 20-2** Visitor Center, Dinosaur National Monument. Here visitors may view fossil bones in place *(National Park Service.)*

**Figure 20-3** Naturalist explains how strangling fig kills large trees, Everglades National Park. *(National Park Service.)*

activities in common and all require the same basic philosophy in planning and maintaining park facilities.

A park organization like the National Park Service or many of the state park services is large in size, spends considerable sums of money, and serves the needs of millions of people. For this reason park administration becomes an important activity, requiring the same type of managerial skills demanded of forest administrators. The superintendent of the national park has many functions. He must plan the acquisition of land, the development of road and trail systems, the organization of a fire-protection system, the erection of campgrounds and picnic grounds, and the division of the park into ranger districts. He assigns duties and inspects the performance of the rangers and naturalists. He must maintain good relations with the public, both the people who visit the park and those who live in the surrounding territory. He must perform all other duties that are necessary to maintain the park in good condition and provide for economic expenditure of public funds dedicated to this purpose.

## EDUCATION FOR PARK AND RECREATION MANAGEMENT

The tasks of the modern park planner and administrator call for a variety of technical skills. Among the professional people employed by the

National Park Service are engineers, architects, landscape architects, historians, naturalists, archaeologists, and biologists. The Service also employs many foresters. The strictly professional requirements of park planners and administrators have been recognized by special committees of three nationwide park organizations: the National Conference on State Parks, the American Institute of Park Executives, and the National Park Service. The committees pointed out the need for studying the basic sciences such as botany, chemistry, geology, entomology, and zoology. In addition, they urged the inclusion of specific subjects related to forest areas, such as wood technology, wildlife management, pathology, forest entomology, and silviculture. But, most important, they stressed the need for a background in landscape design, an appreciation of architecture, a knowledge of engineering and construction methods, and familiarity with plant materials other than forest trees.

Landscape architecture has been defined as "the fitting of land for human use and enjoyment." Such a definition is especially applicable to the planning of lands for recreational use. Education that fits men for national, state, county, and municipal park work also fits them to think properly about the development of private land as well. The city planner receives his basic education in landscape architecture, as do the men who become camp planners, recreational planners and park planners.

Drawing is to the landscape designer what writing is to most people, a means of expression. Park planners need to use drawings to show others what they plan to do and how they plan to do it.

The park manager also must have the ability to understand the wide range of purposes and ideals that will be found among his clients. He will need to express himself well in writing and speaking as well as in the graphic arts. Skill in dealing with people is an essential trait for the person interested in park management.

Park management therefore is a special professional field closely related to forestry but with its own specialties in landscape design, land-use planning, and recreational planning.

## RECREATION ON PUBLIC AND PRIVATE FORESTS

The broad principles outlined above for park planning and administration apply equally for recreational developments on public and private forests. There are certain differences to be kept in mind. In contrast to parks, in which no taking of wildlife nor disturbance of natural vegetation is permitted, forests are usually managed for multiple-use purposes. This requires that conflicting uses be separated. Generally, recently logged areas should be closed to public use for the safety of the public as well as for the protection of the new forest stand. Timber cutting on and near recreational sites generally should be restricted to removing trees that are a hazard to the public and cutting to enhance the recreational features. Public hunting and trapping on forests are permitted and should be carried out in keeping with good game-management practices. Removal of animals in parks should be handled by the park authorities so that a healthy animal population is maintained. Difference in facility planning is mainly a question of scale and of funds available for development.

## BRINGING FOREST AMENITIES INTO THE METROPOLITAN ENVIRONMENT

Urban man can enjoy the amenities of the forest in two ways: by escaping from the city to the forest for vacations and weekends, and by saving or creating forests in the metropolitan areas. Large parks, such as the Cooke County Forest Preserve in Chicago, are one approach. Our nation is fortunate in that, in the densely settled states of the Atlantic seaboard, some 50 percent of the land remains in forest; in Connecticut 75 percent is forested. At present most such forests are in private holdings. Public forests should be extended. Tax and other incentives may be used to open portions of private forests to restricted public use. Another approach is to plan new subdivisions so as to retain 20 to 40 percent of the land as open space. The Open Space Action Institute of New York advocates a variety of measures that can be taken.[9] Cluster development of new housing subdivisions is one approach that offers many advantages. By leaving some of the land in park or natural forest owned by the public, the householder becomes content with a slightly smaller house lot. Also, by clustering houses the land used and expense required for constructing roads, water, sewer, electric, and telephone lines is diminished. By linking open space in contiguous developments, wooded corridors along streams and steep slopes can be reserved that invite forest birds and other wildlife to the suburban dwellers' doors. It is surprising how attractive small areas of natural woodland can be to wild creatures if the areas are linked together with appropriate hedgerow cover. A white-tailed deer made her 1970 summer home in just such wooded corridors that wound through highly developed residential sections of the city of Syracuse, New York.

An enterprising Florida developer created a uniquely attractive lake shore community by deeding the land back from the lake to the National Audubon Society as a permanent wildlife and bird sanctuary. Each home looks out on the lake on one side and into the sanctuary on the other. In this way the home owners can watch the large cranes, herons, and anhingas as well as the coots, ducks, cormorants, and alligators that such an environment attracts.

Forest amenities can even be brought into the heart of cities by taking advantage of open space along streams, space given up by rerouting railways and highways, and spaces cleared for urban renewal. Wooded corridors in cities can serve to mask noise and unsightly structures, to filter air, and to provide pathways on which men may walk with safety and pleasure. Most cities have a plethora of unsightly low-value land-use on their outskirts, some of which might, with forethought and planning, be developed into amenity forests.

Man may preserve and even help to recreate functioning natural ecosystems near his place of dwelling if he has the imagination and desire to do so. Constructive use of open space, compatible with its protection and preservation, should be encouraged by programs in the schools, scout organizations, and other youth groups. Forest areas need to be complemented by playgrounds and other areas for heavy recreational use.

## THE VACATION HOME

The fruits of our industrial society have become so abundant and so widespread in America that owning a vacation home of some sort can be a goal of almost any full-time employed householder. It may be but a simple shelter in the woods such as Thoreau's cabin on Walden Pond or John Burrough's "Slabsides" retreat in the Catskills. Simplicity has the advantage of minimizing housekeeping so that time for recreation and reflection is ample. Such can suffice for a weekend hideaway, provided land with a suitable setting can be acquired.

But if the vacation home is to be occupied all summer long by the housewife with young children, she will want and need some of the conveniences of a city home: running water, electricity, sewage disposal, and ready access to a shopping center, church, companionable neighbors, and a physician. Also, children want playmates of their own age and interests. A modification of the cluster-development plan outlined for suburbs can enhance the vacation home. Dwellings can be grouped to minimize costs for utility services, custodial care, and playgrounds. The reserved land, which may be as much as 90 percent of the total, is then available for forest recreation pursuits by the members of the vacation community. A number of private clubs and associations operate such vacation communities.

## OUTLOOK FOR FOREST RECREATION

A comprehensive study of future outdoor recreational needs for America was completed by the President's Outdoor Recreation Resources Review

Commission in January, 1962.[6] The Commission's report indicated that by the year 2000 the demand for outdoor recreation would increase at least threefold over that of 1961. This increased recreational demand is expected to arise as a result of increased leisure, along with increased income that will enable people to enjoy their leisure. The Commission found that simple forms of outdoor recreation, such as motoring for pleasure, walking for pleasure, playing outdoor games and sports, swimming, sightseeing, bicycling, and fishing are by far the most popular. Other forms that get considerable attention, such as water skiing, canoeing, sailing, mountain climbing, and snow skiing, are indulged in by a relatively limited proportion of the total population. The survey also revealed that as family income and years of education increase, the participation in outdoor recreation also increases. It found that people participate more in outdoor recreation when facilities are available within reasonable distance of their home. The increased mobility of society, however, has evoked a great increase in demand for resource-oriented outdoor recreation, such as unique natural areas and primitive areas. Forest areas, both public and private, were found to play a key role in public outdoor recreation.

A particularly significant finding of the Commission was that there is plenty of land for outdoor recreation, but the problem is to make it available and fitted for the type of recreation that the people need. As they expressed it, "The need is not for a series of crash programs. Large-scale acquisition and development programs are needed; so is money—lots of it. The essential ingredient, however, is imagination. The effectiveness of land, not sheer quantity, is the key." The Commission stressed especially the need for planning suburban development so that outdoor recreation becomes a part of it. They ask such significant questions as, "Do the children have to be driven to school—or can they walk or cycle to it safely over wooded paths?" It is in the development of our suburban communities and their environs that the greatest opportunity lies for meeting the day-to-day outdoor recreation needs of our expanding population.

Outside the rural areas much can be done to take the concentrated use away from a few specific areas by developing and publicizing historic sites, interesting land-use developments, demonstration forests, nature trails, geologic features, and other natural phenomena of note that can intrigue the recreationist. The Commission recommended a comprehensive approach to future outdoor recreation. Provision should be made for high-density recreation areas; general recreation areas; use of natural environment such as forests; use of unique natural areas that contain spectacular waterfalls, peaks, or other features; primitive areas; and historic sites.[1, 7, 12]

## MAXIMIZING HUMAN SATISFACTIONS

Parks, recreation areas, public and private forests, are but a part, albeit an important one, of man's total environment. How they are developed and used, therefore, affects environment in a significant way. For this reason the public is concerned that park and forest policy embrace environmental protection. Both park and forest managers will avow that this has always been one of their chief concerns. But searching new questions are being asked, such as: What is the long-term objective in national park management? Is it to attract the largest number of annual visitors, or to assure the preservation of the natural features that give unique character to the park? Is there not a maximum number of visitors that can be accommodated in a national park without damage to the natural features and downgrading the satisfactions to those who come? Is it necessary or appropriate to build lodges, campgrounds and play areas inside the park instead of restricting them to the periphery? Should not use of the private automobile be restricted within the park to minimize traffic on roads and the need for service stations and huge parking areas within the park?[2,3] The enjoyment of superb natural features requires concentrated observation and reflection if the experience is to be relived in memory. Some of the essence of the place must flow into the viewer's consciousness and lift his spirit. Distraction must be at a low level if this is to happen (Fig. 20-4).

The arguments for and the means by which peak-season visits to the most popular parks could be kept within the capacity of the park to absorb have been stated by several writers.[2,3,5,6] They include advanced scheduling, diversion of visitors to high-quality but less-frequented points of interest, higher fees for visits to the most popular sites than to others, low off-season entrance fees, and even deliberately making visiting difficult or inconvenient by restricting the construction of paved highways and parking lots. Scheduled bus tours from peripheral stations could also help to distribute the visitor load.

One forecast seems unchallengable: Pressure for outdoor recreation will continue to increase faster than population as long as both leisure time and affluence to enjoy it increase. Wilderness champions frankly believe in restricted visitor use to retain the features of solitude and minimum human disturbance of natural features of plant and animal life.

The Leopold commission on wildlife in the national parks recognized that both vegetation and the animal life it supports change with time. As one goal, the commission report suggested maintenance of the plant cover with its attendant wildlife in as nearly as possible the same condition as when the white man first came to America. They recognized that frequent fires were a major factor in maintaining large areas of grassland, and of other forms of preclimax vegetation favored by certain forms of wildlife.[4]

**Figure 20-4** Crowded campground, Yosemite. *(National Park Service.)*

Fires have long been and in 1970 still were a major factor in making and keeping conditions favorable for the heavy animal populations in East African game parks. Timber harvesting is another means of rejuvenating the forest and of promoting diverse vegetative cover favorable to wildlife.

It is to be hoped that the concern for environmental integrity so widespread in the 1970s will bring about new thinking on the national objectives to be sought in public management of environment and especially of park and recreational lands.

One characteristic of forest recreation is the challenge it poses to man's physical resources. May it always do so! But it requires, in addition to sheer physical vigor, some knowledge on the part of the participants. Probably devotees of forest recreation in the decades ahead will give increasing consideration to developing such skills as winter mountaineering, rock climbing, white-water canoeing, and finding one's way through trackless forest.

It may be hoped also that forest recreation as well as other types of recreation will bring to the participant something in addition to mere fun.

Complete devotion to self-amusement ultimately becomes boring unless it adds to the skills and knowledge of the participant as well as diverts him.

The field of systematic research on what people seek in forest recreation and environment was laboriously trying to develop its methodology in the 1960s.[4,6,7,8,11] National goals are still to be established in other-than-general terms. But should these not include making available to our people the types of experiences that will increase their sensitivity to pleasant surroundings and their zeal to preserve them for all to enjoy? Should not our national park and environment objectives aim to strengthen harmonious living through exposure to inspiring experiences, as well as to preserve natural wonderlands? Our public needs to feel that it is participating in building the type of rich living experiences we all so much desire.

## LITERATURE CITED

1  Clawson, Marion. 1963. "Land for Americans," Rand McNally & Company, Chicago. 141 pp.
2  Clawson, Marion, and Jack L. Knetsch. 1966. "Economics of Outdoor Recreation," The Johns Hopkins Press, Baltimore. 328 pp.
3  Darling, F. Fraser, and Noel D. Eichhorn. 1967. "Man and Nature in the National Parks," The Conservation Foundation, Washington, D.C. 80 pp.
4  Heinselman, M. L. 1965. Vegetative Management in Wilderness Areas and Primitive Parks, *Journal of Forestry,* **63**:440–445.
5  McCurdy, Dwight R. 1970. Recreationists' Attitudes towards User Fees: Management Implications, *Journal of Forestry,* **68**:645–646.
6  Outdoor Recreation Resources Review Commission. 1962. Outdoor Recreation for America, Washington, D.C. 245 pp.
7  *Recreation Symposium Proceedings.* 1971. Northeastern Forest Experiment Station, Forest Service, U.S. Department of Agriculture, Upper Darby, Pa. 211 pp.
8  Rutherford, William, Jr., and Elwood L. Shafer, Jr. 1969. Selection Cuts Increased Natural Beauty in Two Adirondack Forest Stands, *Journal of Forestry,* **67**:415–419.
9  Stewardship. 1965. Open Space Action Committee, New York.
10  Streeby, Larry L. 1970. Scenic Management Impact on Other Forest Activities, *Journal of Forestry,* **68**:430–432.
11  U.S. Department of Agriculture, Forest Service. 1970. Search for Solitude, U.S. Government Printing Office, Washington. 34 pp.
12  Wilm, Harold G. 1969. How Big Does a Wilderness Have to Be? *American Forests,* **75**(9):13.

Chapter 21

# Forest Policy

Worrell in his book *Principles of Forest Policy* defines policy as settled courses of action adopted and followed by society.[7] In this chapter, the term "policy" will also refer to courses of action followed by, or vigorously advocated for adoption by, government agencies, industry groups, and various citizens' associations that purport to be working in the public interest. Policies may or may not be backed by legislation. A policy issue is a course of action advocated but not yet adopted.

**FOREST POLICY ISSUES OF THE PAST**

From 1492 to 1891 the major policies followed by the people of the United States were to clear forests for farms, dispose of the public domain, and tacitly to allow cutting of public forests by timbermen to supply national needs for lumber.

In 1891 public sentiment was strong enough to persuade the Congress to pass a measure authorizing the President to set aside timber lands in national reserves.

In 1905 the major issue was to provide for management of the federal forest lands, renamed national forests.

In 1911 the federal government recognized that it had concern for the protection and management of all forest lands and established the principle of cooperating wtth the states to effect such protection and management.

During the 1920s the major issues up for public action were to increase federal-state involvement in protecting all forest lands from fire, insect, and disease outbreaks; to provide forest extension service to stimulate better practice by farmers; to develop forest research; to block up federal holdings in the national forests through exchange of land and of public timber for land; and to improve forest practices on large private holdings.

In the 1930s public attention was directed to using public forests to spread employment, thereby hastening recovery from the Great Depression and relieving hard-pressed private owners of the burden of carrying cutover lands.

From 1942 to 1945 the major concern was to get out wood to support the war effort.

At the close of 1945 the issue of public control over forest practices on private lands, first raised in 1920, again came to the fore. Some 17 states enacted regulatory laws. By this time, however, the more progressive forest industries embarked of their own will on forestry programs that began to match and in some cases to excel the intensity of management on the national and state forests. Meanwhile, the efforts to bring about better practice on small holdings had borne but limited fruit nationwide, except by heavy subsidies that were provided by various farmland-retirement measures.

By 1952 public regulation of forest practices on large corporate holdings no longer seemed to be a major need. With research results expanding and corporations ready to put them into practice, attention shifted toward building up the productivity of forests through intensive management practices. Recreational use was expanding at a rapid rate and with the report of the President's Outdoor Recreation Review Commission in 1962 major public concern shifted to meeting recreation demands.

## FOREST POLICY ISSUES OF THE 1970s

The major concern of the public in the early 1970s appeared to focus on restoring and maintaining a quality environment throughout the land. This concern was expressed in congressional speeches, news editorials, journals of opinion, books, and Earth Day Teach-in's at universities. New national organizations such as "Friends of the Earth" joined others in proposals for state and national legislation and in court actions to halt steps believed to degrade the environment. The theme of the 1970 national convention of the Society of American Foresters was "Managing Environment—Role of Resource Professionals." Here it was clearly recognized that forests make a great positive contribution to environment, which foresters are obliged to protect. It was also clear that forest industries, because of the sheer bulk of products processed, make their contribution to, and impress on, environment. Professor Frederick E. Smith in his keynote address before the 1970 national convention called attention to the rapid increase in per capita consumption of various commodities and what such increase portended. He pointed out that in 1918 the United States population was half that of 1969, but the dates at which consumption of certain products was at half their 1969 level were as follows:

| Product | Year product was at half 1969-consumption level |
|---|---|
| Food | 1940 |
| Beef | 1953 |
| Registered automobiles | 1953 |
| Gross national product | 1958 |
| Electric power | 1960 |
| Fertilizers | 1960 |
| Pesticides (all kinds) | 1963 |

The great disparity between the time it took to double our population, 42 years, contrasted with the 30 years to double food production, and the 10 and 7 years to double the use of fertilizers and pesticides, points to increased environmental resistance to growth of population and consumption. World population is reported to be increasing at about 1.5 percent per year, whereas the rate at which material is being processed increases at 5 percent per year. Demand on the environment and friction among people seem to increase exponentially with increase in population. The increasing rate per capita of crime, Smith cites as providing an example.[6]

It should be pointed out that most of the increase in the rate of material processing and use is in the affluent societies. From this it clearly follows that environmental resistance becomes operative at much lower population densities in the industrial nations than it does in those with lower consumption per capita. Must man then choose between future penury or environmental cataclysm? This is not an idle question. It can be stated categorically that if population and consumption per capita continue to rise at present rates for a few more decades, environmental resistance will begin to operate at a catastrophic scale.

There are two things man can do. One is to lower his demand on the environment by slowing down population growth and consumption. The second is to focus his scientific and technologic talent on means of minimizing his impact on the environment. In view of the rising expectations of peoples and the crippling economic depression that a sharp curtailment of industrial output would cause, it is unrealistic to expect that major dependence can be placed immediately on option number 1 above. Fortunately, much can be expected of option number 2, provided professional resource scientists and managers can gain and hold the public support they will need to meet the overall task. If they succeed, time will be gained during which man can learn to realize his greatest satisfactions through living creatively in his environment rather than merely exploiting it.

In 1971 it was too early to assess the strength of the public commitment to environmental improvement. It appeared to be nationwide and to be strong in many other nations with advanced technologies. A further difficulty has been lack of data to document the case in precise terms. It is well to keep in mind, though, that public sentiment, once sufficiently aroused to the need for corrective action, seldom waits for quantitative proof before enacting policy. The setting aside of forest reserves in 1891 to avert a threatened timber famine is a case in point. Where safety and health are threatened, action to protect the public is already national policy. It need only be extended to cope with new recognized hazards. But how far the public will go to maintain an esthetically pleasing environment has yet to be determined. It may first be necessary to establish that pleasing surroundings promote the national economy and the health of the nation before the public adopts preservation of a quality environment as a national policy.

## THE IMPACT OF FOREST POLICY ON MAN

As long as man survives on earth, he must look to forests for products and services. From Chap. 18 it is clear that future wood needs of our own nation and of most others will increase substantially by the year 2000. So also will our needs for water, for products and services of wildlife, and for outdoor recreation.

These multiple values and services of forests have given rise to the principle of multiple-use management on both public and private forests. No longer can all needs or desires be supplied for one product or service without impact on the others. Policy involves deciding among the many demands, the allocation of uses that provides the highest total degree of satisfaction both for the present and for the long-term future. In essence, forest policy deals with who gets what, and when, and who pays the costs involved. Forest policy is, therefore, inevitably controversial, and the more acute the demand for forest products and services, the more heated the controversy becomes.

At no time in our nation's history have more questions of forest policy come under challenge than in the 1970s. The forest manager, be he a government or industrial employee, must recognize this concern of everyone with forest policy and learn how to elicit a consensus so that he can get along with his task of resource administration. If he fails to do this it will mean that society rejects his services. Consensus will not be easy to effect, for in the 1970s forest policy was recognized to embrace more than allocating use among timber, wildlife, water, forage, and recreation. It involved protecting the contribution of forests to the total environment, a task so all-embracing that man had still to define it, establish realizable objectives, and develop the essential techniques to achieve the desired results.

**Interrelations Involved**  Forest policy in a body-politic is a blending of policies of the public with respect to public lands, and of private landowners with respect to lands they own or control. But it is also much more than this. The public has a genuine and legitimate interest in how private forest lands are managed, and the private users of public lands have an interest in the regulations governing their use of public land. Government exercises restraints over use of private lands through its powers of escheat, eminent domain, police, assessment and taxation, zoning, permits and licenses, and various regulatory laws. Citizens in turn exercise restraints over practices of government land-managing agencies through petition, appeal to the national Congress, pressures applied by citizen organizations, and actions initiated in courts of law. In these ways stockmen, timber-users, water-users, hunters, fishermen, woodland owners, recreationists, and people concerned about environment and esthetics exercise their rights as citizens to influence government policy and actions. By analogous actions citizens' groups can make their wishes known to owners and managers of corporate and other private land as well. Altruistic, as well as self- and financial interests, motivate private citizens and the organizations they support to speak for them. Conflicting viewpoints are the rule.

But the conflict goes far beyond that between self-interest seekers, altruistic people, and organizations. In essence, the conflict exists within each individual and within the membership of the organizations that speak for him on public issues. For everyone benefits from many products and services of the forest and so is concerned with their totality as well as with the one or two that appeal to him most. It is this totality of use and concern that forms the substrate on which consensus can be built.

For a more complete treatment of the nature of forest policy, its economic basis, how it comes to be adopted, and the effects of its administration on societal objectives, the reader is referred to Worrell's book of 1970 on the subject.[7]

A behind-the-scenes description of how the Multiple Use–Sustained Yield Act was conceived, the pro's and con's debated, the drafts prepared and refined, the support for the bill enlisted, the diversionary amendments defeated, and the ultimate act passed, is given by Dr. Edward C. Crafts, who played a key role in the process.[4]

## PUBLIC LAND POLICY

**The Public Land Law Review Commission**  A key document that attracted widespread attention in 1970 was the report of the Public Land Law Review Commission established by the United States Congress in 1965.[5] The report reviewed in a comprehensive way the status of the various public land laws and how they related to the major forestry and other land issues of the 1970s. It recommended changes that needed to be made and brought constructive thought to bear on the entire task of managing the third of the nation's land owned by the federal government, including the national forests (Fig 21-1).

The Commission found many of the existing federal laws to be no longer needed and recommended their repeal. It found others to be in conflict with present concepts of best use of the land, and most to have poorly defined directives to govern administrative action. Among its 137 major recommendations for legislative action and administrative practices were the following:

Future disposal of federal lands should be only of those that will bring maximum benefits to the general public in nonfederal ownership. Some might be transferred to states, some sold to private individuals and corporations.

Federal land-management agencies should be required to consider all points of view before promulgating comprehensive rules and regulations. Means of appeal from such regulations should be simplified and judicial review generally be available.

Land-use plans should provide for the maximum number of compatible uses with the recognition that where maximum benefit can be had through one particular use it be recognized as the dominant use.

Federal lands should be managed so as to maintain or improve the

**Figure 21-1** A field-study conference of forest officers with forest users, permittees, and interested citizens. Sharp differences of opinion can often be resolved when issues are studied on the ground. *(U.S. Forest Service.)*

quality of the environment, and should be managed to this end in conformance with state laws.

In general the federal government should receive full value for the use of public land resources. Where no consumptive use of land or resources is involved the monetary payment need not represent full value.

Where public lands and resources are made available for private use, firm tenure and security of investment should be provided.

Payments to states and local governments in lieu of taxes should be based on local tax effort and not on revenues from use of resources.

Statutory authority should be provided for sale at full value of public domain lands for certain mining activities, dryland farming, grazing, residential, commercial, or industrial uses, provided important public values will not be sacrificed.

Statutory authority should be granted for limited disposal of lands in national forests and national grassland where such disposal would result in higher use of such lands than if they continued in federal ownership.

Mechanisms should be provided for transfer of federal land not needed for federal purposes to state and local governments at less than full value.

Public lands that are highly productive for timber should be classified for commercial timber production as the dominant use.

Federal programs of timber production should be financed by appropriations from a revolving fund made up of receipts from timber sales.

Dominant timber-production units should be managed primarily so as to maximize net returns to the federal treasury.

Major timber-management decisions including allowable-cut determinations should include specific consideration of economic factors.

Access road construction should be accelerated.

Timber sales procedures should be simplified wherever possible.

Communities and firms dependent on public-land timber should be given consideration in the management and disposal of public timber.

The Forest Service should be merged with the Department of Interior into a new Department of Natural Resources.

Public reaction to the report and its recommendation was mixed. Many people who were well informed on the inadequacies and conflicts in existing land law commended the Commission for the thoroughness of the task it had performed. They recognized that the report was only a source document from which specific legislation would have to be enacted, and they recognized also that each bill prepared would come under public scrutiny when it was introduced into the Congress.

Other individuals endorsed the report as a whole but took vigorous exception to specific recommendations. Those involving disposal of public land, increased timber cutting, and mining rights, were especially subject to criticism. The general consensus of spokesmen seemed to be suspicion of the dominant-use concept as opposed to the multiple-use principle.

Few, if any, spokesmen endorsed the report in its entirety. Some, however, condemned it generally for being western-oriented, too generous to mining, grazing, and timber interests, and oriented to protect private use of public lands at the expense of public values.

A few students of public administration looked askance at its emphasis on congressional specification of detailed administrative procedures and actions rather than concentrating legislation on defining objectives and specifying general procedures. They argued that the administering agency should be granted latitude to fit actions to local needs.

There was the further criticism that policy on federal forests should be related to that on private and state-owned lands.

There appeared to be no commentators who argued that no revision of public-land laws was needed. The confusion, evasion, and corrupt practices that occurred in the disposal of the public domain, as Cameron[1] clearly pointed out, were due to laws inconsistent with the needs of a developing nation. Many such are still on the statutes.

A number of bills were introduced in Congress in 1971 to carry out recommendations of the commission.

**The Beleaguered Forest Service**  The U.S. Forest Service has been under attack in the past for zealous protection of the public interest against corporate and other private interests. Beginning in 1968 and 1969 it came under attack by new conservation advocates that took sharp exception to its programs for use of national forest resources. What had happened to the Forest Service that spearheaded the conservation movement in the early years of this century; that initiated and led all other federal agencies in setting aside natural areas, roadless areas, primitive areas, and wilderness areas; that nurtured such pioneer champions of wilderness as Benton McKaye, Robert Marshall, and Aldo Leopold; that introduced the concept of multiple-use management; that first brought management to public grazing lands; and that has been repeatedly commended for devotion to the public good and to efficient, corruption-free management of public lands? Had the Forest Service failed, as averred, to keep abreast of current conservation thinking, or was it encircled by advocates of various conflicting uses (Fig. 21-2)?

Skiers and ski-facility developers were seeking new areas for public use. The Sierra Club took court action against the Forest Service to stop the most elaborate such development so far conceived. The President and Congress advocated more access roads to open new forest areas for timber harvesting on national forests. Preservation and conservation societies have resorted to court injunctions to oppose building some such roads. The forest industries in general have urged that the allowable cut on national forests be increased so that lumber may be made available for the increased housing construction advocated by the Federal Housing and Urban Development Agency. Conservationists and wilderness exponents have demanded that before increasing timber cut, more lands be set aside for recreational use, roadless areas, primitive areas, and wilderness, arguing with justification that once opened to timber cutting their wild character will be destroyed. They insist that large areas of superb forests of the nation should be perpetuated in natural and wilderness areas.

Industry spokesmen have urged the Forest Service to emulate their own successful methods of clear-cutting mature timber and replacing it with vigorously growing young stands. Ecologists and silviculturists have denounced clear-cutting of climax-type forests. They have also called attention to the hazards of monoculture over large areas because of potential susceptibility to insect or disease outbreaks. Ecologists also question the desirability of establishing an artificial as opposed to perpetuating a natural forest ecosystem. When the Forest Service moved in the direction of even-aged timber management by clear-cutting in the

**Figure 21-2** Robert Marshall, forester, scientist, author, explorer. As Chief, Division of Recreation and Lands, 1937–1939, Marshall was responsible for establishing Forest Service policies that resulted in reservation of natural, roadless, and wilderness areas. Marshall was founder and patron of the Wilderness Society. *(U.S. Forest Service.)*

Bitterroots and in West Virginia, conservationists, joined by some foresters, resorted to court action to halt clear-cutting until the question could be further studied.

For a period in 1970 hardly a week seemed to go by that a new suit was not filed in the courts to halt timber harvesting, leasing for recreational developments, access-road construction, and other activities the Forest Service had been engaged in for years and considered to be routine practice.

Environmentalists were unhappy also about what seems to them a reversal of Forest Service policy on use of fire and clear-cutting. The Forest Service was long an advocate of complete protection against fire and of partial as opposed to clear-cutting of forests. Current thinking among foresters in 1971 is that conditions exist where fire is a useful silvicultural method of reducing fire hazard, favoring forest reproduction and controlling the development of desirable ecosystems. Clear-cutting of overmature timber is also held to be good practice for pioneer and some subclimax types. Neither, though, should be prescribed for wholesale use irrespective of local conditions. It is highly important that they not be used where irreparable damage to the environment would result. To this the Forest Service agrees. It has also admitted that mistakes in application of clear-cutting have occurred.

Part of the difficulty seemed to stem from the strongly centralized administrative control the Forest Service has developed over the years. Such control seemed to work well when use of national forests was relatively light and conflicting uses rare. The intensive use demand of the 1970s was placing increasing strain upon administration all up and down the line.

The essence of the challenge to the Forest Service is the insistence of conservation groups for "pieces of the action"—a chance to be consulted on important policy issues before decisions are reached and commitments made. Many feel that the commercial users of the forest have their means of influencing Forest Service policy, and the conservationists and environmentalists are demanding equal voice. The timber industries and livestock men are not opposed in principle to wilderness areas on national forests, nor are recreationists or wilderness advocates opposed to harvesting timber on lands that have been committed to such use. The controversy arises because the Forest Service, in the interest of national housing, has responded to the Congress and the forest industry urge to open new forest areas to timber cutting. Wilderness advocates in turn have taken the position that new areas should not be opened to timber harvesting until it has been determined that such areas are unsuited or not needed for wilderness preservation or other restricted use. A group whose members had the hardihood to spend 5,842,800 days in national forest wilderness areas in 1970 is one—both in numbers and in spirit—to be treated with respect. The Public Land Law Review Commission has recommended to the President and Congress that public hearings be arranged by land agencies before major policy decisions are reached.

**The Role of Commercial Users of Public Forests** The forest industries frankly recognize that the greatest reserve of high-quality timber now remains on the national forests and they would like to harvest it and convert it to salable products. They have found on their own lands that timber grows rapidly in quantity during early ages. They argue that what is important for future timber supply is not the remaining more or less stagnant stands of old growth and virgin timber but rather the growth potential of thrifty young stands. They therefore urge increased rates of harvesting the old-growth timber and have promoted federal legislation to bring this about.

Stockmen have had their battles with the Forest Service in the past and have gained participation in grazing use on national forests. Resort and facility operators who erect recreation structures on permits and leases are a group with which the Forest Service also must reckon.

The Forest Service therefore has the thorny task of encouraging use of the national forests' renewable resources of water, wood, wildlife, and forage to meet the growing demand for these forest products. It also must strive to satisfy the demands of recreationists. Many want developed campgrounds, bathing beaches, ski slopes, and snowmobile trails. Others want the challenge of wilderness and roadless areas. The time has come when the expressed desires are outgrowing what the land can continually supply. And once a land area is opened to general use, it is seldom possible to reverse the action.

**Financing Forest Expenditures** Clawson[2] makes a strong case for businesslike budgeting of national forest and other land-management activities. He stresses the desirability of adjusting fees for grazing, recreation, and other uses gradually upward to the levels that prevail on comparable privately managed lands. He strongly feels that capital expenditures for planting and timber-stand improvement be made at levels that can be justified on the basis of expected future income. He would like agencies to show in their budgets just what the nation is getting from a given level of management and how future revenue might be affected if the level were decreased or increased by various amounts. Such accounting will require the collection and analyzing of much field data.

## POLICY ISSUES CONFRONTING PRIVATE FORESTRY

Private forest owners may be grouped into three classes:

1 Individuals with small holdings, less than 5,000 acres and generally less than 500 acres.
2 Holders of properties of 5,000 acres or more. These are mostly corporate or large individually owned properties held mainly for timber production.
3 Miscellaneous mostly small ownerships held by resort keepers and others as adjuncts to another business. Timber harvesting for revenue is a secondary objective or is completely rejected.

Collectively class 1 owners hold over one-half of the commercial forest land of the United States. They are made up of farmers, local business and professional people, retired folk, and similar classes of nonresident owners. Few depend solely or even mainly on revenue from their timber lands for a livelihood. Surveys have shown that for more than half of such small-forest owners revenue from timber sale is a secondary consideration. Still, perhaps most would permit timber harvesting if they could get the quality of logging done that would be consistent with their other reasons for owning the land. This has not been easy in the past. Logging on such properties as a group has been below standards on national forests and well-managed corporate lands, and their timber yield reflects this. The land of small owners is potentially among the most productive forest areas of the nation. Collectively such owners could be major contributors to the nation's timber supply. To aid them various federal-state programs have been established. Their total holdings are of such importance to future timber supply that public control of cutting and management practices was being advocated in 1970 by some industrial foresters. That such a position is serious is borne out by the fact that some companies in the South offer free to landowners a timber-marking service. To get the cooperation of such owners the services offered must be consistent with the landowner's motives for owning the land. No foresters, either public or private, have yet developed a nationwide program that has the appeal essential to elicit the full cooperation of these owners.

Nationwide the task of bringing good forest practice to the lands held by class-1 owners was still unresolved as of 1971, although new legislation was pending (Fig. 21-3).

Leaders among the forest industries, class 2, were managing their lands well in 1970. A significant number of companies kept their forests open to public hunting and other recreational use.[3] Some maintained campgrounds, picnic areas, and scenic trails. Restricted timber cutting along streams and roadsides was often observed. In the South, timber harvesting is often followed by slash cleanup and replanting. One company even established a small wilderness area accessible only by a 6-mile trail over rugged terrain.

Owners in class 3, though growing in terms of area owned, still held in 1970 relatively minor areas of forest lands. Some harvested timber using good forestry methods. Many had holdings of such small size as to be of little importance from the logger's standpoint. For the decade of the 1970s such owners are expected to contribute but a minor percentage of total harvest of logs and pulpwood.

**Policy Needs of Forest Industries** The forest industries themselves need to develop an aggressive policy for increasing their contribution to the national well-being and widening the usefulness of their products. They also need to regain the public esteem they lost during the early

**Figure 21-3** Abandoned farm in upstate New York on land no longer fertile enough for agricultural use. Much of such land has returned to forest by natural reseeding or planting.

forest exploitation in the country. This can be done in part by improving and diversifying their output and tailoring it more closely to consumer needs. But of even more immediate importance to their public esteem is to lessen the impact of their forest operations on the environment.

Logging operations need to be made far less destructive than they appear to be. So far loggers have sought equipment that would increase output of logs, rather than that which causes minimal disturbance to the forest environment. Much could be done to reduce logging impact by better planning and layout of work and more careful operation of machinery in the forest. Closer utilization of product and postlogging cleanup could reduce unsightliness and damage to soil and water.

Logging machinery needs to be developed that can be used for thinning young stands, for making intermediate cuttings, and for group-selection cutting in uneven-aged stands and other partial harvest cuts. The use of such machinery should cause but insignificant damage to the trees left for a future cut or to their environment. Similar equipment should be developed that is suitable for use on small-sized forest properties.

Forest industries should be prepared to restore the land they cut over to productive conditions, whether this be their own land or land on which they have exercised cutting rights. The above actions may add to logging costs.

The forest industrtes seem long to have been on the defensive. They have engaged in little imaginative promotion of their product, yet this is needed to retain and enlarge wood use in the interest of the public and of environmental protection. This will have to be done largely through using wood to make products that are more attractive and serviceable than those now available. Such improvement is a constant need and should be provided for by well-organized financial support.

## EXPECTATIONS OF THE 1970s

**Wilderness** It is expected that by the close of the 1970s reservations of federal lands for wilderness—roadless, primitive, and natural areas—will be largely completed except perhaps in Alaska. Such a program is provided for under the Wilderness Act, has been urged by conservation groups, and endorsed by the Public Land Law Review Commission.

**Environmental Improvement** Substantial progress is anticipated in improvement of the environment, yet the growing pressure upon it may largely offset the progress. This is not an issue that can be resolved once and for all but must be continually pursued if retrogression is to be avoided. A hopeful feature is the growing recognition of the necessity for restrictions on private developments that jar esthetic sensibilities and degrade environment. New York State in 1971 established by legislative action an agency to perform this function in the Adirondack Park.

Forests occupy almost one-half the land area of some of our most populous eastern seaboard states. Managing these forests for enhancing the total environment becomes a new task for foresters that they should seek to discharge in collaboration with forest owners, planners, landscape architects, rural developers, highway builders, and local governments.

A new look at forestry practices, broader than is implied by multiple-use for wood, water, wildlife, forage, and recreation, should be expected. These will still remain important, all of them. But esthetics and integration of forests into the suburban and rural ecology are a more comprehensive task, and one requiring new sensibilities to human yearnings and what forests can contribute toward their satisfaction.

Fortunately in many European nations examples exist of harmony between urban and rural land-use. With our nation's far greater land resource and affluence, little real reason exists for our not making a satisfying adjustment between man and his environment.

**Public Controls over Private Forests** The issue of public control over management of private forests that had been more or less dormant during the 1950s and 1960s appeared likely to be revived in the 1970s and 1980s. A bill in Congress in 1971 would impose strong regulatory authority over private forests to protect both environment and future timber supplies. Another bill was introduced to achieve the same objectives through a program of incentives. Industrial foresters, who have as a group opposed public controls, found in 1970 that there were those in their ranks who felt that such controls would be needed to upgrade the output from small ownerships. Foresters, conservationists, and many small owners themselves had come to recognize by 1970 that public control of some kind may be necessary. If 1969 Forest Service projections of future timber demands prove correct, measures to improve management of small-forest holdings will become urgent during the 1970s and 1980s. Public sentiment for effective measures to improve practices on such forests could then increase rapidly. Applying improved practices in the woodlands will be a major task for foresters. To be effective, the latter must recognize the objectives of modern owners and provide a means whereby small holdings can be grouped through cooperative or other means into economically sized operating units. A strong positive approach is needed.

Additional questions should be faced if steps are to be taken in time to meet the needs of the year 2000 and beyond. How much timber can be grown within the United States and at what cost? How much should be exported as paper and other products? How much should be imported from Canada and the tropics? How far should the United States go in restricting timber exports to preserve the local supply? Which of the many choices available will best serve the long-run interests of the people of the United States?

**Land-Use Adjustments** Important shifts in land use were occurring in many parts of the United States in 1970. These were most in evidence near existing population centers. But new areas also were being opened to intensified use. Some of the shifts in use will have impact on forests and timber production. New questions thus arise. How much commercial forest land must be given up to agricultural expansion, to urban and suburban development, to industrial sites, and to transportation? How much commercial forest land should be devoted to nontimber use for such purposes as national and state parks, roadside strips, scenic drives and vistas, wilderness areas, intensive and extensive recreational use, protection forests, and small private holdings uneconomic to manage and from which to harvest the timber?

How shall the costs of such reserved areas be distributed among user fees, local government support, and federal government support?

How can logging operations be improved, or, by contrast, how can the public attitude toward logging unsightliness be softened so that intensive forestry is accepted as necessary and desirable, just as intensive agriculture is accepted?

A rapidly growing issue is the need to make opportunities available to more people to participate in forest-based outdoor recreation and to enjoy forest amenities. Additional public forests and parks are to be expected. But pressures seem likely to develop for public use of private forests where these lie close to metropolitan areas.

If real progress is to be made in enhancing the esthetic and recreation attractions of the rural landscapes, such measures as zoning, scenic easements, taxation based on current use rather than so-called highest use, county planning and related measures, will need to be brought into play. Laws in themselves will not get action; positive administration is essential.

## LITERATURE CITED

1. Cameron, Jenks. 1928. "The Development of Governmental Forest Control in the United States," The Johns Hopkins Press, Baltimore. 471 pp.
2. Clawson, Marion. 1967. "The Federal Lands Since 1956: Recent Trends in Use and Management," The Johns Hopkins Press, Baltimore. 113 pp.
3. Cordell, Harold K., and Stephen J. Maddock. 1969. Recreational Policies of the Major Pulp and Paper Companies in the South, *Journal of Forestry,* **67**:229–231.
4. Crafts, Edward C. 1970. The Saga of a Law, *American Forests,* **76**(6):12–19, 52, 54; (7):29–35.
5. Public Land Law Review Commission. 1970. "One-Third of the Nation's Land," Government Printing Office, Washington. 342 pp.
6. Smith, Frederick E. 1970. Ecological Demand and Environmental Response, *Journal of Forestry,* **68**:752–755.
7. Worrell, Albert C. 1970. "Principles of Forest Policy," McGraw-Hill Book Company, New York. 243 pp.

Part Seven

# Forestry Education, Research, and Employment

Chapter 22

# Education in Forestry

The art and science of managing for human benefit the resources on and associated with forest land draws upon many fields of science, technology, and art. Biology in its broadest sense forms the scientific base, for forestry is the management of forest ecosystems. Mathematics, physical science, and engineering are drawn upon to open the forest to use, and to harvest, transport, and process its products. They are also used to quantify the output of wood and water, and to some extent of other products and services. Economics and business administration come into play in maximizing the returns from expenditures for forest improvements. Behavioral science and public administration are drawn upon to develop firm policies for forest use based upon consideration of diverse human needs and demands, some of which are conflicting. To use the above fields of science and technology so as to achieve a high level of utility and satisfaction from forest resources is in itself an art. To preserve and augment forest amenities is a fine art in the broad meaning of the term.

## A SYSTEM OF EDUCATION IN FORESTRY

Everyone who enjoys being in a forest, or viewing it from a car window, or who uses wood or paper in any form, is affected to some degree by how forests are cared for and managed. The practice of forestry is basically a publicly sponsored activity in most nations of the world and so enjoys the financial backing of governments. It is therefore important that many people in a modern nation have an awareness of forests and the role they play in environmental protection as well as in the economic welfare of the nation. Consumer groups, teachers, clergymen, journalists, and community leaders help to create public awareness and to formulate public opinion. All need access to scientifically sound information on the basic nature of forest care and use. Engineers and architects who design wooden structures need technical information on wood and its availability. Legislators, budget officers, executives, and financiers are in a sense trustees for the public in channeling funds and giving overall direction to those who manage forests. They especially need unbiased sources of information if a nation's forests are to be managed in the long-term public interest. Obviously workmen who harvest and process forest products need specialized skills. Those who plan and supervise their work require technical knowledge. Those who develop and apply the technology of forestry require professional competence. Meeting the educational and training needs of professionals, technicians, and skilled workmen who make a life career of forestry constitutes a major element in a national system of forestry education. Informing the lay public and those who are involved in formulating, administering, financing, and adjudicating forest policies and practices is the other function to be performed.[7]

A national system of education in forestry has three broad responsibilities:

1  To accumulate, preserve, and disseminate knowledge of forests that has been acquired and tested over the years by successful application.
2  To use current developments in supporting fields of science, technology, and art to synthesize new concepts of forest care and use and to test such concepts for soundness and utility.
3  Upon request to make available to legislators, jurists, administrators, and others the present state of the art and science of forestry as it may be applied to enhance the welfare of man.

Every active forester has a role to play in such a national system of education in forestry. The professional college is its keystone.

## THE PROFESSIONAL COLLEGE OF FORESTRY

The first more or less formal education in forestry was in the master schools of Germany and France. A few outstanding *Forstmeisters* gathered about them a small group of neophytes whom they instructed in the art of managing forests and game. These evolved ultimately into state-supported forestry schools, usually located in or near a forest, with instructors assigned by the state forest service, generally for periods of 1 to 4 years. Students were recruited by the state service on a competitive basis and given a service appointment with student stipend. This type of school was operating in 1970 in India, Pakistan, Spain, France, and Sweden.

A second type was open to state appointees and also to "free" students, those who received no state appointment or stipend and had no assurance of state employment upon graduation. The schools of this type were mostly located at universities. They were operating in Germany, Turkey, and the U.S.S.R. in 1970 and in a few developing countries.

The third type of forestry school had few if any students with state appointments. Some might have scholarships but were under no obligations to work for any specific public or private agency upon graduation except for the rare individual financed by his corporation. Nor were any graduates assured of a public position upon completing their studies. Schools of this type were the common pattern in Japan, Finland, the United Kingdom, the Philippines, Canada, and the United States. All had university connections in 1970.

The first type has the obvious advantage of efficiency in that no more men need be educated than are to be placed in the public service. Selection of students can be rigid because competition for admission is generally keen. Rarely do students withdraw before completing their studies. Instruction is kept closely geared to operating needs by bringing in fresh officers as instructors at least as often as with each new class. Esprit de corps is generally high. The major weakness is that it tends to become provincial as it has no university connection or at best a rather tenuous one. Stimulation from nonforestry disciplines tends to be slight.

The second type has the stimulus of a university environment but the disadvantage of a divided student body.

The third type has the advantage of being well integrated into the university of which it is a part. Competition among students is high as only those with good academic records are likely to be selected for the choice positions upon graduation. Course and even curricular offerings tend to proliferate to widen employment opportunities for students and to afford each faculty member a chance to offer instruction in his specialized field. Such schools tend to be innovative and progressive. Many sponsor stimulative research programs that enrich instruction and bring prestige to faculty members. Many also offer postgraduate instruction, a level in which both professors and students are alike teachers and learners (Fig. 22-1).

The chief disadvantage of this type is that the pace of instruction causes many students to fall behind and drop out. Also, more are usually graduated than can find desirable employment in the profession. Many, however, find employment in peripheral fields such as photogrammetry, land-surveying, various engineering and construction fields. Others have become successful bankers, resort managers, and business men. With additional education a few have become lawyers, schoolteachers, journalists, engineers, physicians, and clergymen. Among these, some have avowed that their forestry education provided a good general education for their new profession. The forestry profession also has benefited by the interest and understanding of forest resource management, with such people spread among nonforesters.

**Forestry Curriculums**  All forestry schools require a general grounding in mathematics, physics, chemistry, biology, and social sciences. The art of expression, both orally and in writing, is normally stressed because

Figure 22-1  The Forestry College, University of Ibadan, Nigeria.

reporting and communicating forms an important part of a forester's duties. A speaking acquaintance with French and Spanish is important for those who would work for the UN Food and Agriculture Organization. On such a foundation are built the various professional curriculums in forestry, wood technology, wildlife management, and other fields. Subjects that Dana suggests as indispensable for a general forestry curriculum are dendrology, forest ecology, silviculture, forest protection, forest measurements, forest management, forest economics, forest policy, and forest administration.[2]

Major undergraduate curriculums offered by colleges of forestry in 1970 covered the following subject-matter fields:

General forestry with or without options in biological science, social science, engineering, management, and industrial forestry.
Range management.
Wood utilization with options in chemical technology and engineering.
Wildlife management.
Wood technology.
Recreation management.
Landscape architecture.
Park management.
Lumber, merchandising, and light construction.
Forest environmental science.
Furniture technology.
Logging engineering.
Conservation of natural resources.
Paper science and engineering.
Forest chemistry.
Packaging technology.
Other curriculums have been urged upon the schools by various industry associations.

Important as curriculums are in higher education, other concerns were demanding the attention of both forestry educators and men in the profession in 1971. It was being recognized that managing the resources of the forest demanded a variety of skills and educational competencies. The profession could ill afford to neglect any of them. Furthermore, to attempt to educate each man to become a competent broad-scope practitioner upon graduation was beyond the time limit of an undergraduate curriculum. What could be hoped for was to give a man a reasonable competence in one or more major forestry activities and to stimulate him to acquire competence in additional subjects as he progressed in his career. It was found that few employers of foresters inquired much into the subjects a man had pursued in college. Rather, they sought to ascertain whether or not he had the personal drive, breadth of interest, and curiosity to continue to be an effective learner and performer throughout life.

Forestry schools in 1971 were using a variety of techniques to stimulate individual learning and to develop a thirst for broad education among their students. In fact it seemed that the broader the programs within a forestry school the more alert the faculty became to urge its students to study ancillary subjects in the university as a whole.

As forestry grew in complexity, specialization became inescapable.

A study of the tasks performed by foresters graduated from the State University College of Forestry at Syracuse University revealed some interesting facts. Supervision and administration occupied 41 percent of their time, research 13 percent, sales and promotion 11 percent, and technical forestry work only 8 percent. Knowledge of technical forestry, though, formed the base for many of their supervisory, administrative, and other duties.

Many foresters have expressed the conviction that forestry colleges should increase instruction in subjects that prepare men for administration and other high-level posts. This point received much attention at the Roanoke Conference of forestry educators and employers, held in 1969. Some men felt that the entire program of forestry education needed reorientation to stress the subjects currently in the foreground for executive education, such as decision theory, organization and management theory, economic theory, simulation, and operations research and systems analysis. They also emphasized the need of orientation toward the needs and wishes of the general public as the consumer of forest products and services.[3]

While not denying the desirability of such studies, others thought that somehow the modern foresters had not done so badly in competition with associates of different educational background. They contended that a large number of foresters did display leadership in public affairs; they did have access to public officials, congressmen, and state legislators; and they were respected for the quality of their contributions to general-resource management.[6] Cases could be cited of men with forestry backgrounds who became vice-presidents and presidents of colleges and universities, congressmen, governors, corporation executives, and bank presidents. A few had developed giant industrial enterprises. It needs to be borne in mind that educating men to manage complex ecosystems in itself requires a broad outlook, scientific awareness, social understanding, and an appreciation that much of man's work must deal with uncertainties. There is little place for reliance on dogma in a forester's career.

It is a healthy sign that forestry educators and employers of graduates both engage in such debate. It assures examination of educational objectives, curriculums, and instructional methods. It helps foresters to avoid becoming provincial and should alert students to the need for continued learning. It may also be stimulating for forestry educators to get a "dig" now and then from outsiders. One writer, incensed by clear-cutting practices on certain national forests, stated in testimony before a Senate Committee in April 1971, "This kind of abomination takes place because foresters are narrow in their training, limited in their contacts and reading, and sheltered by the syndromes of their profession."[1] It should be recognized by students, faculty, and employers alike that a man with a bachelor's degree from a university has really just begun, not finished his professional education. Postgraduate and continuing education lie ahead. But most important is for each professional to become a self-learner throughout life (Fig. 22-2).

Professional success comes to those who recognize the limitations of their education and study assiduously to gain new competencies when

**Figure 22-2** Students in ecology studying winter forest vegetation. *(State College of Forestry at Syracuse.)*

faced with opportunities they otherwise could not master. It is important that self-learning be purposely directed lest it but dissipate energies on fields having little to contribute to a man's life career.

The special needs of education in wood-science and technology are dealt with by Ellis.[4]

**Postgraduate Education**   Postgraduate education leading to the master's and doctor's degrees is offered by many forestry schools in the United States, Europe, and elsewhere. The purpose of such instruction is twofold: to broaden a man's general knowledge of forestry, and to teach him to become a creative learner in some restricted subject area. To achieve this dual goal, each postgraduate student in consultation with his major professor plans his own study program. A major decision is choosing a subject for his personal investigation and report.

The postgraduate seminar is a valuable and widely used instructional method. In it, each student and often each faculty participant is asked to review the literature on a particular subject and present an oral report before the assembled seminar group in a clear and convincing manner. A question period follows in which the speaker is queried in depth on the subject to uncover what is and what is not yet well established by investigators in the assigned area of study. The participant thus performs the type of task he is likely to be called upon to perform by his superior in research, administrative, and academic employment.

It is through postgraduate study that young foresters can gain critical understanding and competence in such specialized fields as chemical ecology, ethology, operations research, systems analysis, mathematical simulation, management theory, free radical chemistry, polymer synthesis, rheology of wood, ultrastructure of wood, and others. Effective study in many such fields requires expensive specialized equipment and library resources to be found only in universities and advanced research institutions.

## EDUCATING FOREST TECHNICIANS

The UN Food and Agriculture Organization, in projecting the forestry educational systems needed in developing nations, first projected the contribution that forestry might make to a future gross national product. This took the form of making projections of land to be planted, natural forest areas to be developed and harvested, and forest products to be processed. To these were added ancillary activities in water, wildlife, range, and recreation management. The next step was to determine the labor and technical supervisory force needed for these tasks. The final step was to project the number of professionals needed to plan and give direction to the technical staff. The point is that UN-FAO recognized the technician as the key man in translating a large forestry task into its manpower components. It found that in most developing countries the ratio of technicians to professionals was out of balance with the roughly 5 to 1 ratio that had become long established in European nations. But nowhere was the ratio wider of the mark than in the United States of America.

Fernow, Pinchot, Schenck, Graves, and others vigorously urged the establishment of forest technician schools in the early 1900s. By 1910 some 10 or more schools were operating on a short-course basis. Demand for graduates failed to materialize so support was withdrawn.

The first school to survive and prosper was established at Wanakena, New York, in 1912, under the sponsorship of the New York State College of Forestry at Syracuse University, later to become a part of the State University of New York system. Understanding support was supplied by the parent college. A well-designed and vigorously administered program of instruction was established. The New York State Ranger School soon gained the reputation of graduating men who could perform well on the job (Fig. 22-3). Other forest technician schools to follow patterned their programs after it. It was not, however, until around 1960 that forest

**Figure 22-3** Professor Philip T. Coolidge, first director of New York State Ranger School (1912–1913), addressing alumni, and Professor James F. Dubuar (director 1921–1958), at extreme right. *(State College of Forestry at Syracuse.)*

technician schools began to gain prominence in the United States. Meanwhile, from 1900 until 1960 and beyond, much of the type of work performed in Europe by forest technicians was assigned to recent four-year graduates. With the pressure on for much higher performance by professional foresters, the day for the forest technician seems to have arrived. In 1970, 51 forest technician schools were operating in the United States.[8] Yet the ratio of technicians to professionals of 4 to 1 that prevailed in Canada was inverted in the United States where it stood at 1 technician to 2 professionals.

The forest and forest-industry technicians carry out the plans and policies formulated by the professional foresters. They perform such functions as training and instructing foremen, assigning work to them and to skilled workmen, and inspecting performance for quality and quantity of output. They perform a variety of tasks requiring technical knowledge and skills such as surveying, map making, timber cruising, timber marking, together with the office work these entail. Forest industry technicians may operate dry kilns, or supervise other operating units such as wood machining, finishing, packaging, and shipping. Forest wildlife technicians may organize game censuses, check on use of browse and forage, cull game herds, and engage in various wildlife improvement measures.

All are obliged to keep records, summarize these, and make reports to superiors. Most have at least some responsibility for dealing with the public and forest-users such as recreationists, sportsmen, loggers, livestock graziers, and various organized groups.

In short the technician is the man to perform technical and supervisory tasks requiring technical knowledge and skills, to establish and maintain work standards, and to keep his superior officer informed on the progress of work.

**The Technician School Curriculum** The curriculum of a forest, wood products, or wildlife-technician school will probably include several of the same subjects that appear in the professional man's education. The difference is in depth of treatment and emphasis. In general, the professional school concentrates on principles and fundamental understanding; the technician school on applications and performance. The professional is employed to be an analyzer, synthesizer, and creator. The technician is taught to be a performer of technical work to high standards of output and quality.

Instructors in technician schools will, therefore, seek not only to cover the subject matter of their courses but also to require each man to meet high standards of performance on his assigned tasks. It is important that technicians know from personal experience the standards of output to be expected of others. Experience in crew leadership and foreman duties are generally included in the program of instruction. A well-educated technician is expected to follow instructions closely and to perform at a high level with a minimum of on-the-job training, and with little supervision.

The difference in education of the professional and of the technician is stressed because of the difference in the tasks each is expected to perform. The nation needs far more technicians in forestry and could do with fewer professionally educated men who perform largely technician-level work. Does this mean that an ambitious and able technician cannot aspire to become a professional? In some countries he will find the professional status closed to him. A few men of extraordinary energy and ability have in the past risen to full professional stature by individual study and high performance on the job. The quicker and easier way, however, is to transfer to a professional-level school after completing the technician program. Among those who have followed the latter course have been heads of forestry schools, high-level government administrators, and vice-presidents of large corporations. No stigma is attached to a man in America for having graduated from a technician school.

## SCHOOLS FOR FOREST GUARDS, FOREMEN, AND MACHINE OPERATORS

Vocational schools to train forest guards, foremen, and forest-industry skilled workmen were operating in France, Pakistan, U.S.S.R., Kenya, Chile, and a few other countries in 1970. In Latin America there were some 10 schools operating in 1968 at a level between that of the vocational and that of the technician school. In addition, UN-FAO had teams of specialists that traveled from country to country giving instruction to forest officers in how to organize and instruct in such schools.

In the United States a number of training courses have been organized from time to time for loggers, head sawyers in sawmills, dry-kiln operators, lumber graders, wood-machining operators, gluing and finishing workmen, and forest guards. Instruction was also given in logging safety by safety engineers of forest industries and insurance companies. Nicolet College at Rhinelander, Wisconsin, was offering vocational training for woods workers on a sustained full-time basis in 1971.

Instruction of mechanics, electricians, machine-operators, and repairmen was offered in vocational high schools and in some community colleges.

Most of the workmen in these categories learned on the job under the tutelage of experienced men. Probably the most widely known vocational school in the forestry field up to the year 1968 was the lumber graders' school at Memphis, Tennessee. A course for retail lumber salesmen and managers operated vigorously for several years in New York and New England under the sponsorship of the Northeastern Retail Lumber Dealers Association in cooperation with two forestry colleges.

There appears to be sustained need in America for vocational education for workers in forestry and forest industries if we are to quickly bring into responsible positions young men with interests and ability in this work. Much time is wasted and frustration involved in depending solely on the long slow rise of men on the job. For example, it has taken some 15 to 20 years of on-the-job experience for a man to qualify as a paper mill machine-tender in the United States. A company in Finland has trained young men 21 years of age to tend the huge complex high-speed Beloit paper machines in 1 year's time.

## EDUCATION OF THE CITIZEN

Acquainting the resource-conscious citizen, schoolteachers and community leaders with concepts and objectives in forest care and use and in environmental protection, is a task for conservation organizations, extension foresters, and public-information specialists in federal and state conservation agencies and universities. Various lectures, slide shows, exhibits, paid advertisements, leaflets, news briefs, spot announcements, and full-length films are used. Pamphlets, lectures, leaflets, and films are generally available to responsible groups. Also, from time to time forestry agencies organize field excursions and special field days to acquaint community leaders and others with forest-management activities. Serious-minded civic organizations such as the League of Women Voters occasionally study specific conservation and environmental subjects in depth and bring their findings to the attention of lawmakers, administrators, and the public in general.

Many conservation associations publish attractive journals that carry conservation messages to the reading public. They also are not shy about urging citizens to write their congressmen or state legislators urging support for or protest against certain bills.

The broad concern of citizens in the late 1960s and 1970s with environmental protection and enhancement is encouraging. It bids fair to result ultimately in a far better informed general public on conservation issues than our nation has had since the turn of the century.

When the general public becomes sufficiently aroused and evinces its willingness to pay the costs for environmental improvement, it can be confidently predicted that lawmakers, budget officers, corporation executives, and financiers will find means to get the job under way.

## CONTINUING EDUCATION

Continuing education is a responsibility of the individual, of his employing agency, of his profession, and of the universities.

A man enters a profession not because it offers a good life, which is important, but because it offers an opportunity for self-fulfillment through a life of service. The individual himself must be interested in the subject of his life's work and in its peripheral fields. Keeping abreast of developments in a fast-moving field such as environmental protection and enhancement will require plenty of study and thought. Generally as much as an hour's time each working day should be spent by the professional to keep up with new developments in his specialty. Those who do not do this will soon find that the half-life of their education has expired. Professional journals and especially attendance at professional meetings keep a man in touch with things he will want to know more about. Pertinent training sessions and short courses should also be participated in.

Employers of foresters seldom have a more valuable asset than the competence of their professional men. They should, therefore, encourage such men to attend professional meetings and special short courses that are pertinent to their work assignments. In-service training, and lectures and seminars by invited speakers, have done much to stimulate a desire among professional men to keep abreast of developments. Encouragement of personal investigation and publication also puts a man on his mettle.

The professional society serves continuing education by publishing stimulating articles in its journals, by holding inspiring national and sectional meetings, by recognition of outstanding achievements, and by appointing committees to investigate and report on subject matter fields. The Society of American Foresters can point to outstanding performance in all the above ways. The professional society can also keep before the membership the obligation to keep informed. By referring pertinent matters to the membership for referendum it assures that policy issues as well as professional achievements are given constructive thought.

Universities have an obligation to hold continuing educational sessions on rapidly developing subject areas. To be successful, these sessions need to be planned for a specific clientele and the instruction should be carefully fitted to their background and needs. This is not difficult in scientific areas because the workers themselves are generally well informed and may serve both as participants and instructors. Stimulating discussion is usually essential to fix points in the participants' minds. The art of directing such discussions is being developed in many universities.

Continuing education should not be thought of as for the professional man only. Technicians must be brought up to date in new methods and techniques as well. Equipment operators, foremen, and other skilled workmen will need to be trained to operate new models of machines, perform new processes, and meet new performance standards in their work. And education of laymen in a rapidly developing field is a never-ending responsibility.

## STATUS AND OUTLOOK FOR FORESTRY EDUCATION IN AMERICA

The Society of American Foresters grants accreditation to forestry schools that meet professional standards that the Society prescribes. An accreditation visit by the Society is generally conducted in cooperation with a regional college and university-accrediting association. Separate standards have been drawn up for schools of wood technology, wildlife, and range management by societies in these fields. Accreditation of professional schools of landscape architecture is provided by the American Society of Landscape Architects.

Accredited schools of forestry numbered 35 in 1971. Affiliated degree-granting but nonaccredited schools numbered 13, and other institutions offering instruction but not degrees in forestry numbered 6, for a total of 54.[9] The enrollment and degrees granted by the forestry schools in 1970 were as follows:

|  | Bachelor's | Master's | Doctor's |
|---|---|---|---|
| Total enrollment | 17,178 | 1,926 | 960 |
| Women students | 696 | 115 | 24 |
| Degrees awarded | 2,771 | 513 | 154 |

Of the bachelor's degrees, 55 percent were awarded in general forestry, 19 percent in wildlife management, 8 percent in forest recreation, and 6 percent in wood technology and forest utilization.[5]

Comparable data on forest technician and vocational schools were not available in 1970.

**Outlook** Forestry educators in 1971 faced a number of challenges. Among them were the following:

1  The growing insistence of students that their education fit them to participate effectively in changing our society in the direction toward greater harmony with the environment, greater humanity toward people and nations, and toward maximum creativity of individuals.

2  The rapid increase in education and employment of forestry technicians. This means that the standard of professional education must be changed.

3  The growing participation of planners, ecologists, and others in decision making in areas where formerly foresters occupied first place.

4  The need for sharpened vision to formulate and define important new national needs in forest-resource leadership, research, and administration, and to educate young men to meet such needs.

5  A rapidly growing demand for all products and services of forests as population and economic activity expand.

6  Complicating the educators' task will be the insistence of growing numbers of people that the environment be as little impaired as possible and that natural ecosystems be largely unmodified. Moreover, conservation spokesmen are insisting that they be permitted a voice at the council table where decisions are made on where and how the timber may be harvested.

7  Not only will the forestry profession be expected to meet the timber demands of 1980 but it will be expected to provide vigorously growing young timber to meet the still higher demands anticipated for 1990, 2000, and beyond. This can be done most satisfactorily only by viewing the timber outlook on a worldwide rather than a national basis.

8  The need to develop integrated use of forest land and integrated use of the timber resource in manufacturing plants.

9  The forestry profession has advocated multiple use, but has not clarified this so that it meets with complete public satisfaction and confidence. What the public needs to understand is how the several uses can be integrated in a forest so as to realize optimum public benefits. The same principle applies to use of timber in the processing plant. Integrated use is highly developed in Scandinavian and Finnish timber-processing plants. It could be much more widespread in the United States.

10  Perhaps the greatest challenge of all to forestry education comes from the realization that the success foresters achieve in meeting the demands on forests in the year 2000 will depend on the education, stimulus, and inspiration that students of the 1970s receive. That this must be imparted during a period when American youth is groping for new values in life and public institutions are undergoing challenge as never before only adds to the forest educator's difficulties and responsibilities.

The forestry profession in the past has been noted not only for vigorous leadership in the protection, development, and use of forests and timber, but of watersheds, rangelands, wildlife resources, and outdoor recreation. The profession should be a leader in environmental improvement insofar as it involves forest land and resources. This is a task for the profession as a whole in which the forestry schools are in a favored position to exercise important influence.

Fortunately, many examples of creative planning and innovation in forestry schools were in evidence in 1970. There seems to be growing awareness that a policy of continued upgrading is necessary. Duerr expresses the responsibility of forestry educators as follows: "The professional school can strive to help the student cultivate his creativity, question goals and means, understand and use the decision process, and work as handily with the social and institutional elements of his system as with the technical."[3]

It will not be easy for the forestry profession to regain the leadership it once enjoyed in the renewable resource-management field. Perhaps it is best that this function be shared with men educated in other disciplines that have much to contribute. Leadership comes from deep insight into major societal needs matched by creative thought directed toward devising workable and feasible means by which such needs can be met. To gain such insight requires deep concern for human welfare. To develop practical plans of action requires technological competence. To bring technical plans to fruition requires vigorous and understanding administration. The forester has a role to play in each of these.

**LITERATURE CITED**

1. Brooks, Fred C. 1971. Senate Hears Clearcut Concerns, *Journal of Forestry,* **69**:299–302.
2. Dana, Samuel T., and Evert W. Johnson. 1963. Forestry Education in America Today and Tomorrow, Society of American Foresters, Washington. 402 pp.
3. Duerr, William A. 1970. Report on Professional Forestry Education Policy, *Journal of Forestry,* **68**:650–653.
4. Ellis, Everett L. 1970. Education and Employment in the Forest Products Field, *Journal of Forestry,* **68**:411–414.
5. Marckworth, Gordon D. 1971. Statistics from Schools of Forestry for 1970: Degrees Granted and Enrollments, *Journal of Forestry,* **69**:503.
6. Nelson, Thomas C. 1970. A Look at SAF's Educational Policy for the 1980's, *Journal of Forestry,* **68**:653–654.
7. Shirley, Hardy L. 1966. Forestry Education in a Changing World, *Proceedings of the Sixth World Forestry Congress,* **1**:895–902, Madrid.
8. Society of American Foresters. 1970. Institutions in the U.S. Offering Forest Technician Training, *Journal of Forestry,* **68**:249.
9. Society of American Foresters. 1971. Institutions in the United States Offering Professional Education in Forestry, *Journal of Forestry,* **69**:103.

Chapter 23

# Research in Forestry

Forest road building in the 1920s required hard, grueling work. After staking out, the right-of-way had to be cleared of trees and undergrowth with ax, crosscut saw, and scythe. Rocks and stumps were blasted and dragged away by a tractor. Surface litter and debris was scraped aside by a heavy road grader drawn by a crawler tractor. The grading could then be begun using the same grader and tractor. Behind the grader two men with axes followed to cut and remove the roots. The tractor with its grader in tow required considerable clear space to turn around. A forest officer responsible for building narrow roads in mountainous country reasoned, Why could not a grader blade be mounted on the front of the highly maneuverable crawler tractor? It could then smooth its own path and both cut and fill as it moved along the contour. Could it not also uproot small stumps and shove them and boulders off the travel surface? The officer set about attaching a movable blade to the tractor frame and so constructed the first bulldozer.

Logs in the 1920s were skidded from the stump to the landing by horses. To lighten the draft, high-wheeled logging arches were used to raise the forward end of the log off the ground. For logs too heavy for horses, a crawler tractor with a track-laying arch was devised. Another forest officer thought the arch and tractor could be built as one machine, thereby increasing maneuverability. He widened the tractor frame and its treads. He mounted a fairlead and buffer plate over the winch and so built the first integrated arch-and-tractor skidding machine, which he called the "tomcat." Lacking patent protection it was not built commercially in the United States, but a similar model was built and widely used in the U.S.S.R.

Still, it was heavy, slow, and clumsy. An ingenious equipment manufacturer thought, Why not build a high-wheeled tractor mounted on large pneumatic tires with articulated frame, four-wheel drive, locking axles, and elevated fairlead? Would this not gain the traction needed to maneuver in the forest without bogging down in soft earth? Thus the modern, high-speed, rubber-mounted skidder was evolved.

It was a much more complicated task to develop the tree harvester. An early model was built in the U.S.S.R. that could sever stems from the stump and drag full trees to a landing. The first United States–built harvester could shear the stem near ground level, delimb the stem, shear off pulpwood-sized bolts, and carry these to the landing for loading onto a truck.

A corporate vice-president was asked by a fiscal officer why he must keep more than $1 million tied up in reserve pulpwood supplies. "That is to assure no stoppage of paper manufacturing at seven company mills," he replied. Upon reflection he thought, "Maybe that budget man raised a good question. Perhaps it would be less costly to hold lower reserves and ship wood from one plant to another when emergencies arise." A consulting firm was called in to work with local management on the problem. All factors of pulpwood flow and of costs were quantified and their relationship mathematically simulated. The problem was programmed for a computer, feeding the necessary data into the mathematical model. The answer was positive. Inventories could be substantially reduced. The costs of the small amount of transshipping likely to be required would be generously covered by the saving in interest on capital otherwise frozen in inventories.

An organic chemist thought that wood fabrication could be immensely improved if wood could be temporarily plasticized for bending and shaping. He reasoned that this would require a substance that would soften lignin and lubricate the interfaces of cellulose molecules. What would do the job at reasonable cost? A brief literature search plus some chemical calculations indicated that ammonia might serve. It did. He thus discovered not only a better means for bending wood, but a new method of preparing paper pulp, a new wood-plastic material, and a new art medium. All these awaited development and industrial application as of 1971.

## THE NATURE OF FOREST RESEARCH

The construction of the first bulldozer, integrated skidder, and timber harvester are examples of developmental research. Solving the pulpwood-inventory problem involved devising new mathematical tools to minimize costs. Plasticizing wood can be considered as basic research, for although the chemist who did this was fully aware of its potential uses, a long road of developmental research still lay ahead in 1971. Each of the five examples have certain aspects in common:

1 They arise from a significant question in the mind of a knowledgeable individual.
2 The question is elaborated to outline what work will probably be required to find an answer.
3 The effort and cost to solve the problem are weighed against the probable benefits that may flow from the solution.
4 Search is made of literature and other sources to discover what has already been done that may be used to find a solution.
5 The problem is viewed in detail and the various steps required for a solution outlined.
6 A budget is drawn up to provide for the expected costs entailed. Often this may have to be revised if unforeseen difficulties arise.
7 Final decision to proceed and allocation of necessary funds are made.
8 Qualified personnel are assigned to the several steps involved.
9 As each step is completed a determination of its credibility or reliability is made.
10 All stages are integrated into the complete solution and its reliability is determined. This may require construction and testing of a prototype, as in the case of the bulldozer and log-skidder. In the case of the pulpwood inventory, it takes the form of a probability ratio of say 99 in 100 cases reshipping will prove cheaper than holding a complete wood reserve in each mill. In the case of plasticizing wood with ammonia, a complete pilot plant test will be required before the process can be placed into commercial use.

For the chemist, plasticizing wood was but a side issue in his investigation of cellulose. He set about the task of polymerizing sugars to synthesize cellulose precursors: one of these turned out to be a precursor of blood sugar. Medical biochemists are deeply interested in his work because of its possible relation to how the human body synthesizes, stores, and uses glycogen. For plant biochemists, it is of equal concern for it relates to how green plants build cellulose and starch from simple sugars and in turn break them down for use in growth.

A fascinating feature of forest research is that often the answer, though difficult to discover, lies in the forest itself. This is because of the intricate interrelationship among organisms. In the 1930s it was demonstrated that both nutritional difficulties and seedling diseases could be minimized in forest nurseries by introduction of composted forest litter or by establishing a nursery in a small forest clearing made for the purpose. It is now known that soil mycorrhizae may aid in nutrition and that they and other fungi produce antibiotics that can inhibit fungal diseases. Another case is insect control. The creature's own hormones and pheromones probably can be used to keep it in check. But to discover these, determine their chemical nature, synthesize them in amounts for control use, and find a method of using them requires costly and highly sophisticated research. It is equally fascinating that solutions to other forest problems may be arrived at through abstruse mathematical approaches and computer technology.

The man with his mind in the stratosphere used to be a subject for derision. In 1970, the first timber inventory was made in which space photos taken during the flight of Apollo 9 were used for delineating forest types.[2] This was but an extension of the use of aerial photos in timber inventories and forest-mapping that has been common practice since 1946.

The above cited cases alone serve to indicate that forest research involves a threefold approach. One is through the practical difficulties encountered and observations made by the field administrators; a second is through ecology and the functioning of the ecosystem; a third is through basic sciences and engineering.

Literally hundreds of small problems and major ones also have been dealt with effectively through refined silvicultural practices. Northern forest soils can be improved in fertility by admitting enough sunlight to the forest floor to promote rapid litter decay and recycling of nitrogen. Prairie soils inoculated with mycorrhizae may enhance shelterbelt growth. Small mammals and other predators of the forest floor keep many insects in check. Thinned, vigorously growing young stands tend to overcome many of the disease and insect problems of crowded and overmature stands. The list of examples could be greatly expanded.

The ecological approach is but a somewhat more sophisticated approach to that of the silviculturist and manager. In fact, foresters helped greatly to develop it. The forester's concepts of tolerance, his experiments on root competition, his discovery of root inhibition, of climatic races and the values of mixed stands, have all helped to understand ecology. So have the concepts of plant succession, climax types, and ecosystems by the ecologist added to the forester's working tools for research and management.

Basic science has afforded the broadest and perhaps the most fruitful approaches to forest research. It avoids needless floundering, testing everything empirically as the investigator proceeds. It also offers many possibilities of new and more productive approaches to problems. A Lake states forester wanted to find out the best time of year to weed aspen root suckers from conifer plantations. He carried out elaborate field tests, cutting the sprouts each month of the year and recording the number and growth rates of the resulting new sprouts. A plant physiologist learning of his objective gave him the results offhand. "Sprouting will be least vigorous," he said, "if sprouts are cut in the spring immediately after the new flush of stem and leaf growth has passed, for it is then that food reserves in the roots are at their lowest level."

But foresters as such, though given introductory courses in physics, chemistry, geology, botany, and zoology, are still not research physicists, chemists, earth scientists, nor biologists. Hence the U.S. Forest Service and other forest-research institutes do employ men with research competence in the above and additional fields. Many of these work in teams made up of men expert in several different disciplines. A number of research foresters hold advanced degrees in such subjects as meteorology, plant physiology, genetics, soil science, chemistry, pathology, entomology, operations research, statistics, economics, and other sciences.

## ORGANIZATION AND FINANCING OF RESEARCH

Though a short splint of wood was first plasticized with liquid ammonia in a test tube, forest research and development in general is no test-tube operation. It requires sophisticated laboratories, expensive and delicate instruments, electronic data processing, and above all highly educated, imaginative, resourceful, and dedicated workers. To obtain the space, equipment, library resources, and personnel required for creative work is costly. Moreover, forest responses are often available only after some five or more growing seasons have elapsed. Time thereby adds to the cost. Hence, only well-established and well-financed research institutions can engage effectively in comprehensive forest research programs. These are found in the U.S. Forest Service Experiment Stations, the U.S. Forest Products Laboratory, the universities, the larger forest industries, chiefly the pulp and paper and plywood corporations, and large private consulting firms.

The expenditures for forest research by the above agencies have shown a gratifying increase since the first nationwide study was made in 1926. The estimated annual expenditures by study years have been as follows:

| Year | Estimated expenditure, in millions of dollars |
|------|------|
| 1926 | 4 |
| 1937 | 14 |
| 1953 | 45 |
| 1961 | 92 |
| 1971 | 140 |

The 1971 expenditures for research by the U.S. Department of Agriculture and the universities combined were expected to be about $68 million. Expenditures by the forest industries for research and development, including those of manufacturers of logging and other forestry equipment, have been somewhat larger than those of all other agencies. The total expenditure for 1971 was therefore expected to be in the neighborhood of $140 million. This projection, if correct, would mean that expenditures were generally in line with the 1955 projection made by Kaufert of $200 million by 1978.[3]

Those who bear these costs—the taxpayers, universities, and industries—expect that the expenditures will result in benefits commensurate with the costs and risks entailed. This means in most cases savings in dollar costs of forest and forest-industry operations. But it also means returns through such social benefits as pure water supplies, an esthetically satisfying countryside, an improved environment, recreational opportunities, stabilized rural economy, and remunerative employment.

Most forest research is, therefore, problem-oriented with each project weighed in terms of costs and expected benefits. And it is supported by well-established research institutions. Some workers may delve deeply into fundamental science, if that is needed to solve the problems. But the solution will have its benefits in direct application or in opening new fields of investigation that do result in significant human satisfactions.

Federal forest research is concentrated in eight regional forest experiment stations, the Forest Products Laboratory, the Institute of Tropical Forestry in Puerto Rico, and the Washington Office of the Forest

Service. The Forest Service cooperates on research with various universities and private industries. It also supports research under Public Law 480 at a number of foreign universities and research institutes (Fig. 23-1).

The U.S. Department of Agriculture, Cooperative State Research Service administers the McIntire–Stennis Cooperative Research Program that supplies money on a matching-fund basis to finance research projects in forestry at state-supported universities. The Department also supports research in forestry at land-grant universities through various funds. The U.S. Department of Agriculture, Office of Science and Education administers the Current Research Information Service. This Service enables individual workers to maintain contact with others working in areas of mutual interest so that coordination of effort is possible. It also enables research administrators to keep in touch with the total research programs of the states and the Department of Agriculture.

**Figure 23-1** Raphael Zon, early advocate of forest research, and planner of the system of U.S. Forest Experiment Stations and the Forest Products Laboratory. *(U.S. Forest Service.)*

A rough measure of the emphasis given to various broad areas of forest research is the number of publications issued annually in the major subject fields. The 20 broad areas listed in the 1970 Forest Service report on research have been grouped into 7 and are listed below with the number of publications pertaining to each:[5]

| General areas reported upon | Number of publications pertaining to each |
|---|---|
| Improving the environment | 147 |
| Protecting forests from fire, including remote sensing and weather studies | 99 |
| Protection from insects and diseases | 259 |
| Forest surveys, engineering, economics, and marketing | 125 |
| Forest products and housing | 214 |
| Improving timber production | 155 |
| Water quality and yield | 102 |
| Total | 1,101 |

The McIntire–Stennis Cooperative Forest Research Program sponsored in 1969 a total of 507 projects at 60 universities. Working on these were 521 scientists and 366 graduate students, and 290 publications resulted.[4] Projects under this program also covered a wide range of forestry subjects paralleling many of those listed for the Forest Service. Universities also financed research in forestry and related subjects from their own research funds and from those available to them from other granting agencies, notably the National Science Foundation and forest industries (Fig. 23-2).

Forest-industry research was conducted at the Institute of Paper Chemistry, other group-sponsored institutions, and in the individual company forests and laboratories. It has been common practice for industries to form associations for sponsoring research of fundamental nature in which the industry as a whole is interested. Tree improvement and genetics has been one such field and chemistry of pulp manufacture another.

Is there then no place for the hobbyist in forest research? Fortunately there is, provided he undertakes something he is competent to solve in his spare time. Much imaginative work has been so carried out. The priest, Gregor Mendel, discovered laws of inheritance during time he could spare from his duties as prelate of the monastery. Phenological observations can be recorded and analyzed with but modest expenditure of time. Ecosystems can be observed and described on vacation trips.

**Figure 23-2** Testing a laminated wooden arch. *(State College of Forestry at Syracuse.)*

Simple fertilizing trials can be established and responses measured. New designs for articles to be made of wood can be prepared and tested in the home workshop. A host of others could be suggested.

## ACHIEVEMENTS OF FOREST RESEARCH

Research results tend to come in bits and pieces. For this reason the uninitiated may think that the accomplishments of a worker during a year consist of only a few unrelated facts of little or no use. Even the annual reports of large research organizations may list many accomplishments that in themselves appear but a plethora of unrelated facts. For example, the 1970 report of accomplishments of Forest Service contains such research findings as these:[5]

Losses through chemical degradation during kraft pulping can be partially prevented by treatments to modify wood carbohydrates.

Stream flow from two watersheds increased from 4 to 8 inches following clearing the lower half of one watershed and the upper half of a second.

The amount of evapotranspiration by grass cover on a deforested 22-acre catchment area in the southern Appalachians was closely related to the amount of grass produced.

Certain eastern white pine trees are extremely sensitive to low levels of sulphur dioxide in the atmosphere.

Generally it is only after accumulating and interrelating other results to such findings as the above that significant modifications of practice come about.

Kaufert,[3] in his study of 1953, asked whether the $45.4 million spent that year had resulted in accomplishments sufficient to warrant the expenditure and the increased expenditures estimated to be needed in the future. Opinions were sought from some 200 leaders of considerable stature in the industries and the public agencies. After discussion with the research investigator, these leaders invariably came to the conclusion that the products, practices, and processes with which they dealt had been greatly affected by research accomplishments.

Specific accomplishments are legion. By developing gum-flow stimulants and mechanization of operations, research workers have demonstrated that the daily output of workers in naval stores harvesting can be increased by 12-fold.

Male shoot moths have been completely sterilized by topical application of tepa, without adversely affecting mating ability or longevity. This is an additional finding in the widespread search for new biological methods to control insect pests.

A high-capacity water pumping and distribution system has been developed that creates a water curtain that proved effective in stopping a 50-acre head fire in jack pine logging slash. Developing new gluing techniques for resinous southern pines together with pine timber inventories made possible the development of the southern pine plywood industry.

Genetic research has given us hybrid poplars that are superior in growth rate to other poplars. It has given us superior strains of pines, spruces, and hardwoods. Statistical research has enabled foresters to design forest inventories to attain desired levels of accuracy at least cost. They have added immensely to the reliability of projections of forest growth and of timber requirements. There is hardly a chapter in this book that does not refer to actions and products that came about as a result of forest research.

Forest research has also made contributions to other fields. Its contribution to ecology, to blood chemistry, and to metabolism have been mentioned. Foresters' development of equipment and methods for smoke-jumpers has been adapted for parachute troops. Laminated wooden arches and trusses have made possible wide unsupported spans and new architectural shapes and structures. New house construction methods have been devised that reduce on-site work, lessen material costs and improve performance in service. Treatment of wood with preservatives and fire retardants have greatly prolonged its life in service. These are a substantial contribution to the construction industry.

Research has been basic to the greatly improved condition of the western range and to increase beef production therefrom. It has taught us much of wildlife behavior and of how to improve game-carrying capacity.

These are but a few illustrations of the contributions of forest research to society. The 1970 annual list of new research contributions to forestry, forest industries, consumers, and rural environment would make a book of approximately 200 pages.

**REWARDS TO THE INDIVIDUAL**

Forest research requires a large amount of meticulous, mundane work. Research problems must be formulated and broken down into component phases. Search of literature must be thorough. Experiments must be designed so that conclusive evidence will be obtained. Statistical and other controls must be provided. Instruments and equipment must often be devised, constructed, and standardized. Workers must be trained to carry out the installation, routine maintenance, and record keeping. Data must be transferred to cards or tapes for processing. The processing must be programmed and carried out by electronic computers or other means. Results must be studied, compared, and weighed against indicated experimental errors. Finally, the results must be tested for consistency and reliability against phenomena known to be related. Additional experiments may then be indicated or extensive field or pilot-plant testing provided for. When well-established conclusions can be drawn, a report for publication or patent application is prepared.

The above reads like a lot of drudgery and so it is likely to be. The thrill and excitement of research is often hidden by the mass of detail. But the excitement should be there. First the problem, if well conceived, presents its own challenge to the investigator. How can it best be attacked? The search of literature itself often arouses speculation as to new methods to try. The need to devise instruments and equipment stimulates creativity. Planning an elegant design that will result in conclusive results with minimum extraneous work is a scientific art to be cultivated. Measurements and recording seem mundane but one needs to be on the alert not to overlook something out of the ordinary. It may be highly important. And finally, as the data are being processed, the suspense as to how it actually will turn out heightens. "Have I proved something or have I not?" Or even, "Have I opened a pathway to something heretofore unseen?" The anticipation grows as the results are studied to exhaust and to ponder their full meaning. So research offers its thrills and intellectual excitement.

**EDUCATION FOR RESEARCH**

Research is a highly specialized activity. It operates on the fringes of knowledge. Unless one knows where these frontiers are he may be wandering about aimlessly and achieve nothing of significance. Research also has its methodology of investigation and criteria for scientific evidence and proof. These must be understood and applied. The basic education, therefore, is the Doctor of Philosophy degree or its equivalent. This concentrates on teaching one how to recognize a frontier in science and how to go about exploring it.

To make truly significant contributions through research, however, requires a great deal more than painstaking study of literature, rigorously planned experimentation, and meticulous analyses of results. It takes a flash of insight—the genius to see beyond the evident to possible new relationships. It takes daring and faith to stake one's time and energy on a new approach that may lead nowhere. Imagination and insight are to a degree inborn traits. But they can be immensely stimulated by association with men of fertile mind and febrile intellectual activity. Such are to be found in outstanding universities and research institutions. An ambitious young Ph.D. can therefore well afford taking a postdoctoral fellowship where such stimulus can be experienced.

Firsthand acquaintance with conditions at the operating level—the forest, the wood-processing plant, or the pulp and paper mill—is essential for effective leadership in project research.

It is well for the forest research worker also to know the natural area and the wilderness where man-made disturbance is at a minimum. Study of these may sometimes suggest a new approach to a problem.

## OUTLOOK FOR RESEARCH IN FORESTRY

Many problems that currently thwart satisfaction of deeply felt human needs probably will be solved eventually. Forestry has its share of these. In the forefront are stream and air pollution by pulp mills. Vast steps have already been made in reducing both. Two new process modifications were announced in 1970. Much more remains to be done.

The unsightliness of recent logging operations troubles foresters and environmentalists alike. It will take a change in attitude and lots of pressure on loggers, equipment manufacturers, and woods labor before logged areas in America become as neat as those in Europe. Close utilization of limbwood accounts for tidy logged areas there and is helping in America also.

An important problem is to improve design of many articles and structures made of wood, particularly furniture and dwellings. Neither art nor technology have kept up with what wood science has to offer.

Finding means acceptable to owners for increasing the timber output and use of the 60 percent of forests in small holdings is a broad human and social problem still to be solved.

Related to the above problem is managing the forests associated with developing suburbia. Planners and foresters should be able to achieve widespread human satisfactions through making extensive forest areas available for man's recreational use. This has been done by the city of Zurich, Switzerland, in the Sihlwald, and by the Austrians in the Wienerwald. Both of these forests are managed for commercial products, recreation, and amenities.

The above involve planning and development, but they also involve finding new approaches to people if plans are to be effected with harmony.

Research in ecology should increase our knowledge of how to maintain and perpetuate healthy, high-yielding, esthetically pleasing forests with their associated animals and subordinate plants. A final solution is unlikely for man can always learn more and manage better.

It may be expected that future pest-control problems will be solved through use of pheromones, hormones, naturally occurring antibiosis, pest diseases, and other means of attacking specific organisms. Much research lies ahead to realize this.

Protection of forests against fire was one of the first major tasks of foresters in America and will remain a major task for years to come. Research underway in 1970 that may be expected to aid in future fire control includes seeding clouds with silver iodide to minimize lightning fires; a systems approach to fire prevention during critical periods such as occur in southern California when Santa Ana winds are blowing; determining the damage to environment that results from prescribed fires and wildfires; use of infrared scanners to detect, locate, and map going fires; studies of forest fire meteorology and fire physics that make possible mathematical simulation of fire spread and intensity against which control techniques can be tested.[1]

One can hopefully expect also that much of the danger to life and limb will be lifted from the forest workmen as logging is made more tidy and cheap logging costs become no longer the sole measure of efficient woods operations. With this should come added dignity and economic security to the workmen themselves.

The above are only a few illustrations of problems to which future research may be addressed. As one tough problem is solved, it makes possible progress on others. The growth of research should be limited only by what it can be expected to return in significant benefits to man and his environment.

The trend in recent years toward increasing specialization, increased use of expensive equipment, and increased competence of investigators, seems certain to continue in the future.

Inherent in problem-solving research has been narrowness of approach. It makes it possible to solve one problem but sometimes creates others. The approach should be to cast the problem in a broader setting that integrates the action of all major factors. This is the essence of the systems approach.

Foresters need to provide for all human satisfactions that flow from forests and use of their products. Research has a large role to play in bringing this about. A shift from major concentration on analytic to integrative research should add new dimensions to usefulness of results. It should also play a significant role in realizing an inspiring environment.

The little-used forests of tropical Latin America, Africa, Asia, and Oceania present baffling difficulties for forest management because of their complexity in species composition and ecological relationships. Ultimately they offer the potential for furnishing many high-quality products for world commerce. Research, though active in all four areas, has progressed little beyond coping with some of the major commercial species and developing means of replacing native vegetation with high-yielding plantations. What effect the latter may have on long-term site productivity is still to be determined. Much developmental and fundamental research lie ahead.

## LITERATURE CITED

1. Barrows, Jack S. 1971. Forest Fire Research for Environmental Protection, *Journal of Forestry,* **69**:17-20.
2. Hay, Edwards. 1971. An Eye in the Sky, *American Forests,* **77**(1):20-23.
3. Kaufert, Frank H., and William H. Cummings. 1955. Forestry and Related Research in North America, Society of American Foresters, Washington D.C. 280 pp.
4. U.S. Department of Agriculture, Cooperative State Research Service. 1970. Forest Research Progress in 1969 McIntire-Stennis Cooperative Forest Research Program, Division of Information, Office of Management Services, U.S. Department of Agriculture, Washington. 108 pp.
5. U.S. Department of Agriculture, Forest Service. 1970. Research Accomplishments, Washington. 174 pp.

Chapter 24

# Employment in Forestry

Gifford Pinchot told how as a young man he sought the advice of Dr. B. E. Fernow, then director of the Division of Forestry in the Department of Agriculture, on forestry as a life career. Fernow encouraged Pinchot but suggested that he study horticulture also in case no forestry positions opened in America. After succeeding to Fernow's position and building up the Forest Service to an organization of over 300 employees, Pinchot himself advised young men not to become foresters unless they were so attracted that they could not resist following it. Employment of professional foresters in the United States had grown from 1 in 1886 when Fernow was appointed, to over 23,000 in 1971. Still, uncertainty remained as to future employment opportunities.

We can, however, take a look at employment in forestry as of 1971 by agencies and activities, assess annual replacements needed, and consider how the various forestry activities may expand or contract in the future. Many of the foregoing chapters contain a section dealing with future outlook in specialized fields. Chap. 18 contains projections of future use of various timber products. In this chapter consideration will be given to four overall factors that will affect demands on the forest and the forestry effort that will be required to meet them. These factors are population growth, rising expectations of people, environmental resistance, and the initiative of foresters themselves in opening new areas for service.

In conclusion we shall consider the type of life forestry offers and the satisfactions it provides those who follow it as a life career.

## EMPLOYERS OF FORESTERS AND FOREST TECHNICIANS IN THE UNITED STATES

Any organization that owns, uses, manages, or carries out operations on forest land may employ foresters and forest technicians. So also may agencies that use the resources of forest land. Included among them are federal, state, and local governments; hydroelectric companies, water companies, railways, mining companies; loggers and timber processors, range and wildlife managing agencies, operators of private recreation facilities and camps; national defense agencies, consultants to industries involved in forest products, land-planning agencies, timber importing and exporting corporations, and many others that find the knowledge and skills of foresters useful to them.

The major employing agencies are listed by groups in Table 24-1,

Table 24-1 Agencies in the United States Employing Professionals and Technicians in Forestry and Related Work as of 1971

| | Number of foresters employed | |
|---|---|---|
| Agency | Professional foresters | Forest technicians |
| USDA Forest Service | 5,732[1] | 4,309[1] |
| U.S. Department of Interior | 2,000[2] | 1,000[2] |
| Other federal agencies | 500[2] | — |
| State conservation agencies | 2,870[3] | 3,500[2] |
| Universities | 1,013[4] | 75[2] |
| Forest technician schools | 145[5] | 25[2] |
| Consulting firms | 500[5] | 200[2] |
| Forest industries | 8,500[2] | 2,000[2] |
| Miscellaneous and self-employed | 2,000[2] | 150[2] |
| Total | 23,260 | 11,259 |

*Sources:*
[1]U.S. Department of Agriculture, 1971 figures.
[2]Estimates made by Donald R. Theoe of the Society of American Foresters, or by the author.
[3]National Association of State Foresters, 1970 report.
[4]Compiled by Donald R. Theoe for 1969–1970 academic year.
[5]Compiled by Donald Theoe.

together with the reported or estimated numbers of men employed in 1971.

Over the 20-year period from 1951 to 1970, employment of foresters in education, industry, the federal government, and of all foresters, approximately doubled. The increase in state-employed foresters for the period was apparently in the order of 15 percent. Much of the increase occurred during the first decade of the period. The employment of forest technicians, on the other hand, must have increased in the order of 10- to 11-fold during the two decades, most of which occurred since 1960 when many new schools began to turn out graduates.

The activities of the several agency groups involved in forestry and related resource management together with the types of professionals and technicians employed are listed below:

*Federal agencies*
Department of Agriculture, Forest Service, and other bureaus
*Activities:* National forest administration, forest and forest-products research, world forestry. Cooperates with other federal agencies, states, and private forest owners and industries in forestry activities.
*Employ:* Foresters, wood technologists, pulp and paper technologists, engineers, landscape architects, range managers, wildlife managers, forest pathologists, entomologists, ecologists, and hydrologists, at the professional and technician level.

Department of Interior: Bureau of Land Management, Bureau of Indian Affairs, National Park Service, Bureau of Outdoor Recreation, Grazing Service, Fish and Wildlife Service
*Activities:* Administer public-domain lands, grazing lands, Indian lands, national parks, wildlife refuges. Cooperate with other federal agencies and states in outdoor recreation planning and coordination, cooperate with states on fish and wildlife administration and research.
*Employ:* Foresters, wildlife managers, fishery biologists, range managers, landscape architects, and recreation specialists at the professional and technician level.

*Other federal agencies*
Other federal agencies concerned with managing land, reviewing budgets, and reviewing land policies employ foresters on a full-time or consulting basis. Among them are the Department of Defense, the Agency for International Development, the Office of Management and Budget, the Environmental Protection Agency, the Congress, the General Accounting Office, the Geological Survey, and the Tennessee Valley Authority.

As of 1971 such agencies employed mainly professional foresters, in a variety of capacities.

*State conservation departments*
*Activities:* Administer state forests and parks, game refuges and manage-

ment areas, fire, insect and disease control on state and private lands, administer state game laws, license motor boats and off-road motor vehicles, administer in-state waters and their use, administer state forest-practice laws affecting private lands, furnish services to private landowners and industries, operate state fish hatcheries and game-rearing stations and distribute the fish and reared game, operate forest nurseries and distribute the stock. Cooperate with sportsmen's organizations in fish- and wildlife-betterment activities. Furnish advice to forest landowners and industries. Conduct research on forestry, fish, and wildlife.

*Employ:* Foresters, fish and wildlife biologists, law-enforcement officers, forest pest-control officers, landscape architects, and wood technologists at professional and technician levels.

*Universities and colleges*

*Activities:* Offer education at the professional and technician levels in forestry, wood technology, pulp and paper technology, wildlife and range management, forest protection, architecture and landscape architecture, hydrology, engineering, biological science including ecology and environmental science, physical education and outdoor recreation, and many related subject areas. Conduct research and offer postgraduate education in the above fields. Conduct continuing education and sponsor symposiums and conferences in various conservation subjects. Sponsor publications in above fields. Operate college forests and summer camps in conservation.

*Employ:* Professors, research specialists, and technicians in above-mentioned fields of science and technology.

*Consulting firms*

*Activities:* Perform consulting services for industries, investors, associations, and governments.

*Employ:* Foresters, wood technologists, pulp and paper technologists, forest hydrologists, and experts on environmental sciences at professional and technician level.

*Forest-products industries, primary manufacturing*

*Activities:* Manage corporate forests, purchase and harvest timber from public and private forest owners, conduct research in forestry, wood technology, and pulp and paper technology. Manufacture lumber, veneer, plywood, particle board, and paper products. Some operate land-development subsidiaries.

*Employ:* Foresters, logging engineers, mechanical engineers, pulp and paper technologists, and, for land development, landscape architects. Also employ technicians in above specialties.

*Forest-products industries, secondary manufacturing*

*Activities:* Manufacture prefabricated houses, mobile homes, millwork, sash and doors, pallets, furniture and fixtures, athletic equipment, turned products, crates and boxes, musical instruments, and numerous other products; also includes lumber dealers and the light construction industry.

*Employ:* Wood technologists and technicians.

*Suppliers to foresters and forest industries*
  *Activities:* Supply forestry instruments, logging machinery, sawmill and wood-processing machinery, pulp and paper machinery, and chemicals, dry kilns, adhesives, and other equipment and products to the wood-processors.
  *Employ:* Foresters and wood technologists at professional and technician level in product development, and as salesmen and customer-relations men.

*Associations of forest industries and conservationists*
  *Activities:* Provide timely information and promote constructive cooperation among forest industries and users and laymen interested in conservation of nature and environmental improvement.
  *Employ:* Foresters, wood technologists, wildlife biologists, forest ecologists, range experts, landscape architects, and recreation specialists in executive and staff work. Employment limited mainly to professionals and men with writing skills.

*Miscellaneous employers*
  Various agencies with jurisdiction over land, and budgets of forestry agencies, highway departments, power companies, city and county park boards, youth conservation groups, and a few land-development agencies employ foresters, landscape architects, hydrologists, and wildlife biologists on a full-time, part-time, or consulting basis. Some may employ forest technicians also.

*Self-employment*
  A few imaginative foresters and wildlife biologists support themselves at least in part as writers, lecturers, managers of land, and as landowners and operators. Others manage boys' and girls' camps and camps for adults and families. Some are wilderness outfitters and managers of other outdoor activities. Success in such ventures calls for talents and experiences going beyond those included in curricula of forestry colleges.
  Other foresters and technicians have become independent loggers, sawmill operators, nurserymen, house builders, and wood fabricators.

## OUTLOOK FOR FUTURE EMPLOYMENT

**Population Growth** Aroused by the drain that industrial societies place upon natural resources and the wastes they discharge into the environment, a vigorous movement arose in the late 1960s advocating zero population growth. In 1971 it was too early to assess what influence such advocacy might have on future population. U.S. Census and other population projections by demographers were in the order of a 40-percent increase over the 1971 level by the year 2000. The projection was accepted by the U.S. Forest Service in its 1969 study of future timber needs. To house and supply 80 million more people even at 1971 standards of consumption will place a heavy burden on timber resources. It will certainly require more housing, furniture, consumption of paper and paperboard, and use of other wood-based products. As of 1971 the best-documented projection was that made by the U.S. Forest Service and referred to in Chap. 18.

**Rising Human Expectations** Over the years, per capita consumption of most goods has increased along with growth of population. This is expected to continue. But again, people concerned about environment were advocating decreased consumption in addition to zero population growth. An apparent decline in luxury consumption by the well-to-do has occurred over the years which is reflected in less emphasis on elaborate private dwellings, dress, and entertainment. Consumption by the middle- and lower-income groups, on the other hand, has increased markedly, and this increase seems likely to continue for some time. Meanwhile many young people of affluent families seem to be eschewing elaborate dress and extravagant habits in their search for meaningful goals in life. What effect this is likely to have on future consumption of goods by the nation as a whole is something yet to be determined. Strong doubt exists concerning the extent to which this movement may offset the rising expectations of people whose levels of consumption were still low to moderate in 1971.

The measure used for total consumption in the United States is the gross national product (GNP), which includes both commodities and services. The U.S. Forest Service, in accepting the projection of a threefold increase in GNP between 1970 and 2000, obviously anticipates a substantial rise in living standards.

In considering projections of future timber needs and the impact on employment of foresters it should be kept in mind that wide and rapid changes occur in both birth rates and total consumption. Future projections of population and GNP are likely to differ widely from those made in 1970. The reader is cautioned therefore to review the future employment outlook for foresters in light of events that will have occurred since 1971.

**Environmental Resistance** Environmental resistance as mentioned in Chap. 10 may cause animal populations to stop increasing and even to decline sharply by inducing migration, causing high death rates, or declining birth rates. Humans are affected by environmental resistance also, although they are generally able to exercise some control over both environment and death rate. What we are concerned with here, however, is the increased forestry effort that may be required as man presses for yields that approach the biologic potential of the forest site to produce timber, other products, forest amenities, and services. If our nation is to consume twice as much timber in the year 2000 as it did in 1970, it may be expected that considerably more than twice the forestry effort expanded in 1970 will be required. In Europe, for example, timber yields per unit area are somewhat less than double those of the United States, but input of forestry effort as measured by foresters employed per million hectares is close to four times that in the United States. It is conceivable that we may need a similar level of management intensity in the United States to meet anticipated timber needs by the year 2000. Not only will timber output need to be increased, but also forest services and amenities. All three will require increased input of forestry effort.

**Future Role of the Profession** The major factor in growth of forestry employment in the past has been the imaginative leadership exerted by such men as Fernow, Pinchot, Graves, Greeley, and Clapp of the United States; McMillan of Canada; Saari of Finland; and a great many others in Germany, France, Scandinavia, the United Kingdom, and elsewhere. It is foresters working with concerned people that make jobs for foresters. They have done this by recognizing socially important tasks that needed to be done that foresters were competent to perform. They explored these tasks, prepared feasible plans for their accomplishment, drafted proposals for action by the Congress, state legislatures, or corporate boards of directors, and set forth in convincing terms the benefits to be expected from the proposed action. Such steps by foresters and their supporters led to the management and use of national forests, to federal-state cooperation in fire control on all forest land, to various programs for promoting forestry on farm and other small woodlands, to passage of the McSweeney–McNary Forest Research Act and its funding, to incorporating forestry measures into comprehensive flood-control projects, to the several state programs for improved management of private forests, and a host of other forestry programs and actions.

Many such proposals did not originate in the mind of the chief forester, but instead in the minds of others both inside and outside the Forest Service. Among men whose ideas and planning have led to new opportunities for foresters were such individuals as Overton Price, Ralph Hosmer, Raphael Zon, Hugh Baker, Aldo Leopold, Edward Kotok, Carlos Bates, Carl Hartley, Harry Gisborne, and many, many others still living whose fertile minds, devotion to social good, and energetic follow-through helped to mold the course of forestry in America. Many of these people began to make significant contributions when in their twenties and early thirties. The vigor of forestry in the future and the employment opportunities available will depend upon the young men of vision and dedication of today who prepare the way.

**Annual Replacement** Precise information is not available on the numbers of graduates required each year to replace the men currently employed. The working period of men who follow a life career in forestry is believed to approach 30 years. Retirement age is generally 60 or 65 years, though many men remain active much longer. It is known, however, that a substantial number of men who graduate in forestry do not continue long in the profession. Some accept military careers, others get into peripheral fields of employment, and still others shift to entirely different lines of work. It therefore seems reasonable to accept 20 years as the average service period for a graduate in forestry. The same probably holds for technicians. Accepting this figure, the number of graduates needed annually to replace the 23,260 foresters employed in 1971 would be 1,163, and to replace the 11,259 technicians, 563. The number of bachelor's degrees granted in 1970 was 2,771.

What then are some of the tasks ahead on which foresters could make socially valuable contributions, thereby widening the outlook for their profession?

## PROMISING FIELDS FOR FUTURE EMPLOYMENT

Reference has been made in preceding chapters to a number of both old and new fields to which foresters could make significant contributions in the future. Those with the richest promise are the ones requiring large expenditures of money by government, corporations, or private individuals.

**Housing** U.S. Forest Service 1969 estimates of annual new housing construction by future years were as follows:

| Year | Thousands of units |
|------|-------------------|
| 1980 | 2,785 |
| 1990 | 2,930 |
| 2000 | 3,510 |

*Source:* U.S. Department of Agriculture, Forest Service, August 1, 1969, Possibilities for Meeting Future Demands for Softwood Timber in the United States.

Actual new construction for the year 1968 was 1,833 thousand units. House construction in itself is a task for builders rather than foresters. However, foresters manage the land on which timber grows and they supervise its harvesting. Wood technologists and forest-utilization experts can help in design and fabrication of house furnishings and in service to the construction industry (Fig. 24-1).

**Figure 24-1** Students gain inspiration from superbly executed wood sculpture. *(Ivan Mestrovic Studio, Syracuse University.)*

**Pulp and Paper Industry** The pulp and paper industry has embraced modern chemical engineering and technology and actively supports research in corporate laboratories, the Institute of Paper Chemistry, and universities. As a result it has been one of the most vigorous of forest industries. It faces growing public pressure in two areas: stream and air pollution, and disposal of waste paper and cartons. More can be done about both.

The major employment opportunities offered by the industry are for technicians and technologists to keep its highly complex manufacturing processes working smoothly and to develop and introduce still more complex operations for the future. Since 1955 the pulp and paper industry has experienced difficulty in recruiting the numbers of university-prepared pulp and paper technologists it has needed. This has persisted even after generous scholarships have been offered by the industry to induce young people to enter the field. The opportunities are well worth a young man's consideration.

The pulp and paper industry also employs foresters to manage their forest lands. If increased output is to be expected from land tributary to company mills, considerable additional forestry effort will be required over that expended in 1971.

**Rejuvenation of Rural America** Much of rural America has undergone decline in population, social institutions, and economic base with the retreat of farmers from the land and migration of their young people to the cities. Much enterprise can be conducted outside large urban centers, as many corporations have demonstrated.

Each major move from city to small community offers opportunity for developing a satisfying environment for living. The task of the forester is to offer his plans for environmental improvement before all lands suitable for forests get committed to other use.

Attractive rural communities are unlikely to spring up of their own accord. They must be planned, promoted, and developed. Imagination, innovative design, and skill in presenting the charms of small dispersed communities will be needed to bring them to reality. Landscape architects more than foresters are fitted for such a task and it appears that they will be in increasing demand in the years ahead.

**Increasing Productivity of Small Woodlands** The 300 million acres in small woodland holdings are potentially capable of producing as much as three-fifths of the nation's future timber supply. To do so they will need much more intensive management than they were receiving in 1971. The magnitude of the task in the South has been outlined in *The South's Third Forest*, referred to in earlier chapters. To achieve success in improving forest practice on small ownerships, foresters must demonstrate a clear understanding of owner objectives and display the competence to help him realize them. Many such owners are willing to respond and to encourage their neighbors to do likewise. The rapid growth of such societies as the New York Forest Owners' Association and the enthusiasm displayed in their meetings and publications are evidence of the interest these owners have in their properties. Their joint holdings are far too important to ignore (Fig. 24-2).

In 1971 the Chief of the Forest Service announced a new plan in the process of formulation to extend additional federal aids to nonindustrial private forest owners for improved practices.[1]

**Meeting Water Needs** A great increase in water use is to be expected in the years ahead. The forester's task is to make sure that watersheds are kept in condition to continue to deliver high-quality supplies. It may also be a task on key watersheds to attempt measures

**Figure 24-2** Inspecting back country for possible public acquisition. *(U.S. Department of Interior, Bureau of Outdoor Recreation.)*

that might lead to increased flows. This will probably mean some loss in timber growth and also in recreational use of the land (Fig. 24-2).

**Range Management** The western rangelands have been improved substantially as sheep grazing has diminished and as cattle grazing has been more carefully managed. The problems that seemed to lie ahead in the 1970s were: how to reverse the decline in sheep numbers; how to maintain continued favorable water flow from the land; how to integrate livestock use with wildlife use of the same range; how to minimize conflict between recreational users and livestock users; and how to integrate range use with feedlot management. A further and never ending problem is how to bring enduring prosperity to the range-livestock industry. Hanging over it is always the threat of a prolonged dry cycle that can nullify long-term improvement.

Range managers will certainly be called upon to work with wildlife managers to determine the respective roles of livestock and large wild herbivores in range use. This will also involve the role of carnivores in maintaining the health of both and in keeping range rodents in check.

**Outdoor Education and Recreation** Almost all managers of large areas of natural landscape will be obliged to provide opportunities for people to engage in outdoor recreation. Foresters are heavily involved. Forest interpretation in the outdoors where ecological principles can be pointed out and explained needs to be expanded and made available for a host of people if they are to understand and exercise the protective care our environment will need against growing intensity of use. It is people who degrade environment. It is people who must be taught to appreciate and protect it (Fig. 24-3).

**Forest Protection** Foresters long ago learned the importance of working constructively with the decomposing organisms of the soil. Gradually they are learning also how to use fire for constructive purposes, including control of the composition of forest stands.

At the beginning of the 1970s chemists working with foresters were opening up the new field of chemical ecology that offers promise of new approaches to forest-insect control. Might it be conceivable that man should some day use pheromones to concentrate wood borers and the wood-destroying fungi they carry to clear up logging slash? And after they had completed their task might he lure them to traps to prevent their attacking healthy trees? Could man learn to use defoliators to eliminate

Figure 24-3 Forest engineer instructing Job Corps trainees in map-reading. *(U.S. Forest Service.)*

hardwood competition from conifers on soils poorly suited to hardwoods? Such were far-out speculations in 1971 but not entirely fantasy.

## IMPROVING MAN'S ENVIRONMENT

The 760 million acres of forests within our nation protect the environment whether man manages them or not. Fitting forests around the environment of cities, suburbs, factories, and busy transportation lines will require careful planning and maintenance if maximum benefits are to be achieved. Tree zones should be wide enough to form closed canopies with leaf litter and associated organisms. They can then operate as a sanitizing and decomposing agent for the dust, debris, and city flotsam that they trap. To compensate for the heat absorption of asphalt pavement and blackened roof tops, the transpiring leaf surface should be greater than the black surfaces.

The entire concept still requires further development before a workable scheme will evolve for bringing into the city some of the amenities and environmental benefits of a forest.

An interesting example of bringing a wild environment to the city's edge exists in Africa. The city of Nairobi, Kenya, has just outside its limits a game park where citizens and visitors may watch gazelle, impala, wildebeeste, rhinoceros, eland, zebra, giraffe, cheetah, lions, and leopards as they browse, hunt, select mates, and train their young—unfettered by enclosure.

**Intercity Forest Belts** Green belts around cities were long ago proposed as a means of enhancing urban and suburban living. The original concepts in the United States were that such green areas be rather modest in size. The concept of greenbelts in the U.S.S.R. as exemplified at Kiev is a forest area of some 30,000 hectares, almost 75,000 acres. The Vienna Woods represents more nearly what it may be possible to aspire to in America. The Wienerwald, its German name, encompasses all the forest and woodland that lies within some 15 miles of the city. In it are farms, vineyards, sawmills, villages, and meadows. The Wienerwald is in fact essentially self-sufficient economically, yet it provides a vast wooded area to give quality to the environment of the city. It is some such concept as this that needs development for the heavily populated eastern seaboard area of the United States.

## WORLD FORESTRY

The indicative world plans for agriculture and forestry prepared by the UN Food and Agriculture Organization call for expansion by some twofold or more of the forestry work of developing nations. Plans involve doubling the 1969 out-turn of professional-level graduates and quadrupling that of forest technicians. Outside help will be required to effect this. These nations will continue to need outside help to develop their forest industries and to export forest products.

Some United States forest-industry corporations engage in overseas manufacturing and sales. These, together with importing and exporting companies that deal in forest products, or in logging and timber-processing machinery, afford overseas employment opportunities for forestry graduates.

The UN Food and Agriculture Organization employs a limited number of foresters, most of whom come from countries other than the United States. There are two reasons for this. A speaking and writing knowledge of the three FAO languages, English, French, and Spanish, is expected of full-time employees. Those who are sent on assignment to developing nations are also expected to gain competence in the language of their host nation. Few young American foresters have such language competence, but many European foresters have it. The second reason is that FAO salaries have not been as attractive to Americans as they were to Europeans and men from developing nations.

Effective work in any country requires, in addition to local language competence, the capacity to develop quickly an understanding with one's counterpart and with the local people as a whole to elicit their cooperation. This means not only having and showing respect for people of backgrounds and cultures other than our own, but also having a sensitivity to individuals and their feelings so that harmonious coworking becomes possible.[2]

## WORKING CONDITIONS AND SATISFACTIONS

Working and living conditions for foresters have changed markedly since the days when a forest-ranger examination included a demonstration of how to throw a diamond hitch on a pack horse, and the ranger's wife carried water to a cabin from the nearest spring. Most foresters and forest technicians now live in modern homes in villages or small cities. Here they may find stimulating friends, good schools, library facilities, and cultural programs on radio or television. They may take an active part in community affairs. Their work brings them into contact with important people of the community because it affects community welfare.

Foresters can expect hard work, responsibility, reasonably long working hours, modest salary scales for professionals and technicians, competition for promotions, and the opportunity to exercise considerable initiative. They may be obliged to move frequently if they serve in a large organization. This is one of the means taken to prepare young men for increased responsibilities. In a physical sense the environment is likely to be pleasant and may be inspiring. The forester is expected to be largely self-sufficient in discharging his professional responsibilities, subject to

**Figure 24-4** Someone must provide the light to guide us through the mist of conflicting desires if forests are to serve man's total needs. *(U.S. Forest Service.)*

the direction of his supervising officer. He must adopt a habit of study to keep abreast of the many new developments in forestry and related technologies.

Most foresters find their work rewarding in terms of satisfactions, opportunities for self-development, pleasant working conditions, and fulfillment through service. This holds true whether the forester be a professional or a forest technician. Working conditions. and responsibilities for wildlife managers, range managers, watershed managers, and forest recreation specialists offer similar opportunities and satisfactions.

The wood-products engineer and forest-products technician, as well as the pulp and paper engineers and technicians, will work mainly in manufacturing plants. Many of these also are located in small communities where living conditions may closely parallel those of the forester. Their work too can be highly rewarding, because it is important, rapidly changing, and especially in the case of wood-processing industries offers many opportunities for innovation and initiative.

## SUMMARY AND CONCLUSIONS

The long-term goal of forest management is to make optimum use of the greatest synthetic factory on earth, the forest. Production of wood as a commercial commodity of great utility, renewable through growth, will continue to rank at or near the top among the human benefits of forests (Fig. 24-4). Environmental protection will rank almost equally high. Helping those who gain their livelihood from forest-based work to benefit from their labors in proportion to what those otherwise employed receive is a task to which foresters should give increasing attention in the future. Mechanization is affording a way to achieve this.

To assure mankind a future of affluence and ease need not be the goal of human life. Rather, it should be to extend the creative power of man's intellect to make his tenure on earth harmonious, ennobling, and enduring. The reward sought can be to press eagerly forward toward the receding horizons of greater knowledge, justice, and self-fulfillment.

## LITERATURE CITED

1 Cliff, Edward P. 1971. More Trees for People, *American Forests,* 77(4):16–19, 50, 52–54.
2 Shirley, Hardy L. 1968. Working in Far Away Places, *Journal of Forestry,* 66:894–897.

# Index

Accredited schools, 418
Adams, John, 32
Administration, 251–264
   directing, 257, 258
   organizing, and staffing, 253, 254
   outside talent, 258, 259
   planning and budgeting, 252, 253, 257
   reporting, 256, 257
   setting goals, 251
   staff, 259, 260
   supervision, 254–256
Administrative theory, 260–262
African game parks, 174, 175
Air pollution, 43
Alaskan forests, 72, 73
Allelochemics, 148–151
   bacteria responses, 150
   insect induced response, 150
   suppressants, 149, 150

Allelochemics:
   use of, 150–151
Amenities:
   metropolitan areas, 380
   social benefits of, 368
America, early forestry, 19
American Association for the Advancement of Science, 32, 49
American Fisheries Society, 50
American Forestry Association, 33, 50
American Forestry Congress, 33, 34, 50
Animal behavior, 179, 180
Animal population, 179
   regulation of, 179
   survival of young, 179
Aphids, 128
Appraising, 207–223
   investment value, 219
   liquidation value, 219

Appraising:
  sale value, 219, 220
Aptitudes for administration, 262–263
Asia:
  forestry development, 18
  land use in East, 367
Aspen-birch type, 61, 62
Avalanches, 167

Baker, Hugh, 442
Bangladesh, 335
Bates, Carlos, 442
Biologic potential, 180
Bitterroot National Forest, 395
Board foot, 209, 212
Brazil, forest rehabilitation, 358, 359
Broome County, N.Y., forests of, 361
Bulldozer, invention of, 421
Business, 238–249
  acumen, 246–248
  guides, 239
  information sources, 249
  organizing the forest, 242–244
  record keeping, 245
  return on, 239, 248
  timber growing, 238–249

Canada:
  forests of, 340, 349, 350
  timber export, 345
Carbon cycle, 84
Career, development of administrative, 263–264
Cedars of Lebanon, history of exploitation, 16, 17
Cellulose, 308–310
  chemical nature, 309
  derivatives of, 310
  properties of, 310
Cellulose plastics, 323, 324
Census of wildlife, 182
Central forest formations, 63–65
Chain, Gunter's, 208, 209
Chemical defenses, 151, 152
Chemical ecology, 143–155
  background of, 144–146
  methods of, 153–154
Chemical treatments of wood, 292–294, 324–325

Chemistry of wood, 307–326
  ash content, 308
  cellulose, 308–310
  extractives, 308
  lignin, 310–312
China, timber resources, 337, 368
Chippewa National Forest, 363
Christmas trees, 301
Citizen education, 416
Citizen involvement, 390–393
Citizen organizations, 50
City forestry, employment in, 448
Civilian Conservation Corps, 39
Clapp, Earle H., 40, 442
Clarke-McNary Law, 38, 45, 46
Clear-cutting, 106–108, 235, 394–396
Cliff, Edward P., 261
Climate influence on forests (*see* Site factors)
Climatic factor (*see* Site factors)
Climax forest, 103
  silviculture of, 104–106
Cloud seeding, 125, 167
Coniferous forests, cool, 53, 61
Connor Lumber Company, 360
Continuing education, 417, 418
  role of professional society, 417
"Continuous forest inventory," 214
Cooperative Forest Management Law, 46
Cooperative State Research Service, 427
Cord, 209, 212
Crafts, Edward C., 42
Credit, 245–246
Cropland, 4
Cruising radius, 180
Cunit, 212
Cutting restrictions, 16, 30
Cutting timber (*see* Harvesting)

DDT, 130, 144, 145
Decision making, 260–262
Defense mechanisms, 151–152
Developing nations, forest practices, 21–22
Developmental research, 421–424
Dinosaur National Monument, 377
Directing, 257–258
  outside talent, 258–259

Diseases, 131–135
  Dutch elm, 133
  heart rot, 133, 134
    on sugar maple, 134
  mistletoe, 134, 135
  nonparasitic, 131–132
  parasitic, 132
  rusts, 132
  types of, 131
Dominant use, 392, 393
Douglas fir types, 68–71
Dry forests, 58


Earth, surface features of, 4
Ecologic succession, 90–93
Ecological Society of America, 49
Ecosystem, 3–4
  complexity of, 93
  description of, 79–80
  endurance of, 98
  energy flow, 81–83
  fragility of, 93
  functioning, of, 80
  man in, 94–95
  use by man, 98
Education, 405–420
  objectives of, 406
  outlook for, 418–420
  planning for, 420
  postgraduate, 412
  professional colleges, 407–412
    curriculums, 408–412
    types of, 407
  status of, 418
  system of, 406
  technician schools, 412–415
Egypt, early forestry in, 19
Elm-ash-cottonwood type, 63, 65
Employing agencies, 437–440
  associations, 440
  colleges and universities, 439
  consulting firms, 439
  federal, 438
  forest industries, 439
  self-, 440
  state, 438
  suppliers, 440
Employment, 436–451
  agencies (*see* Employing agencies)

Employment:
  factors affecting, 440–443
    environmental resistance, 441
    future role of forestry, 442
    population growth, 440
    replacement, 443
    rising expectations, 441
  foreign, 448–449
  in forests, 359
  future fields of: environment, 448
    housing, 443
    intercity forests, 448
    paper, 444
    protection, 447
    range, 446
    recreation, 446
    rural America, 445
    small woodlands, 445
    watersheds, 445
    world forestry, 445
  number of professionals, 436
  outlook, 440–451
  satisfactions, 449–451
  working conditions, 449–451
Energy cycle, 3, 81–83
English and metric units, 208–209
Environment, protection of, 10
Environmental impacts, 388
  minimizing, 389
  reduce demands, 389
Environmental improvement, 114, 115, 400
Erosion:
  water, 165, 168–169
  wind, 169–170
European forestry, development of, 19–21
Evapotranspiration, 162–167
  measurement of, 163–166
Everglades National Park, 378


Farm forestry, 46
Farm land abandonment, 44, 339
Federal forestry, beginnings of, 32
Federal-State Cooperative Programs, 37–38
Fernow, Bernhard E., 436, 442
Field-study conference, 392
Fir-spruce type, 68, 70

Fire:
 control, 30, 121–126
  organization, 126
  use of aircraft, 125
 danger meters, 123
 plow, 124
 review, 126
Fires:
 attitude towards, 121
 causes of, 122
 crown, 124
 detection of, 123
 disastrous, 120–121
 effects of, 121
 ground, 123
 historic, 120–121
 Indian caused, 121
 kinds of, 123
 lightning caused, 121
 lives lost, 120–121
 man caused, 122
 prescribed, 122
 presuppression of, 122
 prevention of, 122
 protection from, 120–126
 suppression of, 123
 surface, 123
Fish and Wildlife Service, 183, 184, 438
Flood damage, 168–169
Floods, lives lost, 169
Food and Agriculture Acts, 46
Food and Agriculture Organization of the United Nations, 114, 330, 368, 449
Forage, 9
Forest:
 amenities, 6
  metropolitan, 380, 381
 destruction of, 16–18
 services of, 6
 use of, 8
  by primitive men, 12–15
Forest Preserve, New York State, 44
Forest products, 6
 increased output, 347–348
 need for, 41
 secondary: charcoal, 300
  extractives and derivatives, 298–300

Forest products:
 secondary: fruits and nuts, 302
  greenery, 301
  maple products, 301
  medicinal herbs, 302
  miscellaneous, 298
  naval stores, 298–300
  pharmaceuticals, 302
  tree seed, 301–302
 world, 327–339
Forest Products Laboratory, 426
Forest Products Research Society, 50
Forest recreation, 9, 373–380
Forest regions:
 U.S., 53–76
  Alaska, 72–73
  Central, 63–65
  Northern, 61–63
  Pacific Coast, 70–73
  Rocky Mountain, 68–70
  Southern, 65–68
  Tropical, 73–76
 world: cool coniferous, 53–56
  dry, 58
  equatorial rain, 56–57
  moist tropical, 57
  temperate mixed, 56
  warm moist, 56
Forest renewal, 224–225
Forest Reserves, 32–33
Forest resources, world, 5, 53–55
Forest, restoration of, 18
 China, 18–19
 Mediterranean region, 18–19
Forest Service:
 establishment of, 35
 suits against, 394–397, 426
Foresters:
 education of, 405–420
 functions of, 6, 7
Forestry:
 associations, 49–51
 colonial times, 28
 definition of, 5
 development of: Europe, 19–21
  United States, 27–51
 practice in new countries, 21–22
 scope of coverage, 10–11
 social benefits of, 357–370

Forests:
  area of, 5
  decadence of, 92
  definition of, 5
  function of, 5
  man-made, 339
  nontimber use, 339, 340
  property rights in, 15
  reservation of, 32–33
  stages in use of, 15
Franklin, Benjamin, 30

Game management, 176
  concept of, 176
Genetics, Forest Institute, 348
Gisborne, Harry, 442
Graduate education, 412
Graduates in forestry, 418
Grasslands, 5, 193–196
Graves, Henry Solon, 37, 47, 442
Grazing:
  beneficial effects, 136
  damage to forests, 135, 136
  damage to watersheds, 9
Grazing areas:
  United States, 193–196
  world, 188
Grazing Service, 191
Greek ruins, Aspendos, 366
Greeley, William B., 37, 442
Growth, determination of, 213, 214
Gullies, 168

Habitat management, 177, 181–183
Hartig, Theodor, 20
Hartley, Carl, 442
Harvesting timber, 225–237
  bucking, 226
  delimbing, 226
  experimental, 230
  felling, 225
  hauling, 231, 232
  loading, 231–232
  machines, 229–230
  planning and layout, 232–234
  skidding, 226
Hemicellulose, 308
Hemlock-sitka spruce type, 68, 70, 72

Hepting, George H., 135
Hewitt Law, 45
Homestead Act, 31, 43
Hoover, Herbert C., 39
Hormones, 152–153
  use in insect control, 152–153
Hosmer, Ralph, 442
Hough, Franklin B., 32–33
House construction, 8
Housing, 303–304
  factory-built, 304
Husbandry, 97

Improvement of forests, 99–100
Increment, determination of, 213–214
India, timber resources, 335
Indian Affairs, Bureau of, 438
Indicative world plan, 330, 368
Industrial forestry, 41, 47–48
Industry associations, 51
Insect pheromones, 130, 146–148
Insecticides, DDT, 130
Insects, 126–131
  bark beetles, 129
  control of, 129–131
  defoliators, 128
  outbreaks of, 126
  selective feeding of, 151
  sucking, 128
  wood destroying, 129
Instructing Job Corps trainees, 447
Insurance:
  forest, 246
  workmen's compensation, 233, 246
International Union of Forest Research Organizations, 50
Inventory:
  for continents, 218
  continuous forest, 214
  for counties, states, and nations, 218
  data processing, 217
  field data collecting, 216
  planning, 214
  state of the art, 218–219
Investment:
  case studies, 248–249
  credit, 245–246

Investment:
  forest land, 242
  returns on, 239

Kaingineros, 23
Kalambo Falls, 13
Kaufert, Frank, 426, 430

Labor, forest, 244
Lake states, land use, 362–364
Laminated arch, 429
Laminated wood, 291
Land:
  ownership of, 27–28
  use of, 360–369
    adjustments, 401
    by American Indians, 28
    Broome County, N.Y., 361
    by early colonists, 28
    East Asia, 367
    Lake states, 362–364
    Pakistan, 367
    Philippines, 366
    Western North America, 364–365
    worldwide examples, 365–368
Land acquisition, 446
Land area:
  forest, 4
  world, 4
Land Law Review Commission, 391–394
Land Management, Bureau of, 438
Land and Water Conservation Fund, 42
Landslides, 168
Legislation:
  Clarke-McNary Act, 38, 45, 46
  Cooperative Forest Management Act, 46
  Food and Agriculture Act, 46
  Forest Reserve Act, 32
  Hewitt Act, 45
  McIntire-Stennis, 427, 428
  Multiple Use-Sustained Yield Act, 42
  Norris-Doxey Farm Forestry Act, 46
  Pitman-Robertson Act, 46

Legislation:
  Smith-Lever Act, 46
  Statehood Act of Alaska, 72
  Taylor Grazing Act, 40
  Weeks Law, 38, 45
  Wilderness Preservation Act, 43
Leopold, Aldo, 176–178, 394, 442
Leopold, Starker, Commission, 383
Life, cycle of, 3
Lighting the future, 450
Lignin, 308, 310–312
Litigation, 394
Lodgepole pine type, 68–71
Log rule, 212
Log skidders, invention of, 421, 422
Logging, 224–237
  animal, 226–227
  balloon, 230
  bucking, 226
  cable, 227
    high-lead, 228
    skyline, 227
  damage to water, 165
  delimbing, 226
  felling, 225
  hand, 226
  helicopter, 230
  integration with forest practice, 235–237
  owner, 245
  planning and layout, 232–234
  tractor, 228–229
  waste, 235
Longevity:
  of timber stands, 120
  of trees, 120
Lowdermilk, Walter C., 18
Lumber:
  finishing, 288
  handling, 289
  manufacture of, 285–288
  prices, 247
  seasoning, 288
Lumber Code, 39
Lumber manufacture in colonial times, 29
Lysimeters, 164–165

McIntire-Stennis Research, 427–428

McKaye, Benton, 394
McMillan, Harvey, 442
Magsaysay, Ramon, 22
Man made forests, silviculture of, 109, 339
Management Plan, Sihlwald, 20
Management theory, 260–262
Manufacturing activity, world, 331
Maple-beech-birch type, 61–62
Maple products, 301
Marketing, 244
Marshall, Robert, 394–395
Mathematical simulation, 422
   growth of timber, 213
Measurement:
   of forage, 221
   of land, 208–210
   of recreation use, 222
   of timber, 211–220
   of trees, 211
   of water, 220
   of wildlife, 221
Mediterranean forests, rehabilitation of, 358
Mencius, 16
Metric-English units, 208–209
Michaux, Francois-Andre, 32
Mistletoe, 134–135
Multiple-use, 98, 389–391
   citizen involvement, 390
   incompatible demands, 389
Multiple Use-Sustained Yield Act, 42

National Academy of Sciences, 33, 49
National emergencies, timber reserve for, 359
National forests:
   commercial users of, 396, 397
   financing, 393, 397
   grazing use, 190
   use policy, 35
   users of, 396
National Grasslands, 191–192
National Parks, 374–380
National Wildlife Refuges, 183, 184
Naval stores, 298–300
New York State Ranger School, 413
Nitrogen cycle, 84–85
Norris-Doxey Farm Forestry Act, 46

North America, timber potential of, 350
North American Conservation Congress, 37
Northern forest formations, 61–63
Nursery practice, 110–111
Nutrient cycle, 85

Oak-hickory type, 63–64
Olduvai Gorge, 12–13
Outdoor recreation, 371–385
   conflicting use, 375
   planning for, 372–376
Outdoor Recreation, Bureau of, 42
Outdoor Recreation Resources Review Commission, 42, 381–382

Pakistan:
   land use, 367
   timber resources, 335, 336
Pakistan Forest Institute, 336
Paper, consumption of, 322
Paper machine, 319–321
Paper manufacture, 318–323
   coating, 321
   importance of, 321–323
Paper-plastics combinations, 323, 324
Paper pulp, 312–318
   chemigroundwood, 316–317
   cold soda, 316
   consumption of, 313
   groundwood, 312–314
   neutral sulfite, 316–317
   pollution, 317
   soda, 315
   sulfate, 315–316
   sulfite, 314
   whole wood fiber, 317–318
Paper pulp refiner, 315
Park:
   building, 376
   maintenance, 376
   management, 377–379
      education for, 378
   planning, 374–377
   promoting understanding, 377
Particle board, 291–292
Penn, William, 29

Pest control, 126–136
  research, 135
Pheromones, 130, 146–148, 178
  complex behavior, 148
  use in insect control, 146–148
  use by wildlife, 178
Philippines, 366
Pinchot, Gifford, 33–37, 47, 436, 442
Pingree, David, 47
Pioneer types, silviculture of, 107–108
Plantation care, 112
Planting, 109–114
  arid zone, 113, 114
  for environmental improvement, 114–115
  field, 112
  in tropics, 112–113
Plywood, 290
Policy, 386–402
  citizen involvement in, 389–393
  definition of, 386
  forest industries, 398
  issues of, 386–389
  outlook, 400–402
    environmental constraints, 400
    land use adjustments, 401–402
    wilderness, 400
  private forestry, 397–400
  public controls, 387, 400, 401
Pollution, 171
  abatement, 171
  air, 43
  by pulp and paper mills, 317–319
  water, 43
  water treatment, 317–319
Ponderosa pine types, 68–72
Population, world growth in, 329
Prairie-Plains Shelterbelt Program, 40, 170
Predators, 199
  control of, 199
Price, Overton, 442
Prices:
  lumber, 247
  stumpage, 247
Primitive men:
  knowledge of tree names, 14
  use of forest, 12–15
  use of wood, 13–14
Pristine forest, 6

Private forestry, 46–49
Private forests, public control of, 387, 400
Professional foresters:
  continuing education, 411, 417–418
  criticism of, 411
  education needs, 410
  work of, 410
Protecting:
  soil, 167–171
  water, 156–173
Protection, 98, 119
  from animals, 135–138
  from diastrophism, 140
  from diseases, 131–135
  from fire, 30, 119–126
  from insect attack, 126
  from mistletoes, 134–135
  from weather, 138–140
  from wildlife, 136–138
Public domain, disposal of, 31
Public Land Law Review Commission, 184, 391–394
Puerto Rican forests, 73–76

Rain:
  infiltration in soil, 160–162
  interception of, 158, 159
Rain forests:
  Equatorial, 56
  Pacific, 70–73
Range:
  economics of, 200
  in other lands, 200–201
  protection of, 197–199
Range essentials, 174
Range management, 188–202
  forage improvement, 196–197
  herd management, 196
  importance of, 192
  outlook for, 201–202
  techniques of, 196
Range regions, 193–197
Ranger schools, 412–415
  curriculum of, 414–415
  New York State, 413
Rangers:
  attaining professional status, 415
  work of technicians, 414

Recreation, 371–385
  human satisfactions with, 383–385
  outlook for, 381–382
  planning, 371–375
Recreation lands:
  private, 379–380
  public, 373–378
  use of, 374
Recycling wood and paper, 346–347
Redwood type, 71
Research, 421–435
  achievements of, 430–431
  areas of, 428
  contributions to other fields, 431
  developmental, 422
  ecosystem approach, 424
  education for, 432–433
  expenditures for, 426
  financing, 425–429
  nature of, 423–425
  organizing, 425–429
  outlook for, 433–435
  part time, 428
  publications, 428
  results of, 424
  rewards of, 431–432
  role of profession, 442
  scientific approach, 425
  systems approach, 434
  tropical, 426, 434
  U.S. Department of Agriculture, 426–427
    Current Research Information Service, 427
Resettlement Administration, 45
Resource management, 37
River basin planning, 171–172
Roosevelt, Franklin D., 39, 170
Roosevelt, Theodore, 34, 36–37


Saari, Eino, 442
Sargent, Charles Sprague, 33
Sawmill, 285–288
  automated, 287
  chip-and-saw, 287
Schurz, Carl, 32–33
Scope of treatment, 10–11
Shelterbelts, 170
Shifting cultivation, 15, 21–23, 367

Sierra Club, 43, 51, 394
Silcox, Ferdinand, 40
Silting, 168
Silvics, 98
Silviculture, 98–117
  climax types, 104–106
  mechanization of, 115–117
  natural forest ecosystems, 98, 104
  pioneer types, 107
  potential of, 99–100
  subclimax types, 107
Site factors, 85–90
  biotic, 89–90
  climatic, 86–88
    length of day, 87
    precipitation, 86
    solar radiation, 86
    temperature, 86
    wind, 87–88
  limiting, 90
  soils, 88–89
  life in, 89
Site index, 207
Skyfire, 125
Smith-Lever Act, 46
Snow:
  accumulation of, 159–160
  interception of, 159
Social benefits, 352–369
  employment of people, 359
  national emergencies, 359
  nature of, 359
  nontimber, 368–369
Society of American Foresters, 49–50
Society for Range Management, 49
South, the:
  forests of, 358
  third forest, 348–349, 358
Southern forest formations, 65–68
  hardwood forests, 68
  pine forest, 66–67
Spain, forest rehabilitation in, 358
Spruce-fir type, 61–62
Staff:
  organizing, 253
  use of, 259–260
State College of Forestry, Syracuse, N.Y., 411, 429
State Conservation Departments, 184–185

State forest management, Germany, 20
State forestry, 43–46
 administrations, 44
 commissions, 44
 forest practice laws, 46
State forests, 45
Storm damage, 167–170
 avalanches, 167
 flood, 168
 gullies, 168
 landslides, 168
 silting, 168
Stuart, Robert Y., 40
Subclimax types, silviculture of, 107
Succession, ecologic, 90–93
Supervision, 254–256
Survey System, Public Land, 210
Surveying, land, 210

Taxes, capital gains, 248
Taylor Grazing Act, 40
Technicians:
 education of, 412–415
 relation to professionals, 414
 work of, 414
Temperate forests, 56
 mixed, 56
 warm moist, 56
Timber:
 cutting restrictions, 16–20, 30
 shortages of, 15–16
Timber growing:
 basic factors, 239–249
  price trends, 241
 business of, 238–249
 case studies, 248
 information on, 249
Timber resource:
 China, 337
 Indian subcontinent, 335, 336
 regional balance, 334
 world demand, 338
 world outlook, 333
Timber Resource Review, 42
Tolerance, 103
Training, 254–256
Tree:
 age of, 120

Tree:
 development of, 100–101, 268, 270–272
 hazards to survival, 101–102
 life of, 100–103
 maturity and death, 102–103
 measurement of, 211–212
 nature of, 267–268
 seed, 102, 110
Tropical forests:
 Hawaii, 74
 moist deciduous, 57
 Puerto Rico, 73–76
 rain, 57

Union of Societies of Forestry, 49
U.S.S.R., forests of, 338, 340
United Fruit Company, 341
United Nations, Food and Agriculture Organization, 114, 327–337, 368, 449
United States:
 forests of, 58–76
  areas by regions, 60
  growing stock by regions, 60
  harvest by regions, 60
  increment by regions, 60
  inventories by regions, 60
  map, 59
 increase forest output, 347
 projected timber demands, 344
 reduce timber use, 345
 timber growth and cut, 343
 timber imports, 345
 timber outlook, 342–349
 timber situation 1968, 342–344
U.S. Department of Agriculture:
 Cooperative State Research Service, 427
 Current Research Information Service, 427
 Forest Service, 35, 394–397, 426
 McIntire-Stennis Cooperative Research, 427
U.S. Department of Interior:
 Bureau of Indian Affairs, 438
 Bureau of Land Management, 438
 Bureau of Outdoor Recreation, 42, 438

U.S. Department of Interior:
  Fish and Wildlife Service, 183, 184, 438
  National Park Service, 374–380, 438
University of Ibadan, 408

Vacation homes, 381
Valuation, 207, 240–241
Value to forest owner, 222–223
Veneer, 289–290
Virgin forest, 103
Vocational education, 415–416
Volumes of trees, 212

Water:
  infiltration in soil, 160–162
  interception of, 158
  sources of, 157–158
  use of, 9
    by man, 156–157
    by trees, 162
  yields, 166–167
Water cycle, 83–84
Water measurement, 220, 221
Water pollution, 43
Watershed, damage by logging, 165–166
Watts, Lyle F., 40
Weather, protection from, 138–140
  rain, 138
  snow and ice, 139
  wind damage, 138–139
Weeks Law, 38, 45
Western forest formations, 68
  Pacific Coast, 70–73
  Rocky Mountain, 68–70
  type groups, 68
Western North America, land use, 364–365
Western range:
  area of, 192
  depletion of, 190
  national forest, 191
  regions of, 193–195
  use of, 189–193
  vegetative types, 193–196
White pine-red pine-jack pine type, 61–62

White pine weevil, 127
Wild and scenic river, 372
Wilderness, 103, 104, 400
Wilderness Preservation Act, 43
Wildlife, 174–188
  abundance of, 174–175
  characteristics of, 178
  damage by, 136–138
  forest, 9
  productivity of, 174–175, 189
  propagation rate, 180
  range essentials, 181
  values of, 186–187
Wildlife management, 181
  developing a program, 181–183
Wildlife manager, 187
Wildlife Refuges, National, 183–184
Wildlife Society, 49
Wilson, James, 35
Wood:
  chemical components, 307–309
  design, 296–297
  disadvantages of, 270
  finishing, 296
  formation of, 270–272
  gross structure, 271–272
  identification of, 274
  microscopic structure, 272–274
  physical properties of, 276–283
  properties of, 269–270
  sanding, 296
  ultrastructure of, 274–278
Wood balance by regions, 334
Wood processing:
  administrative control, 298
  lumber manufacture, 285–288
  plant layout, 297
  primary, 284–296
  programming, 297
  quality control, 297
  secondary, 294–298
  significance of, 302–305
Wood processing industries, significance of, 302–305
Wood residues, use of, 346–347
Wood rotting fungi, 134
Wood sculpture, iv, 325, 444
Wood treatment, 292–294, 311, 422
  decay resistance, 292–293
  fastening of, 295

Wood treatment:
　fire-retardant, 293
　gluing, 295
　machining, 295
　seasoning, 295
Wood use:
　ancient Egyptians, 14
　ancient Greeks, 14
　per capita, 333
　early Romans, 14, 15
　in world economy, 330–333
　by world regions, 332
　world trends, 328
Wood uses, 8, 279–283
　athletic equipment, 282
　containers, 282
　esthetic, 282
　miscellaneous, 283
　musical instruments, 282
　relation to properties, 280–283
　structural, 281–282

Woodlands, definition of, 5
Work assignment, 254
Working conditions, 449
World Forestry Congresses, 50
World resources, summary of, 351–353
World timber market, North American role in, 350
World trade, 340–342
World War I, 40, 41
World War II, 40, 41
World's forests, 5, 53–55
　complexity, 58


Yosemite National Park, 384


Zon, Raphael, 41, 427, 442